Electroless copper and nickel–phosphorus plating

AF736839

Related titles:

Nanostructured metals and alloys: Processing, microstructure, mechanical properties and applications
(ISBN 978-1-84569-670-2)
Nanostructured metals and alloys have enhanced tensile strength, fatigue strength and ductility and are suitable for use in applications where strength or strength-to-weight ratios are important. Part I of this important book reviews processing techniques for bulk nanostructured metals and alloys. Parts II and III discuss microstructure and mechanical properties, whilst Part IV outlines applications of this new class of material.

Thermal barrier coatings
(ISBN 978-1-84569-658-0)
Thermal barrier coatings are used to counteract the effects of high-temperature corrosion and degradation of materials exposed to environments with high operating temperatures. The book covers both ceramic and metallic thermal barrier coatings as well as the latest advances in physical vapour deposition and plasma spraying techniques. Advances in nanostructured thermal barrier coatings are also discussed. The book reviews potential failure mechanisms in thermal barrier coatings as well as ways of testing performance and predicting service life. A final chapter reviews emerging materials, processes and technologies in the field.

Surface engineering of light alloys: Aluminium, magnesium and titanium alloys
(ISBN 978-1-84569-537-8)
This authoritative book provides a comprehensive review of the various surface engineering techniques employed to improve the properties of light alloys, focussing on titanium, magnesium and aluminium alloys. It examines surface related degradation of light alloys and covers surface engineering technologies in detail. The book includes chapters on corrosion behaviour of Mg alloys, anodising treatments of Mg alloys, micro-arc oxidation of light alloys, physical vapour deposition of light alloys, PIII/PSII of light alloys; laser surface modification of Ti alloys, plasma nitriding of Ti and Al alloys, duplex surface treatments of light alloys and biomedical devices using Ti alloys.

Details of these and other Woodhead Publishing materials books can be obtained by:

- visiting our web site at www.woodheadpublishing.com
- contacting Customer Services (e-mail: sales@woodheadpublishing.com; fax: +44 (0) 1223 832819; tel.: +44 (0) 1223 499140; address: Woodhead Publishing Limited, 80 High Street, Sawston, Cambridge CB22 3HJ, UK)

If you would like to receive information on forthcoming titles, please send your address details to: Francis Dodds (address, tel. and fax as above; e-mail: francis.dodds@woodheadpublishing.com). Please confirm which subject areas you are interested in.

Electroless copper and nickel–phosphorus plating

Processing, characterisation and modelling

W. Sha, X. Wu and K. G. Keong

WOODHEAD PUBLISHING

Oxford Cambridge Philadelphia New Delhi

Published by Woodhead Publishing Limited,
80 High Street, Sawston, Cambridge CB22 3HJ, UK
www.woodheadpublishing.com

Woodhead Publishing, 1518 Walnut Street, Suite 1100, Philadelphia, PA 19102-3406, USA

Woodhead Publishing India Private Limited, G-2, Vardaan House, 7/28 Ansari Road,
Daryaganj, New Delhi – 110002, India
www.woodheadpublishingindia.com

First published 2011, Woodhead Publishing Limited

British Library Cataloguing in Publication Data
A catalogue record for this book is available from the British Library.

ISBN 978-0-08-101497-4 (print)
ISBN 978-0-85709-096-6 (online)

The publisher's policy is to use permanent paper from mills that operate a sustainable forestry policy, and which has been manufactured from pulp which is processed using acid-free and elemental chlorine-free practices. Furthermore, the publisher ensures that the text paper and cover board used have met acceptable environmental accreditation standards.

Typeset by Replika Press Pvt Ltd, India

Contents

Author contact details

Professor Wei Sha
Professor of Materials Science
Queen's University Belfast
Belfast BT7 1NN
UK

E-mail: w.sha@qub.ac.uk

Dr Xiaomin Wu
Materials innovation institute
Mekelweg 2
2628 CD Delft
The Netherlands

E-mail: x.wu@m2i.nl

Dr Kim Ghee Keong
34 Lorong Jati 1G/1
Taman Jati
Kulim 09000
Kedah
Malaysia

E-mail: kg.keong@hotmail.com

Preface

Electroless deposition uses the chemical reaction between a reducing agent and a metal salt in a liquid solution so that the metal ion can be reduced to a metal atom on a substrate. Compared with electroplating, electroless plating can be plated uniformly over all surfaces, regardless of size, shape and electrical conductivity. The main applications of electroless copper deposits are in the printed circuit board manufacturing, very- or ultra-large scale integrated (VLSI/ULSI) technology, electromagnetic shielding and decorative plating. Electroless nickel–phosphorus deposits are used as protective and functional coatings, benefiting the chemical, oil and gas, automotive, electronics and the aerospace industries. Other metals can also be plated through the electroless route, but they are less widely used.

Electroless deposits are an extremely active field of materials science research. In 2009 alone, included in the Science Citation Index, 529 research papers were published in international journals on the electroless topic, written by 1829 authors, from 489 institutions in 51 countries. Such researchers are targeted readers of this book.

The last books devoted exclusively to electroless plating were all written before 1991:

- *Chemical (electroless) Nickel Plating* by G. G. Gawrilov, Portcullis Press, 1979, 189 pages
- *Electroless Plating: fundamentals and applications* edited by Glenn O. Mallory and Juan B. Hajdu, William Andrew Publishing/Noyes, 1990, 538 pages
- *Electroless Nickel Plating* by Wolfgang Riedel, Finishing Publications Limited and ASM International, 1991, 328 pages.

All three are authoritative books. The third one, by Riedel, was translated from German to English and Chinese. However, notwithstanding they do not include much on electroless copper, even electroless nickel R&D has grown enormously since 1991. Therefore, a book on the electroless subject is very much needed and we hope that this book will be welcomed by academics and industrialists alike.

This book is intended to be a research monograph, cumulating the experience of many years' research by the authors. It will include the more recent results, since 2000, but will also cover relevant, recent work by other researchers around the world.

Since 1999, the authors have completed the following PhD projects:

- Processing and characterisation of electroless copper deposit (funded by the UK Engineering and Physical Sciences Research Council, EPSRC)
- Characterisation and modelling of the evolution of microstructure during thermal processing of electroless nickel–phosphorus depositions.

Other smaller projects include:

- Collaborative research in modelling the evolution of material microstructures (Royal Academy of Engineering, EPSRC, Invest NI and Royal Society)
- Electroplated and electroless copper deposits (Shipley Ronal)
- Friction and wear characteristics of electroless Ni–P–SiC coated aluminium cylinder liner (Royal Society).

The research from these projects has led to many paper publications. The two PhD theses and the research papers are the backbone of this book, but the underlying structure of the book will be based on the physical metallurgical phenomena of the electroless process. In addition, other researchers' work is reviewed and the major results presented and discussed.

Each of the three authors is a world authority in the electroless field. Among the 1153 research papers in electroless nickel published during 2002–2009, Keong and Sha are among the authors for the top ten most cited papers in the Science Citation Index, based on which data the journal impact factors are assembled. During 2006–2009, Sha has refereed, on invitation, 38 manuscripts for the top international archival journal in this subject, *Surface and Coatings Technology*. During 2006–2008, Wu's electroless research has been presented in ten international conferences, symposia and meetings. These demonstrate the international standing of the author team for the book.

Computer-based modelling is a fast growing field in materials science, demonstrated by the large recent literature. The authors of this book are in a unique position, in their combined expertise in electroless research and model development. Though modelling is only a part of it, this book will fill a big gap of the book literature in the application of modelling techniques in electroless depositions, both hot topics in contemporary materials studies, whilst at the same time document latest research in this area. A large chunk of this latest research will be from the authors themselves, based on their world leading position in this area, but, as stated before, the book will also cover important relevant research by others. A lot of the microstructural modelling is about phase transformations and kinetics. There will be coverage of artificial neural network research in the book applied to electroless depositions. Readers should be interested in this.

The book has a clearly defined audience. It is primarily intended for researchers studying electroless deposits. The materials expert, however, will be able to learn modelling and apply this increasingly important technique in their electroless materials research and development. As the statistics given earlier have shown, the research base in electroless plating and its application areas is huge, so we hope that this book will have a big impact.

Combining modelling and electroless deposits into one book is the idea behind this book, which we think, is its strength and selling point. There is not any book covering the combination of electroless deposition and computer modelling at present. Computer modellers have many options on entire books that are devoted to the modelling techniques used in the book, but not applied to electroless depositions.

Wei Sha
Professor of Materials Science, Queen's University Belfast, UK

Xiaomin Wu
Materials innovation institute, The Netherlands

Kim Ghee Keong
Kulim, Kedah, Malaysia

Acknowledgements

We thank:

- Our past and present collaborators including Watriah Sarililah, Zhanli Guo, Z. Xiao Guo, Savko Malinov, Nurzairyani Hj Mohd Zairin, Raj Rajendran.
- Financial support from Hutchison Whampoa/Dorothy Hodgkin Postgraduate Awards.
- Professor A. E. Long and Dr V. Sivakumar from Queen's University Belfast and Dr Simon Gill from the University of Leicester for their encouragements, suggestions and discussions.
- All technicians and staff who have helped in research: Mr Mark Russell, Mr Trevor Foley, Mr Norman Harmon, Mr John Hunter, Mr William MacRoberts, Mr Kenny McDonald, Dr David McNeill, Mr Jim Norney, Mr Norman Cornett, Mr Steven McFarland, Mr James Knox and his team.
- Past and current PhD students in the School of Planning, Architecture and Civil Engineering, Queen's University Belfast: Dr Yu Zheng (Albert), Chunhui Liu (Iris), Xinyao Yang, Kai Yang, Qing Li (Nick), Dr Paul Wilson, Patrice Cairns, Chris Anderson, Dr Jane Boyd (nee Brown), Dr Merlyn Chelliah, Dr Siobhan Cox, Dr Simon Grattan, Jacek Kwasny, Vincent O'Donoghue, Gobithas Tharmarajah, Jason von Meding, David Yeboah for their friendship and support.
- Professor P. A. M. Basheer, former Director of Research, for his valuable suggestions in choosing research programme in October 1999.
- Dr Tatiana Novoselova, Dr Hristo Skulev, Dr G. Georgiev, Dr R. Ivanova, Dr Ivaylo Katzarov and Dr Ani Zhecheva for their friendships and support.
- R. Laughton formerly of Lea Manufacturing (UK) Limited for providing nickel–phosphorus samples.
- Dr John Graves at Shipley Europe Limited for supplying samples.
- London & Scandinavian Metallurgical Co. Limited.
- X-ray unit, Deptartment of Materials, Queen Mary, University of London.

- At Queen's University Belfast: Electron Microscope Unit of School of Mathematics and Physics, X-ray units of School of Geography, Archaeology and Palaeoecology and School of Chemistry and Chemical Engineering, Media Services, Materials Characterisation Laboratory, Environmental Engineering Research Laboratory and Metal Workshop of School of Planning, Architecture and Civil Engineering, and Materials Research Laboratory at School of Mechanical and Aerospace Engineering.
- Dr M. Bououdina (Queen Mary, University of London) for his technical assistance.
- Dr A. Révész for providing the data in numerical format for one of the isothermal DSC curves.
- XW's greatest gratitude goes to his parents, Mr Zhenan Wu and Mrs Haihong Lin, and his lovely wife, Mrs Huaying Jiang, who are always his quiet supporters, inspiring him and giving him confidence to complete his PhD study.
- Mr Mok Yoong Keong (1929–2000), KGK's respectful father, for his inspiration and major financial support that made KGK's PhD study possible.

1
Introduction to electroless copper and nickel–phosphorus (Ni–P) depositions

Abstract: Electroless copper deposition using formaldehyde as a reducing agent at 60 °C is widely used in commercial printed circuit board industries. However, formaldehyde, as a carcinogen, has high potential risk to the environment and the plating operators. Therefore, alternatives to formaldehyde used in electroless copper deposition have been proposed. Electroless nickel–phosphorus (Ni–P) deposits are widely used in various industries, in particular as protective and functional coatings because of their unique combination of material properties. Many of these material properties are adjustable through thermal processing when the amorphous structure of the deposits undergoes a crystalline transition process.

Key words: electroless copper, electroless nickel–phosphorus (Ni–P), formaldehyde, printed circuit board, thermal processing.

1.1 Electroless copper deposition

Electroless copper deposition is widely used in the printed circuit board industry for plating onto bare laminate to make non-conductive substrates conductive for further electroplating, and for plating through holes to connect circuits in different layers. The principle of electroless copper deposition is to use the chemical reaction between a reducing agent and a copper salt in a liquid solution so that the copper ion can be reduced to a copper atom. The reaction should be continuous so that sufficient copper can form a film and cover the substrate.

Currently, commercial electroless copper plating solution uses formaldehyde as the reducing agent. However, formaldehyde is a well-known carcinogen with a strong potential hazard to the environment and the staff working around an electroless copper deposition facility. To address this issue, several alternatives have been proposed, amongst which glyoxylic acid has very high potential. The other way to reduce the hazard of formaldehyde is to reduce the plating temperature so that there will be less formaldehyde fumes released from the plating solution to the air. However, before an alternative or a plating condition can be industrially applied, the properties of the deposits acquired in these conditions need to be tested and analysed.

1.2 Electroless nickel–phosphorus (Ni–P) deposition

Electroless nickel–phosphorus (Ni–P) deposits are produced from the electrochemical reduction of a nickel ion through the autocatalytic plating processes in the plating solution. The electroless plating process that led to the practical application of these deposits was first developed by Brenner and Riddell, and the first patent for it was awarded in 1950. Following this in the next decade, the electroless Ni–P deposits were mainly used for the interiors of railroad tankers carrying concentrated corrosive solutions. Because of various investigations over the years, understanding and enhancement of the unique material properties of electroless Ni–P deposits have become significant, and have broadened the deposits' application and role in varied industries, in particular for the use as protective and functional coatings. There is improvement in material properties, such as solderability, wear and corrosion resistances, and magnetic response through plating and heat-treatment processing. The major industries that have benefited from using electroless Ni–P deposits include the chemical, oil and gas, automotive, electronics and the aerospace industries. Today, most commercially used electroless Ni–P deposits are from a binary system that has a range of 3–14 wt% phosphorus content depending on their application roles. Some properties of the basic elements are given in Table 1.1.

The difference in microstructural characteristics of electroless Ni–P deposits has significant effect on the material properties (Li *et al.*, 2001; Taheri *et al.*, 2001; Ten and Chang, 2001), and is subject to the influence of crystallisation and phase transformation behaviour of deposit during thermal processing (Li *et al.*, 2001; Taheri *et al.*, 2001). Besides, it is evident that

Table 1.1 Elementary properties of phosphorus and nickel

Element	Phosphorus	Nickel
Chemical symbol	P	Ni
Atomic number	15	28
Atomic mass	30.973762	58.6934
Atomic radius (pm)	93	124.6
Electronic configuration*	$[Ne]3s^23p^3$	$[Ar]3d^84s^2$
Most common substance	P_4	Ni
Substance class	Non-metal	Metal
Crystal lattice	Cubic	Face centred cubic
Density at 25 °C (kg m^{-3})	1820	8902
Melting and boiling points (°C)	44 and 280	1453 and 2732
Electrical conductivity at 25 °C ($Ohm^{-1}m^{-1}$)	1.00×10^{-9}	1.46×10^{7}
Thermal conductivity at 25 °C ($Wm^{-1}K^{-1}$)	0.235	90.7

*$[Ne] = 1s^22s^22p^6$, $[Ar] = [Ne]3s^23p^6$.

the crystallisation and phase transformation behaviour is determined by the alloy compositions and thermal processing conditions, such as phosphorus content, heating rate and processing end temperature (Taheri *et al.*, 2001). As a result, the correlation between crystallisation and phase transformation behaviour of electroless Ni–P deposits, and those influencing factors, is of importance to the optimised design of specific material properties for targeted applications. However, despite being the subject matter of many previous investigations, there appear to be conflicting results regarding the microstructural and material properties, and crystallisation and phase transformation behaviour of electroless Ni–P deposits, under those mentioned influencing factors. These variations are likely caused by many other related factors, such as plating condition, material aspect of the substrate, and testing and measuring techniques.

1.3 How to plate the depositions in the laboratory

1.3.1 Electroless copper

Electroless copper deposits can be prepared in the laboratory by using a homemade electroless copper plating line or a mini electroless copper plating line. In the following sections, the ingredients of the plating solutions, procedures of the electroless copper plating, and other relevant essential solutions are described in detail.

Electroless copper plating solutions

In this book, in electroless copper plating solutions, either formaldehyde or glyoxylic acid is used as reducing agent. Each type of plating solution is further categorised into high concentration high temperature (HCHT), low concentration high temperature (LCHT), high concentration low temperature (HCLT) and low concentration low temperature (LCLT), among which, the high temperature plating solutions are operated at 60 °C, the common commercially used temperature, and the low temperature plating solutions are operated at 45 °C. There are eight types of deposits in total:

1. formaldehyde HCHT solutions (FHCHT);
2. formaldehyde LCHT solutions (FLCHT);
3. glyoxylic acid HCHT solutions (GHCHT);
4. glyoxylic acid LCHT solutions (GLCHT);
5. formaldehyde HCLT solutions (FHCLT);
6. formaldehyde LCLT solutions (FLCLT);
7. glyoxylic acid HCLT solutions (GHCLT);
8. glyoxylic acid LCLT solutions (GLCLT).

The detailed ingredients of all of the plating solutions are listed in Table 1.2. The ingredients of the solutions are developed for the purpose of fair comparison among the deposits. In these plating solutions, cupric sulphate is used as the source of the copper ion, and EDTA as the chelate agent or complex agent.

The formaldehyde plating solutions are prepared using the following procedure:

1. Mix EDTA with deionised water. The amount of the deionised water is the same as the amount of the plating solution you want to acquire. EDTA is very hard to dissolve in water, so white precipitates of EDTA will remain in the deionised water.
2. Continuously add sodium hydroxide (NaOH) to the solution, stir, until the white precipitates disappear. The solution is now disodium-EDTA solution.
3. Carefully add cupric sulphate gradually to the solution in several separate goes, making sure that the cupric sulphate is fully dissolved before adding more. Adding cupric sulphate in one go will cause a large amount of blue precipitates (cupric sulphate pentahydrate) to appear in the solution which will take longer to fully dissolve. In this process,

Table 1.2 Ingredients of electroless copper plating solutions

Solution	Cupric sulphate ($CuSO_4$)	EDTA‡	Formaldehyde (HCHO)	Glyoxylic acid (CHOCOOH)
Formaldehyde high concentration high temperature (FHCHT)*	0.047 M**	0.11 M	0.05 M	
Formaldehyde low concentration high temperature (FLCHT)*	0.03 M	0.05 M	0.10 M	
Glyoxylic acid high concentration high temperature (GHCHT)*	0.047 M	0.11 M		0.05 M
Glyoxylic acid low concentration high temperature (GLCHT)*	0.03 M	0.05 M		0.10 M
Formaldehyde high concentration low temperature (FHCLT)+	0.047 M	0.11 M	0.15 M	
Formaldehyde low concentration low temperature (FLCLT)+	0.03 M	0.05 M	0.20 M	
Glyoxylic acid high concentration low temperature (GHCLT)+	0.047 M	0.11 M		0.15 M
Glyoxylic acid low concentration low temperature (GLCLT)+	0.03 M	0.05 M		0.20 M

*Operated at 60 °C.
+Operated at 45 °C.
**1 M = 1 mol L^{-1}.
‡EDTA: ethylenediaminetetraacetate acid.

white precipitates of EDTA might finally appear due to the change of the pH value of the solution after adding cupric sulphate.

4. Add sodium hydroxide until the white precipitates disappear.
5. Use a pH probe and indicator to monitor the pH value of the solution. Add sodium hydroxide gradually; stir until the pH value of the solution reaches 12.5 for high temperature solutions or 13.0 for low temperature solutions. The plating solutions can now be stored and are ready to use.
6. Prior to use, heat the solution up to the operating temperature and add formaldehyde solution into the plating solution and the reaction will initiate on the catalysed surface immersed into the solution. The volume of the formaldehyde solution needed is very small compared to the plating solution, so the change of the concentration of the ingredients in the solution can be ignored.

The glyoxylic acid solutions are prepared with a slightly different procedure:

1. The same first step as for formaldehyde solutions.
2. Continuously add potassium hydroxide (KOH) in the solution; stir until the white precipitates disappear. The solution is now dipotassium-EDTA solution. By using potassium hydroxide in glyoxylic acid solution, the Cannizzaro reaction rate is reduced, and the mechanical properties are improved. The Cannizzaro reaction involved in electroless copper plating using glyoxylic acid as reducing agent consumes additional hydroxyl and glyoxylic acid (Shacham-Diamond, 2000):

$$2HOOCCOH + 2OH^- \rightarrow C_2O_4^{2-} + HOCH_2COOH + H_2O \qquad 1.1$$

3. Carefully add cupric sulphate into the solution in several separate goes.
4. Add potassium hydroxide until the white precipitates disappear.
5. Continue to add potassium hydroxide until the solution reaches the same pH value level as for the formaldehyde solutions described earlier. During this process, glyoxylic acid is added into the plating solution. The timing of adding glyoxylic acid is not critical, but the final pH values of the plating solutions need to be controlled with potassium hydroxide. The reason for adding glyoxylic acid in the solution preparation procedure but not prior to the plating is that even a small amount of glyoxylic acid can dramatically lower the pH value of the plating solutions, due to which the reaction may not be initiated in the setup pH value. The plating solutions need to be stored in a cool dark place due to the presence of the reducing agent, to prevent reaction commencing. It is recommended to use the fresh solution.
6. In case of the immersion heating unit being used in the plating solution,

it is necessary to add glyoxylic acid to the plating solution after it has been heated up to the required temperature and then use condensed potassium hydroxide solution to adjust the pH of the plating solution. It is also needed to make sure that the heating unit is not in heating status while doing so; otherwise the high temperature areas surrounding the heating unit may trigger the reaction and cause decomposition.

For all the plating solutions, air bubbles, in which oxygen is acting as the stabiliser to maintain the stability of the plating solution, are continuously injected into the plating solution through a glass pipe during plating. The stabiliser reduces the chance of the decomposition of the plating solution during plating.

Electroless copper plating procedure

The most effective catalyst solution is palladium–tin colloid solution. It is made from the solutions of palladium chloride, tin chloride and hydrochloric acid, but the procedure of preparing the solution may affect the catalyst efficiency of the final catalyst solution. The most commonly used preparation process can be summarised as below.

In order to get 1 litre condensed catalyst solution, firstly, mix 2 g palladium chloride, 200 ml 37% hydrochloride acid and 400 ml deionised water together and stir until palladium chloride is completely dissolved. Then, add 4 g stannous chloride in the previous solution and stir again until it is completely dissolved. Lastly, add 96 g stannous chloride and 400 ml deionised water into the previous solution and stir until the stannous chloride is completely dissolved. This strong acid solution can be stored for several months. When using, mix 15% of the previous solution (condensed catalyst solution) and 15% hydrochloride acid (37%) together, and balance the rest with deionised water, heating up to 50–60 °C for better catalytic effect.

For all the plating solutions and substrates, the electroless copper plating follows the same procedure. Unless otherwise mentioned, careful rinsing in deionised water is carried out between each step. The whole process includes:

1. *Cleaning and conditioning*: This step removes the oil or grease on the surface of the substrates when they are made, packaged or cut. 5% aqueous sodium hydroxide solution is used in this step. The operating temperature is up to 60 °C for a strong cleaning effect. Some of the substrates, like stainless steel sheet substrates with an oily surface are left in the solution at around 40 °C overnight for thorough cleaning. Substrates after this treatment should be able to be fully wetted with water film on the entire surface, indicating a well-cleaned surface.
2. *Etching*: This step micro-etches the surface of the substrates, especially

epoxy substrates, to prepare the surface of the substrate for better adherence. The chemical used in this step is a solution with 5% sulphuric acid and 5% hydrogen peroxide. For stainless steel substrates, etching is not essential but it helps to remove residues on the substrates that the previous cleaner might not be able to do. The operation temperature for this step is around 60 °C.

3. *Neutralising and catalyst promotion*: This step removes the etchant that might remain on the substrate from the previous step and promotes the surface of the substrates in a better condition for a uniform catalyst film. The chemical and operating temperature used in this step are the same as those in Step 1.
4. *Pre-dipping*: This step protects the expensive catalyst solution to be used in the next step from decomposition. The chemical used in this step is 5% hydrochloric acid solution operated at room temperature.
5. *Catalysing*: Without rinsing, the substrates are then immersed in the catalyst solution The chemical used in this step is a palladium–tin colloidal solution. The ingredients, preparation and use of this solution are presented above. The catalyst solution will cover the surface of the substrate with evenly distributed palladium–tin colloids, each of which has a palladium-rich inner core surrounded by a large amount of hydrous tin(IV) (Sn^{4+}) chlorides. The overall colloid is negatively charged, making it capable of triggering the plating reaction on the surface. The catalyst solution is operated at around 60 °C for a strong catalyst effect. All substrates are immersed in the catalyst solution for 15–20 minutes in this step.
6. *Electroless copper plating*: The plating solutions are heated up to the setup temperature, after which the catalysed substrates, after the final rinsing, are immersed into the plating solutions for the required period.
7. *Anti-tarnish*: After the plating step and deionised water rinsing, the substrates, with copper deposits, are immersed in 5% sulphuric acid solution at room temperature for two minutes. This will enable a very thin layer of protective film to form on the surface of the deposits, preventing them from further decomposition on contact with air. The samples are then finally rinsed and dried.

The procedure described above is summarised in Table 1.3.

1.3.2 Electroless nickel–phosphorus

The deposition is easily done on both sides of a mild steel sheet substrate using the formulation of plating solution; one example as listed in Table 1.4. Surface preparation of the substrate is not carried out if the mild steel sheet is smooth and clean. During the plating process, the mild steel substrate is

Table 1.3 Procedure of electroless copper plating

Step	Function	Chemicals and conditions
Cleaning and conditioning	To clean the surface of the substrates	5% NaOH solution 60 °C for 20 minutes or 40 °C for overnight to days depending on the condition of the substrates
Etching	To prepare the surface of the substrates for better adhesion; to further remove residue on the surface	5% H_2SO_4 and 5% H_2O_2 solution 60 °C for 20 minutes
Neutralising and catalyst promotion	To remove etchant and prepare the surface of the substrates for catalyst	5% NaOH solution 60 °C for 5 minutes
Pre-dipping	To protect the expensive catalyst solution	5% HCl solution room temperature, dip
Catalysing	To catalyse the surface of the substrates	$PdCl_2$ $SnCl_2$ colloidal solution 60 °C for 15–20 minutes
Plating	To electrolessly plate copper on the substrates	Electroless copper plating solution 45 °C or 60 °C for various time intervals
Anti-tarnishing	To protect the deposits	5% H_2SO_4 solution room temperature for 2 minutes
Rinsing	To remove residual solutions on the surface of the substrates; to be carried out between each of the steps above except between pre-dipping and catalysing	Deionised water room temperature for 2–5 minutes

Table 1.4 Solution composition and plating temperature for electroless deposition of Ni–16%P

Constituents of solution	Quantity (g L^{-1})
Nickel sulphate ($NiSO_4 \cdot 6H_2O$)	40
Sodium hypophosphite ($NaH_2PO_2 \cdot H_2O$)	25
Succinic acid ($C_4H_6O_4$)*	50
Temperature	85–95 °C

*Complexing agent used to prevent spontaneous decomposition of plating solution.

totally immersed in the plating solution, which is continuously stirred for better agitation (Fig. 1.1). Evaporation from the plating solution is condensed and directed back to the solution. After deposition, for example for one hour, the sample is lifted out from the plating solution and is rinsed using clean water.

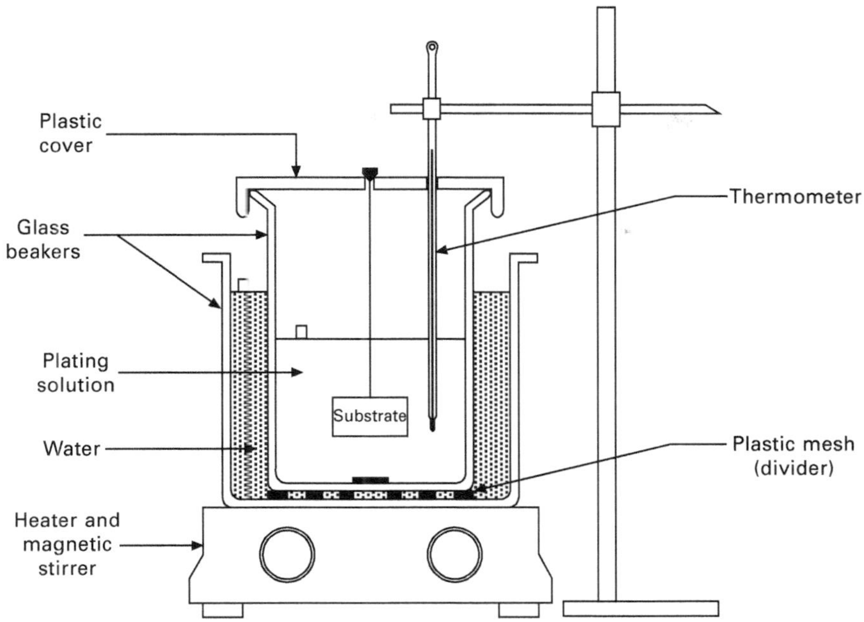

1.1 Apparatus used in the electroless deposition.

1.4 Research objectives

1.4.1 Electroless copper

In Part I of this book, the main objectives will be focused on the development of the plating solutions for glyoxylic acid based on the commercial formaldehyde plating solutions used at both high and low temperatures. Various properties of the deposits acquired from these plating solutions will be examined:

- the electroless copper plating solutions using glyoxylic acid as the alternative to formaldehyde to be used at normal plating temperature (60 °C), and the plating solutions using glyoxylic acid or formaldehyde as reducing agent at lower plating temperature (45 °C);
- the surface morphology evolution and the inner structures of the deposits from different plating solutions;
- the crystal structure and the surface residual stress of the deposits;
- an atomic model for the special surface morphology seen on some of the deposits, and an atomic simulation model to simulate the process of forming such morphology;
- the adhesion strength of the deposits on the epoxy substrates;
- the electrical properties of the deposits.

1.4.2 Electroless Ni–P

Because the amorphous structure of electroless Ni–P deposits is thermally unstable and it could undergo crystalline transition at moderately high temperatures, at about 400 °C, most material properties of the deposits are adjustable through thermal processing. As a result, heat-treatment processes have been commonly used by industries to acquire targeted material properties for meeting specific performance and requirements. Because the material properties of electroless Ni–P deposits could be affected by the crystallisation behaviour during thermal processing, it is important to understand the correlation of the behaviour with its influencing factors, namely the deposit compositions and thermal processing conditions. As the crystallisation behaviour of electroless Ni–P deposits is characterised by the nucleation and growth processes, it is also useful to investigate the crystallisation kinetic of the processes for better understanding of the behaviour. Besides, the evolution of hardness for the electroless Ni–P deposits during thermal processing is of interest, because the hardness parameter has a direct relationship with the wear and abrasion resistance, which is a vital property of the deposits.

The crystallisation behaviour of electroless Ni–P deposits in relation to the effects of deposit compositions and thermal processing conditions can normally be analysed using physical and statistical approaches. However, it would be useful if the analysis work could be performed from a reliable statistical model as this has advantages in terms of the cost and time consumed. Nevertheless, the availability of a reliable database is critical for this option. Because the material properties of electroless Ni–P deposits are dependent on their microstructural characteristics, development of an advanced computer-based model to predict the crystallisation and phase transformation kinetics is highly desirable given that reasonable estimation could be made for the kinetic interpretations, which is of use for the optimisation design of specific material properties.

Therefore, for this book, the following subjects are considered:

- the crystallisation and phase transformation behaviour of electroless Ni–P deposits at different phosphorus content and heating rates;
- the corresponding non-isothermal crystallisation kinetics of electroless Ni–P deposits based on a theoretical approach using a computer-based model;
- a computer-based statistical model for simulation and prediction of the influence of alloy composition, heating rate and processing parameters on the crystallisation temperatures of the Ni–P based amorphous alloys;
- the hardness evolution of the electroless Ni–P deposits in relation to the deposit phosphorus contents and thermal processing temperatures, and empirical model to interpret the kinetic energy of increased hardening effects after the isothermal heat treatment.

1.5 Structure of the book

1.5.1 Part I: Electroless copper depositions

Part I describes the surface morphology of the electroless copper deposit and the atomic scale modelling. This covers the surface morphology evolution, the crystal structure and the surface residual stress, adhesion strength, and electrical property of the deposits. The modelling and simulation include the atomic scale model of the diamond pyramid structures found on the low temperature deposits and the molecular dynamics simulation of the relaxation of the structure based on the atomic model, and the growth of the structure. The Part I chapters are summarised here.

- Chapter 2 describes the surface morphology evolution of the deposit from different plating solutions and shows the high-resolution images of the surface of the deposits. The different evolution processes are compared against the different plating conditions.
- Chapter 3 shows the inner structures of the electroless copper deposits. The image processing method used to calculate the void fraction from the over-etched cross-section images is described. The void fraction of the deposits is compared with the simulation-calculated value.
- Chapter 4 gives an account of the crystal structures of the electroless copper deposits and the surface residual stress of the deposit. The detailed data processing procedure is given.
- Chapter 5 presents the atomic scale model of the diamond pyramid structures that are discovered in some low temperature electroless copper deposits. The mechanism of forming such structures including the effect of the surface stress is discussed.
- Chapter 6 presents the simulation using the molecular dynamics method. The simulation includes the relaxation of the diamond pyramid structures with different sizes and features built based on the atomic model, and the growth process of the diamond pyramid structure under different conditions. The radial distribution functions from the simulation and those calculated from the X-ray diffraction are compared.
- Chapter 7 analyses the adhesion strength between the deposit and the substrate. A computer program is shown to acquire the pull-off parameter from the photos of the fractured surfaces. The possible failure modes when using the direct pull adhesion testing method are discussed.
- Chapter 8 gives the electrical property of the electroless copper deposits, specifically resistivity. Apart from that, a geometric correction method is demonstrated by calculating the length of a digital curve as the cross-section projection of the substrate surface.
- Chapter 9 reviews the application of the electroless copper deposition.

1.5.2 Part II: Electroless nickel–phosphorus depositions

This part describes the characterisation and modelling of the evolution of microstructure during thermal processing of electroless Ni–P depositions. The scope covers the kinetics and behaviour of crystallisation and phase transformation processes, and the evolution of hardness in relation to the deposit phosphorus contents as well as isothermal processing temperatures. Modelling of the kinetics and temperatures for crystallisation and phase transformation processes, and the kinetic energy of increased hardening effects, will also be described. The Part II chapters are summarised below.

- Chapters 10 and 11 detail crystallisation and phase transformation behaviour of the electroless Ni–P deposits with high, medium and low phosphorus contents under continuous heating.
- Chapter 12 details the modelling of non-isothermal crystallisation kinetics of the electroless Ni–P deposits using the Johnson–Mehl–Avrami theory. The degrees of transformation corresponding to the major crystallisation processes of the electroless Ni–P deposits are simulated.
- Chapter 13 is concerned with the artificial neural network modelling of crystallisation temperatures for the Ni–P based amorphous alloys. The computer model is used for simulation and prediction of the influence of alloy composition, heating rate and processing parameters.
- Chapter 14 presents the analyses of hardness evolution for the electroless Ni–P deposits in relation to the effects of phosphorus content and isothermal heat-treated temperature. A model based on the empirical concept of kinetic strength is used to interpret the kinetic energy of increased hardening effects after the isothermal processing.
- Chapter 15 gives a review covering various applications of electroless Ni–P deposits. Other widely used engineering deposits like electroplated nickel and hard chromium are included and briefly described. The advantages, limitations, developments and future challenges of electroless Ni–P deposit are outlined, and the industrial use of the heat-treatment process for the enhancement of material properties is discussed briefly.

1.6 References

Li H X, Wang W J, Chen H Y and Deng J F (2001), 'Surface morphology and electronic state characterization of Ni–P amorphous alloy films', *J Non-Cryst Solids*, **281**, 31–8.

Shacham-Diamond Y Y (2000), 'Electroless copper deposition using glyoxylic acid as reducing agent for ultralarge scale integration metallization', *Electrochem Solid State Lett*, **3**, 279–82.

Taheri R, Oguocha I N A and Yannacopoulos S (2001), 'Effect of heat treatment on age hardening behaviour of electroless nickel–phosphorus coatings', *Mater Sci Technol*, **17**, 278–84.

Ten S S and Chang F Y (2001), 'Electrical resistivity of electroless nickel coated carbon fibers', *Thin Solid Films*, **388**, 143–9.

Part I

Electroless copper depositions

2
Surface morphology evolution of electroless copper deposits

Abstract: This chapter shows the surface morphology evolution of the deposits from different plating solutions, and reveals the different growing process of the deposits. Diamond pyramid structures are discovered on some low temperature deposits. These diamond pyramid structures consist of pure copper. The morphology of the electroplated copper film on the substrate side includes numerous small voids scattering across the surface of the deposit, with relatively large cracks at some portions. No voids are in the deposit on the solution side. In the electroless copper deposits, the granular particles developed on the heat-treated deposit are the most intensive.

Key words: electroless copper, glyoxylic acid, surface morphology, plating, scanning electron microscopy (SEM).

2.1 Introduction and surface morphology of the substrate

The objective of this chapter is to examine the surface morphology evolution of the electroless copper deposits during the plating process. With the changes of the ingredients in the plating solution with different reducing agents and the plating condition, the deposit may grow on the epoxy substrate in a different way, which may further affect the properties of the deposits. The chemical composition of the deposits is analysed.

The scanning electron microscopy image in Fig. 2.1 shows that the substrate before electroless copper plating has a rough, eroded like surface with many dimples, providing good adhesion of the subsequent copper film on the substrate.

2.2 Formaldehyde high temperature solution deposits

This section and the next five will show the surface morphology of eight typical electroless copper deposits. The names for the eight deposits originate from their plating solution compositions as given in Table 1.2.

With 5 minutes plating in the high concentration solution (Fig. 2.2(a)), only a small amount of the deposit appears on the substrate and the surface morphology of the substrate is still clearly presented. After 15 minutes

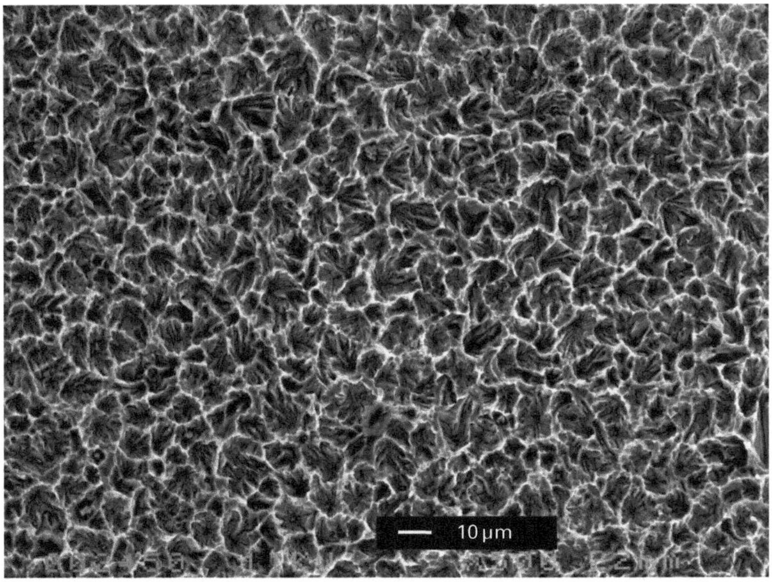

2.1 The surface morphology of FR4 epoxy board substrate. FR4, an abbreviation for Flame Retardant 4, is a type of material used for making a printed circuit board (PCB).

plating (Fig. 2.2(b)), a large amount of the deposit appears both inside and at the edge of the dimples, indicating a similar plating rate in both parts. The deposit starts to cover the dimples but the edges of the dimples can still be distinguished. For the 60-minute deposit (Fig. 2.2(c)), the deposit both inside and at the edge of the dimples continues to build up at a similar rate and forms an uneven surface. The substrate is fully covered with copper deposits, but showing some signs of morphology of the substrate. For the 120-minute deposit (Fig. 2.2(d)), there is a much smoother surface. Copper deposits both inside and on the edge of the dimples fuse together, fully covering the surface of the substrate. Some features of the substrate can still be seen from the surface of the deposit. Some gaps are visible between deposit clusters.

After a circuit is made, a photo resistant coating is required to protect the circuit. This further coating process involves a curing stage at a temperature of around 140 °C. The heat treatment results in no significant change to the surface of the deposit (Fig. 2.2(e)).

After 5 minutes plating (Fig. 2.3(a)), in the lower concentration solution, the deposit starts to build up more from the inside of the dimples and the edge of the dimples is clear and sharp. After 15 minutes plating (Fig. 2.3(b)), the copper deposit is on the edge of the dimples, as well as inside. However, compared to Fig. 2.2(b), less deposit is on the edge of the dimples and copper

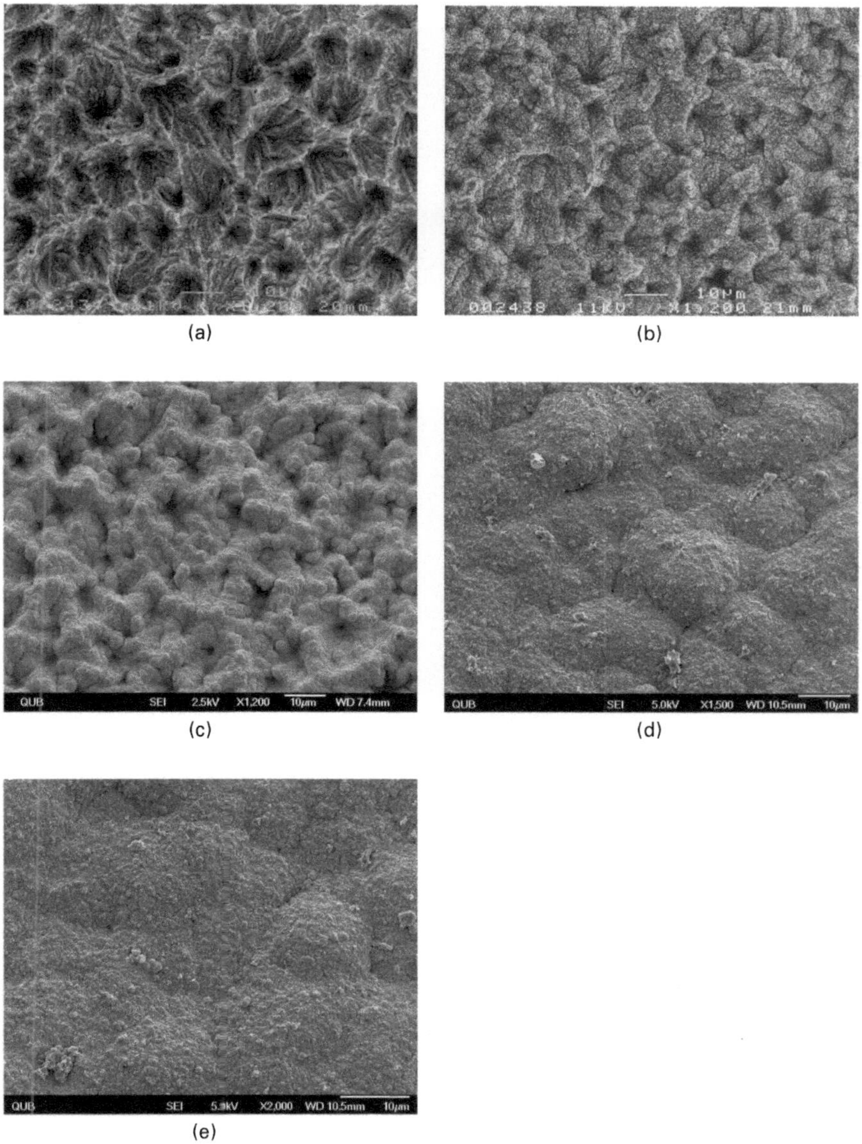

2.2 The surface morphology evolution process of separate electroless copper deposits from the formaldehyde high concentration high temperature solution with (a) 5 minutes, (b) 15 minutes, (c) 60 minutes and (d) 120 minutes plating time, and (e) 120 minutes with 140 °C heat treatment.

tends to build up more from the inside of the dimples. For the 60-minute deposit (Fig. 2.3(c)), the whole surface of the substrate is covered with copper deposit. Less features of the dimples are shown from the surface except for

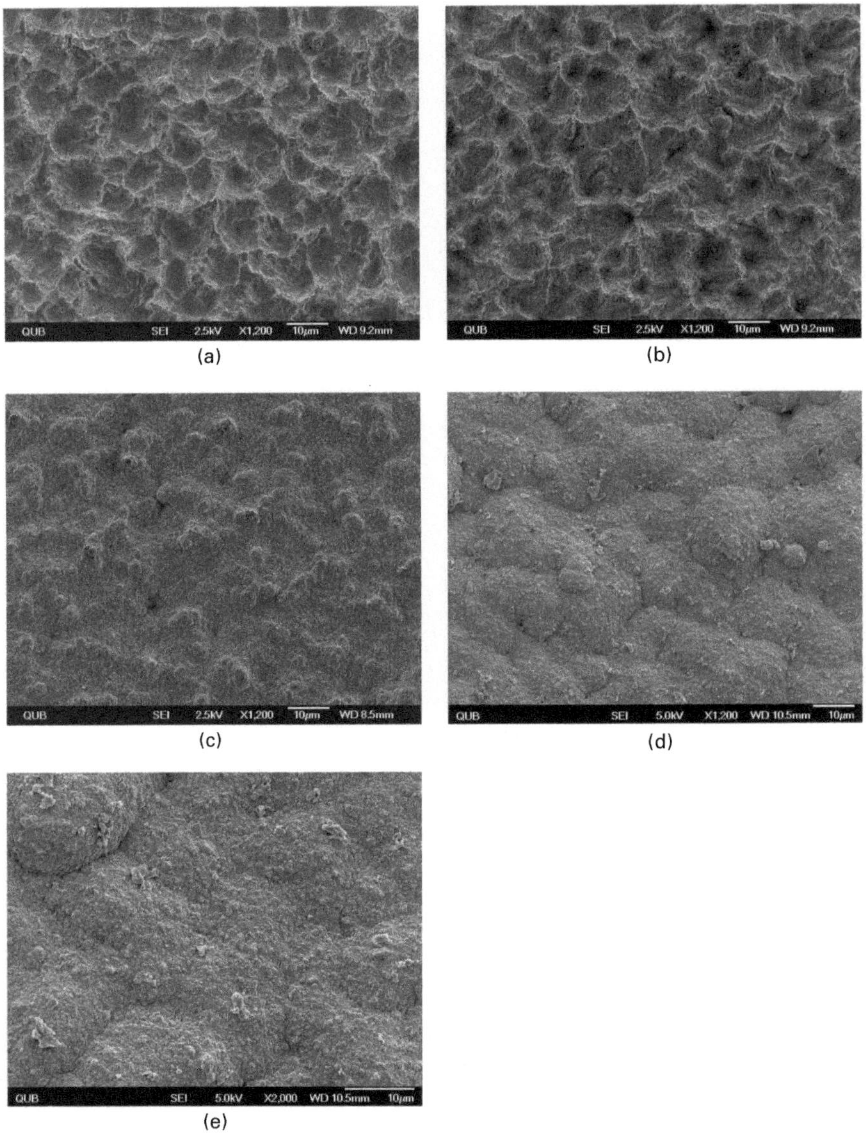

2.3 The surface morphology evolution process of separate electroless copper deposits from the formaldehyde low concentration high temperature solution with (a) 5 minutes, (b) 15 minutes, (c) 60 minutes and (d) 120 minutes plating time, and (e) 120 minutes with 140 °C heat treatment.

some sharp edges. Compared with Fig. 2.2(c), the surface of the deposit is much smoother, when plated using the low concentration plating solution. For the 120-minute deposit (Fig. 2.3(d)), the substrate is fully covered with few features visible on the surface as a result of the substrate morphology. One such feature is some gaps between clusters of copper deposit. The overall deposit surface is smooth and integrated. The 140 °C treatment (Fig. 2.3(e)) does not make a significant difference.

2.3 Glyoxylic acid high temperature solution deposits

For the deposit of 5-minute plating in the high concentration solution (Fig. 2.4(a)), the surface shows basically similar features to those of the surface of the epoxy substrate. Deposit is inside the dimples but little is at the edge of the dimples. For the 15-minute deposit (Fig. 2.4(b)), the copper deposit starts to fill in the dimples. The edge of the dimples is not covered with solid deposit inside. Compared with Fig. 2.2(b), by using glyoxylic acid, the copper deposit tends to grow more from the inside of the dimples. For the 60 minute deposit (Fig. 2.4(c)), the copper deposit has grown further from the inside of the dimples, overwhelming the edge and fusing into large and smooth clusters. Some of the edges of the dimples are covered by these clusters. On the 120-minute deposit (Fig. 2.4(d)), a smooth surface with relatively rough grains is obtained. The round particles are the random copper deposits in the solution attached to the surface of the sample. The large amount of hexagon shaped grains are around 2 microns in diameter. The increased size of grains may be due to the reduction of the concentration of the chemicals in the plating solution in the last stage of plating, which reduces the plating rate and enables the grains to grow more extensively. Some voids on the surface are possibly due to the mismatch at the joint point among clusters during the growth. The heat treatment (Fig. 2.4(e)) does not cause significant change in surface morphology against the as-plated one.

Although the concentration of the copper salt in the solution is low, due to the addition of the reducing agent and high plating temperature, the low concentration plating solution decomposes very easily. Copper is plated on the surface of the substrate but also everywhere in the solution. The consumption of the chemicals due to unwanted deposition greatly affects the plating rate. At 5 minutes, copper is on both the inside and at the edge of the dimples (Fig. 2.5(a)).

After 15 minutes plating (Fig. 2.5(b)), slightly more copper deposit builds up from inside of the dimples providing more coverage on the surface of the substrate. However, due to the reduction of the plating rate mentioned above, the difference between the surface of the 15-minute and 5-minute deposit is not significant. In the 60-minute deposit (Fig. 2.5(c)), more deposit is on the surface but it is not fully covered by copper. Dimples can still be observed. The loose copper deposit debris may come from the individual

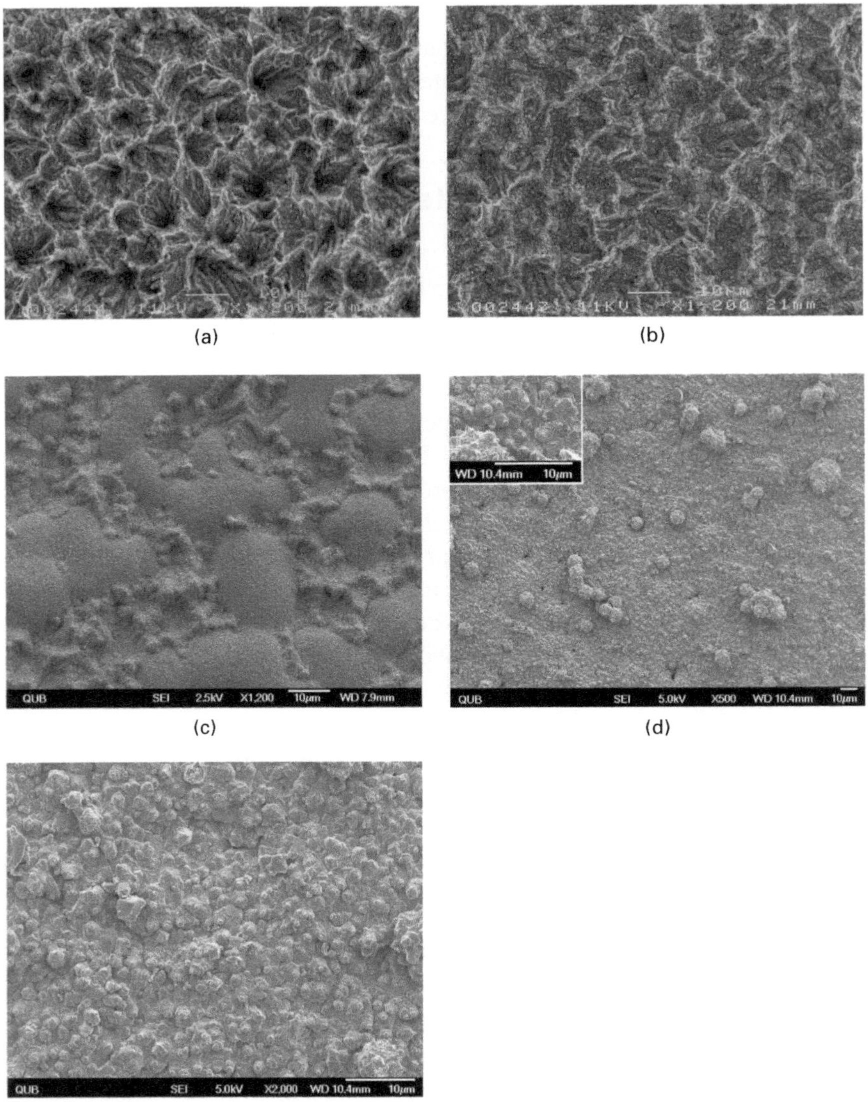

2.4 The surface morphology evolution process of separate electroless copper deposits from the glyoxylic acid high concentration high temperature solution with (a) 5 minutes, (b) 15 minutes, (c) 60 minutes and (d) 120 minutes plating time, and (e) 120 minutes with 140 °C heat treatment.

deposit particles in the solution. The decomposition of the plating solution seriously affects the plating rate, resulting in less deposit plated on the substrate. The plating reaction ends in around 60 minutes.

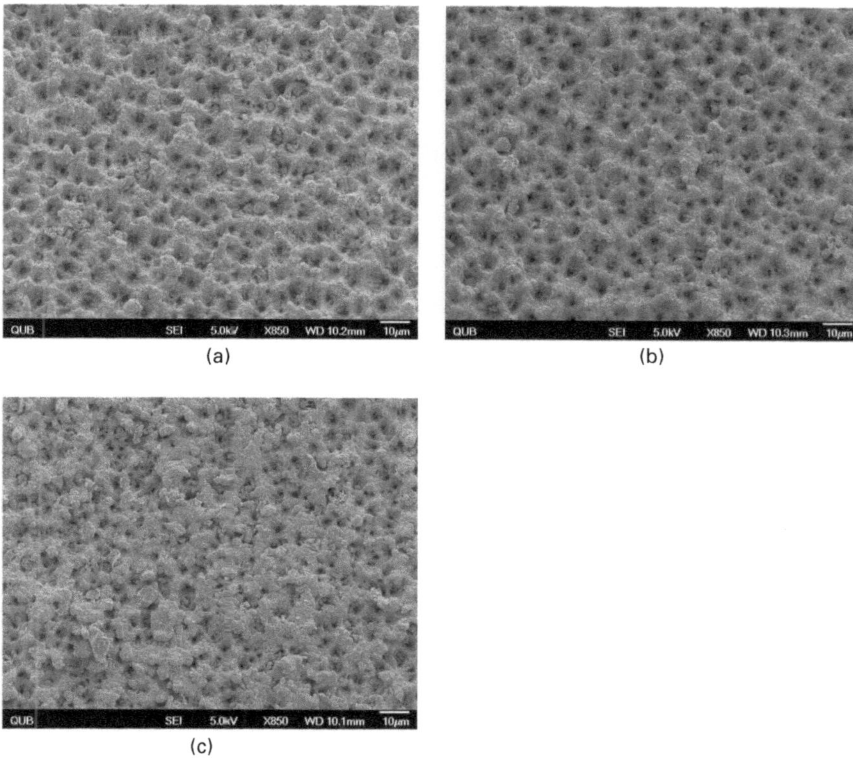

2.5 The surface morphology evolution of separate electroless copper deposits from the glyoxylic acid low concentration high temperature solution with (a) 5 minutes, (b) 15 minutes and (c) 60 minutes plating time.

2.4 Formaldehyde high concentration low temperature (FHCLT) solution deposit

After 5 minutes plating (Fig. 2.6(a)), copper is on both the edge and the inside of the dimples of the substrate, indicating a similar plating rate at both parts of the substrate. The result of this is that deep holes form originating from the dimples cn the surface of the substrate (magnified image in Fig. 2.6(a)).

In Fig. 2.6(b), the dimples of the substrate are nearly covered by the 15-minute copper deposit. Compared to the same plating time from other solutions, Fig. 2.6(b) indicates a fast plating rate in this solution. Spike shaped grain clusters are present (magnified image on the top left of Fig. 2.6(b)), as are some deep voids on the surface of the deposit. The voids may come from the deep holes in Fig. 2.6(a) as a result of the spike shaped clusters. The structures create voids between each other during their growing process

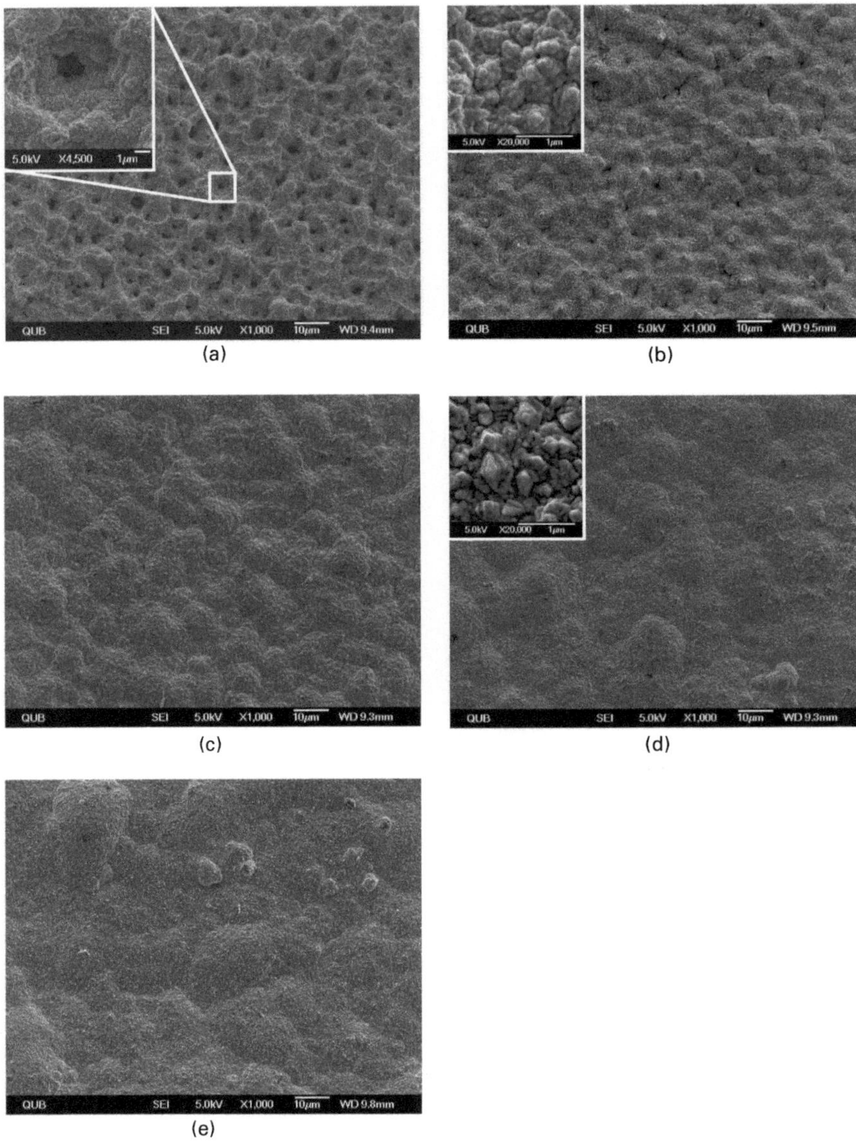

2.6 The surface morphology evolution process of separate electroless copper deposits from the formaldehyde high concentration low temperature solution with (a) 5 minutes, (b) 15 minutes, (c) 60 minutes and (d) 120 minutes plating time, and (e) 120 minutes with 140 °C heat treatment.

and prevent the transportation of the chemicals in the mass solution area to these voids. Once the concentration of the chemicals in these voids reaches a certain low level, the plating reaction is stopped and the voids are left unfilled.

The deposit surface is relatively smooth and covered, with few features from the substrate, after 60 minutes plating time (Fig. 2.6(c)). Voids in Fig. 2.6(b) are filled or covered with electroless copper deposits. However, traces of the substrate are still visible on the rough surface of the deposit.

The deposit surface is very smooth after 120 minutes plating. The morphology features of the substrate are completely covered by copper deposit. There are some highly symmetric multi-layer diamond pyramid structures measuring around 0.2 μm × 0.5 μm on the surface (magnified image on the top left of Fig. 2.6(d)).

Energy dispersive X-ray analysis shows that in the electroless copper deposit with the diamond pyramid structures on the surface, the surface is virtually pure copper. There is a small amount of oxygen apparently present on the surface, which is possibly due to very small amount of oxidation or the residual oxygen in the evacuated microscope chamber. Since no significant oxygen or other element is present, the diamond pyramid structure on the deposits consists of pure copper with no contribution from other elements.

Compared with Fig. 2.6(d), no significant change happens on the heat treated deposit (Fig. 2.6(e)). The diamond pyramid structures are still on the heat treated sample. This means that the heat treatment has no significant effect on the surface morphology of the deposit.

2.5 Formaldehyde low concentration low temperature (FLCLT) solution deposit

After 5 minutes, plating initiates from both inside and the edge of the dimples. A network deposit structure is inside the dimple of the substrate (magnified image on the top left of Fig. 2.7(a)). Pores are left among this network featured deposit and may further lead to a porous structure in the final deposit.

After 15 minutes plating (Fig. 2.7(b)), the copper deposit starts to join together and cover the surface of the substrate. Pores are developed from the hollow structures inside the dimples. Once the voids in between deposits form, it is hard for the chemicals in the bulk plating solution to be transported to the local areas. The plating reaction stops once the concentration of the chemicals reaches a certain level and the voids remain. There is a small amount of small sized multi-layered diamond pyramid structures (magnified image on the top left of Fig. 2.7(b)). This means that the diamond pyramid structures start to form in the very early stage of the plating process in the solution.

After 60 minutes plating (Fig. 2.7(c)), the copper deposit clusters measuring around 10 μm with deep gaps in between each other form on the surface of the deposit. These deep gaps are developed from the pores as shown in Fig. 2.7(b), where the local plating rate is relatively low because of the low

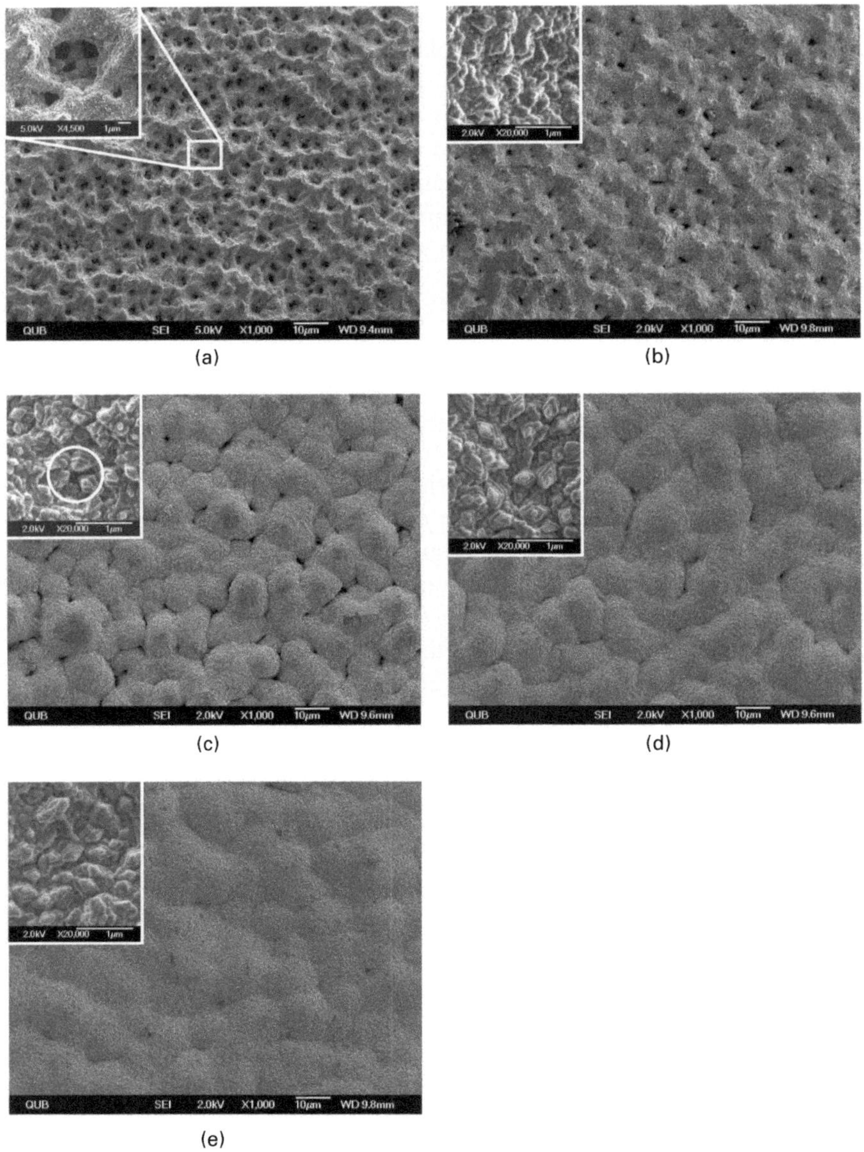

2.7 The surface morphology evolution process of separate electroless copper deposits from the formaldehyde low concentration low temperature solution with (a) 5 minutes, (b) 15 minutes, (c) 60 minutes and (d) 120 minutes plating time, and (e) 120 minutes with 140 °C heat treatment.

chemical concentration and transportaton rate. Diamond pyramid structures are bigger in size and with clearer features (magnified image on the top left of Fig. 2.7(c)) than those in Fig. 2.7(b). A deep void forms among the diamond pyramid structures in the centre, marked with the white circle in the magnified image. The void may develop into a pore inside the deposit with further plating.

After 120 minutes plating (Fig. 2.7(d)), the copper deposit clusters on Fig. 2.7(c) join together into bigger clusters, forming a smoother surface. Fewer deep gaps are on the surface. Fine big multi-layered diamond pyramid structures, measuring around 0.2 µm × 0.4 µm in diagonals, with a clear step feature on their side form on the deposit surface (magnified image on the top left of Fig. 2.7(d)). Closer observation shows that some of the diamond pyramid structures are singles, isolated from others, while others may have two or more growing adjacently, forming diamond pyramid chains. The orientation of the base plane of these diamond pyramid structures is random but their tips are mainly upward.

Compared with Fig. 2.7(d), no significant change happens on the heat treated deposit (Fig. 2.7(e)). A large amount of fine layered diamond pyramid structures still exist on the deposit surface.

2.6 Glyoxylic acid high concentration low temperature (GHCLT) solution deposit

After 5 minutes plating (Fig. 2.8(a)), similar to Fig. 2.7(a), deposits start to grow on both inside and the edge of the dimples of the substrate. The pores on the surface of the deposit are developed from the dimples on the substrate (magnified image in Fig. 2.8(a)). Deposits and voids can be found deep inside of these pores and may lead to porous structure in the deposit after further plating.

After 15 minutes plating (Fig. 2.8(b)), the deposit starts to form big clusters and cover the dimples of the substrate but the feature of the substrate still shows. Deep small pores are developed from the dimples on the surface of the deposit. The plating reaction inside these pores may be greatly restricted due to poor transportation of the chemicals in the narrow local areas. There are very small fine grains, and some layered structures.

The surface becomes very flat and smooth after 60 minutes plating (Fig. 2.8(c)), with few features of the substrate. The dimples are completely covered by the deposit. The magnified image on the top left of Fig. 2.8(c) shows only small grains on the surface of the deposit. There is no obvious diamond pyramid structure similar to those shown in Figs 2.6 and 2.7. It is easy for the small grains to fill the voids and the gaps within deposit, resulting in a flat and smooth coating surface on a rough surface like the epoxy board substrate.

The surface is extremely flat and smooth after 120 minutes plating (Fig. 2.8(d)). The rough surface of the substrate is completely covered. The same as Fig. 2.8(c), small grains are shown in the magnified image on the top left

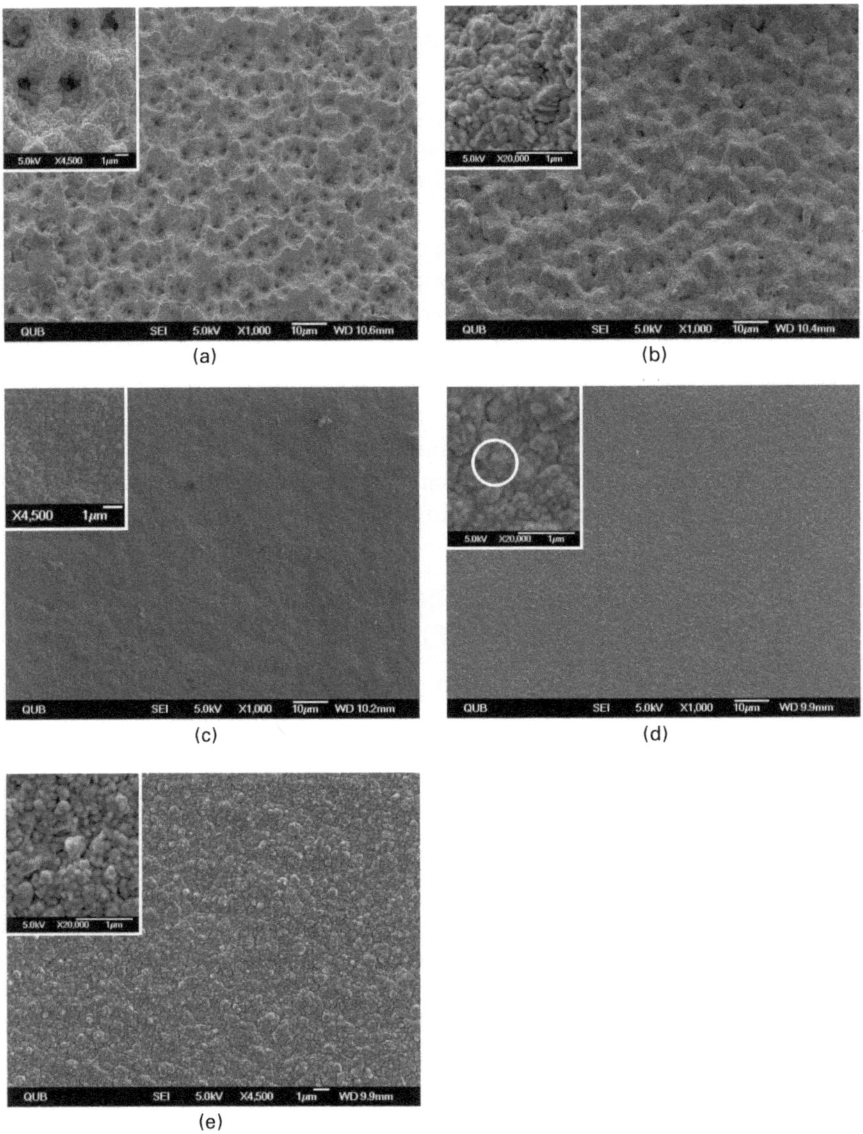

2.8 The surface morphology evolution of separate electroless copper deposits from the glyoxylic acid high concentration low temperature solution with (a) 5 minutes, (b) 15 minutes, (c) 60 minutes and (d) 120 minutes plating time, and (e) 120 minutes with 140 °C heat treatment. The flat, pressed surfaces in the image in (a) are due to the damage to the coated sample by tweezers.

of Fig. 2.8(d), with only a few indistinct diamond pyramid structures, one marked with the white circle in the image.

Compared with Fig. 2.8(d), no significant change happens after heat treatment (Fig. 2.8(e)). The grains remain in small size with no clear diamond pyramid structures, as shown in the magnified image.

2.7 Glyoxylic acid low concentration low temperature (GLCLT) solution deposit

After 5 minutes plating (Fig. 2.9(a)), the copper deposit is on both the edge and the inside of the dimples. Inside the dimple, the copper deposit shows a network shaped feature, similar to the network deposit shown in Fig. 2.7(a). The ball shaped copper clusters at the joints of the dimple edges indicate a relatively high plating rate at the edge of the dimples.

After 15 minutes plating (Fig. 2.9(b)), the features of the substrate are still clearly visible, which means that there is a lower plating rate in the solution than the rest of the low temperature solutions. The magnified image on the top left of Fig. 2.9(b) gives the detail of the deposit around the edge of the dimples and the one on the top right of Fig. 2.9(b) the deposit inside the dimples. Around the edge of the dimples, the deposit shows very fine small particle structures measuring around 0.1 μm. Inside the dimples, the copper deposit shows a totally different band or lath feature. These copper deposits form clear net structures inside the dimples with a large amount of voids. The net structure found inside the dimples can affect the structure of the subsequent deposit.

After 60 minutes plating (Fig. 2.9(c)), the surface is fully covered with the copper deposit. The feature of the substrate is barely visible. However, deep voids on the surface are developed from the dimples of the substrate, which indicates the porous structure inside the deposit. The highly magnified image on the top left of Fig. 2.9(c) shows very fine cubic grains measuring around 0.1 μm on the surface of the deposit. The small grains grown in the solution enable the smooth surface of the deposit to develop.

After 120 minutes plating (Fig. 2.9(d)), the surface is totally covered with the copper deposit with no obvious pores. The surface is slighly more uneven compared with Fig. 2.8(d) but with few features of the substrate. Again, the magnified image on the top left of Fig. 2.9(d) shows very small and fine grains on the surface of the deposit.

Compared with Fig.2.9(d), no significant change happens in the heat treated deposit (Fig. 2.9(e)), which means that the heat treatment has no significant effect on the surface morphology of the deposit from the solution.

2.8 Electroplated and electroless copper deposits

In recent years, there has been a great need for lightweight and more reliable integrated circuits in the electronics industry. For example, plated copper deposits in printed circuit boards (PCBs) have to be considerably thin in

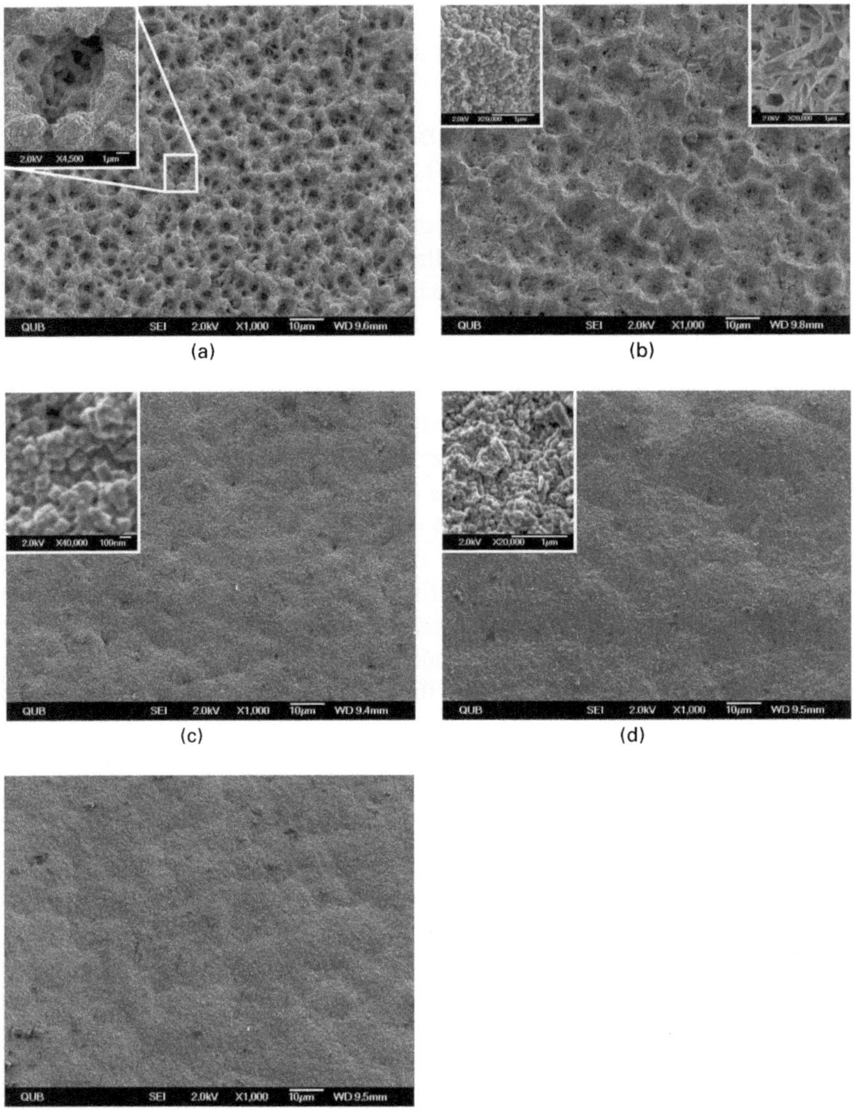

2.9 The surface morphology evolution of separate electroless copper deposits from the glyoxylic acid low concentration low temperature solution with (a) 5 minutes, (b) 15 minutes, (c) 60 minutes and (d) 120 minutes plating time, and (e) 120 minutes with 140 °C heat treatment.

order to reduce the weight and size of electronic components and devices. During processing of certain types of PCB, the mounting or exchanging of components by soldering is critical, as the plated copper deposit is subjected to thermal stresses, which might impair the integrity of the PCB. Consequently, the plated copper deposits used in the electronics industry should withstand physical deformation caused by excessive stresses. Hence, the understanding of the properties and behaviour of copper deposits during processing and application is important. This can be achieved by characterising the plated copper deposits, such as through surface morphology and residual stress measurements.

This section shows the surface microstructural morphology of both the electroplated and electroless copper deposits, and the chemical compositions of the electroplated copper deposits. The catalysts used before the plating of the electroless copper deposits on copper clad/laminated and screen-printed substrates are a Sn–Pd colloid and an organic Pd compound, respectively.

The surface morphology of deposits has been identified as one of the important microstructural factors in dictating their property and behaviour in various applications. A series of micrographs showing the surface morphologies of both the electroplated and electroless copper deposits are shown here. Grain structures of the as-deposited electroplated copper deposit on the substrate side are shown in the optical micrograph in Fig. 2.10. There seem to be relatively large cracks (see arrow on Fig. 2.10), appearing as dark irregular lines of varying width, probably initiated during the peeling of the coating off its substrate. The scanning electron micrograph (SEM) in

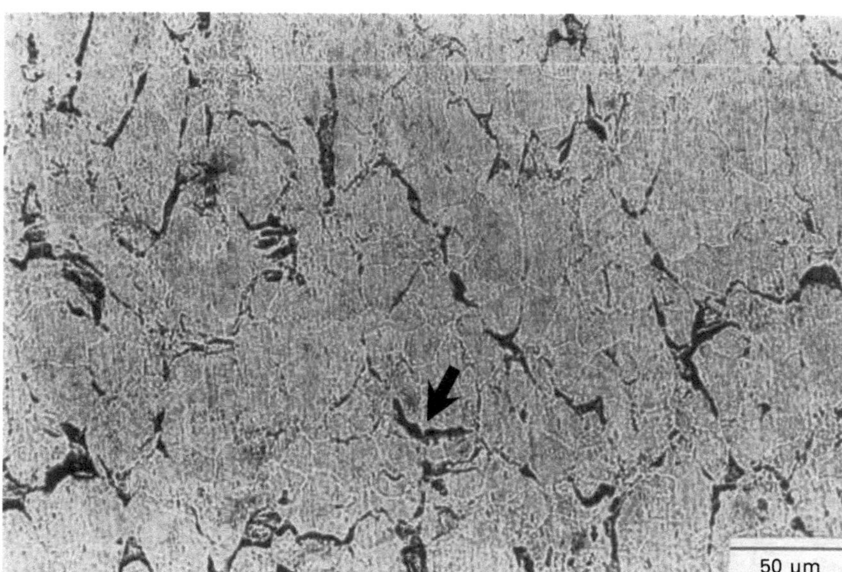

2.10 Optical micrograph of an electroplated copper deposit at substrate side in as-deposited condition. Numerous large cracks exist in the film surface; one example is indicated by the arrow.

Fig. 2.11 shows the small grains from Fig. 2.10. The grain boundaries of these grains are populated with the relatively large cracks. There are series of lamella lines inside the small grains grown into different directions. These lamella appearances might simply be caused by the effect from lamella pearlite in the stainless-steel substrate, rather than from real copper grains in the deposits. Small voids that are probably caused by the hydrogen bubbles during the plating process are also in the deposit, distributed unevenly across the surface.

The surface microstructure of the solution-side of the as-deposited electroplated copper is very different from that of the substrate-side (Fig. 2.12). The fine structure of the solution-side does not show any voids. The overall appearance of the surface structure suggests that the grains are of fine lamella structure. The white spots on the surface of the deposit are likely contaminants, which are attached to the deposit.

For the heat-treated electroplated copper deposit, the microstructural features of the deposit on the substrate and solution sides (Figs 2.13 and 2.14) are similar to that in the as-deposited condition. However, the small voids on the substrate side are slightly more than in the as-deposited condition. The effects of the heat-treatment temperature on the microstructural features of the electroplated copper deposit are apparently not significant. On both the as-deposited and heat-treated electroplated copper deposits, copper was the only element present in the deposits.

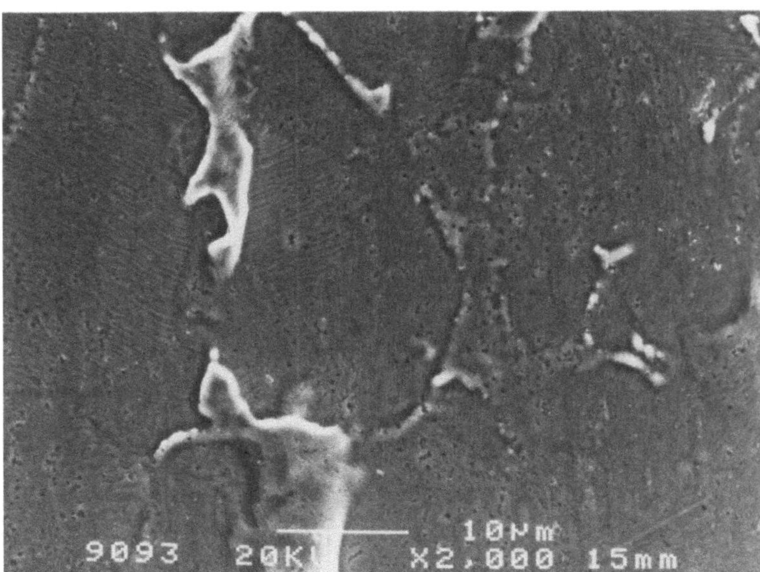

2.11 Scanning electron micrograph of an electroplated copper deposit at substrate side in as-deposited condition.

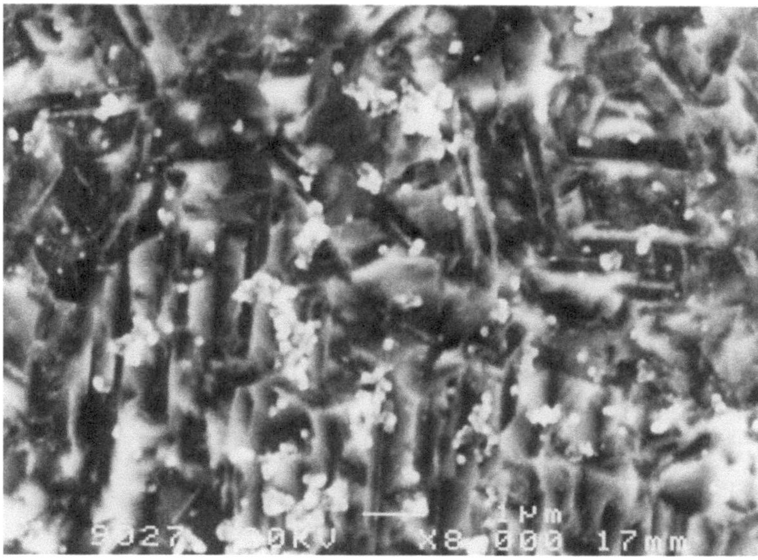

2.12 Scanning electron micrograph of an electroplated copper deposit at solution side in as-deposited condition.

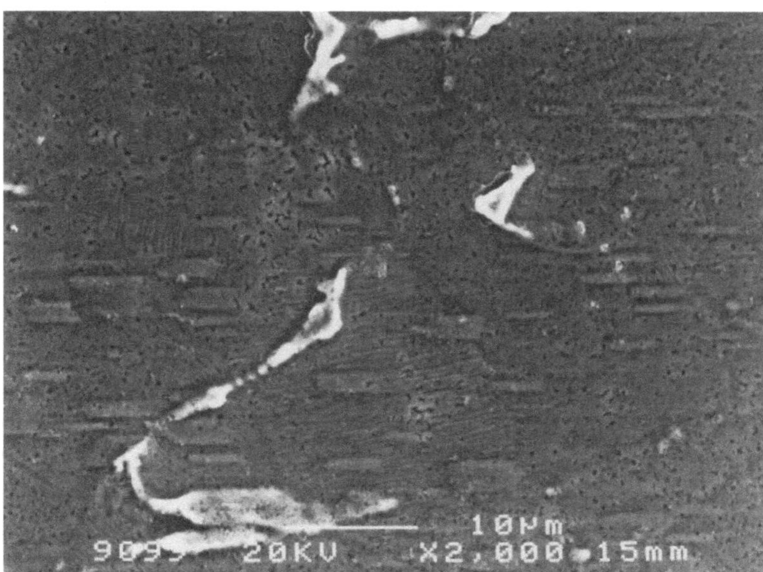

2.13 SEM of heat-treated electroplated copper deposit at substrate side.

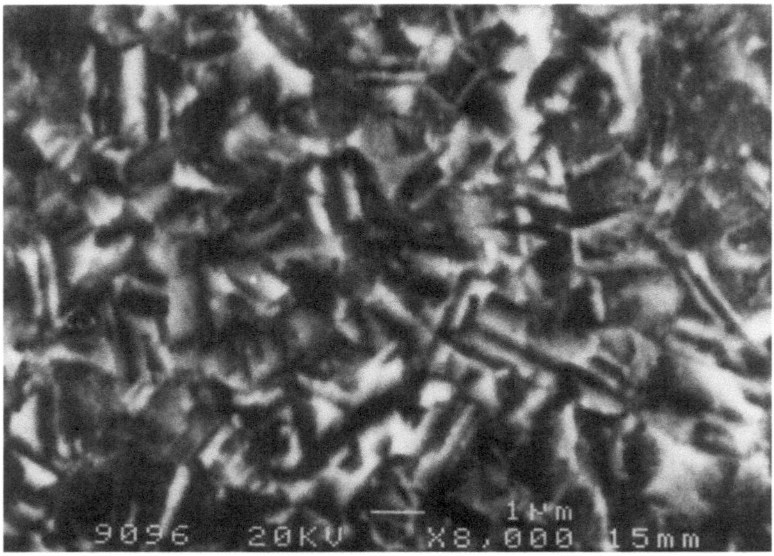

2.14 SEM of heat-treated electroplated copper deposit at solution side.

Compared to the morphology of the electroplated copper deposit on the solution side, the electroless copper deposit has a relatively rough surface structure, large granular particles, and rather faceted grains (Fig. 2.15). The faceted grains are more clearly shown at large magnification (Fig. 2.16). Small voids in the electroplated copper are not in the electroless copper. However, transmission electron microscopy (TEM) analysis shows that small voids are present both within grains and at grain boundaries, and that the electroless deposit contains a significant amount of hydrogen. Furthermore, chemical attack can enlarge the voids and form cracks at grain boundaries, which are populated with relatively large voids. Overall, the surface structure of the electroless copper plated on the copper clad substrate is homogeneous with a small number of relatively large particles scattered randomly on fine faceted grains.

For the solution-side electroless copper deposit plated on the laminated substrate (Fig. 2.17), the surface morphology is very similar to that in the electroless copper deposit plated on the copper clad substrate. The difference is the larger number of granular particles in the copper deposit plated on the laminated substrate.

Apparently, the heat treatment process at a temperature of 140 °C for one hour intensifies the large granular particles (Fig. 2.18). The surface area of the deposit is almost totally covered by these large particles. The large granular particles have well-defined boundaries and valleys are created

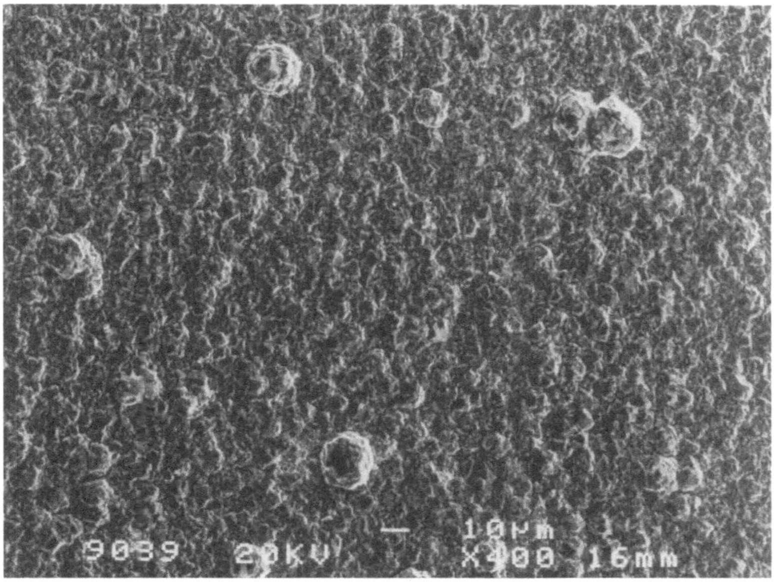

2.15 Low magnification SEM of the surface morphology of electroless copper deposit at solution side on copper clad substrate in as-deposited condition.

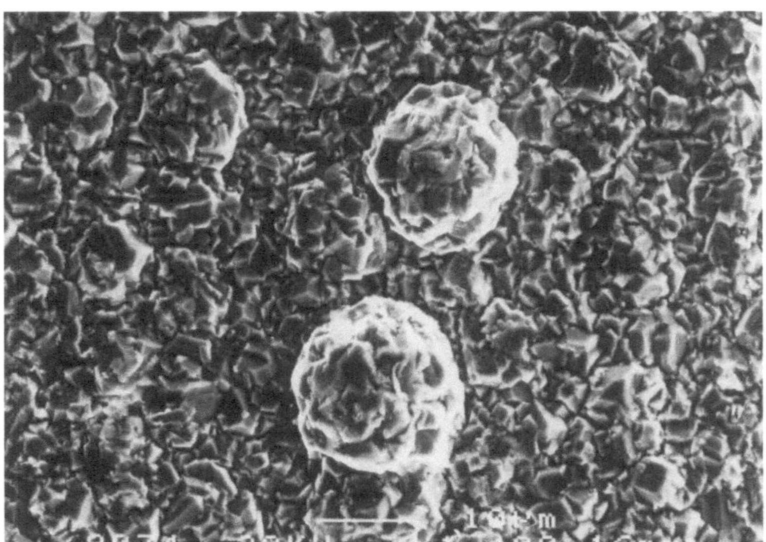

2.16 High magnification SEM of electroless copper deposit at solution side on copper clad substrate in as-deposited condition.

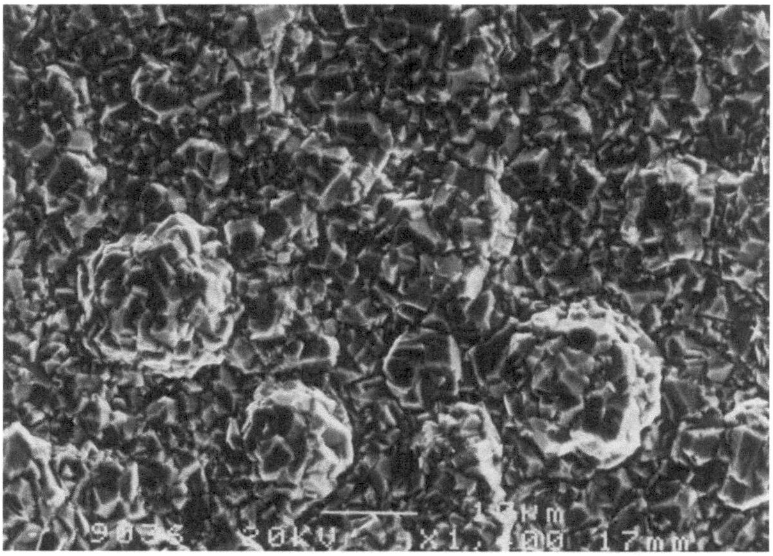

2.17 High magnification SEM of electroless copper deposit at solution side on laminated substrate in as-deposited condition.

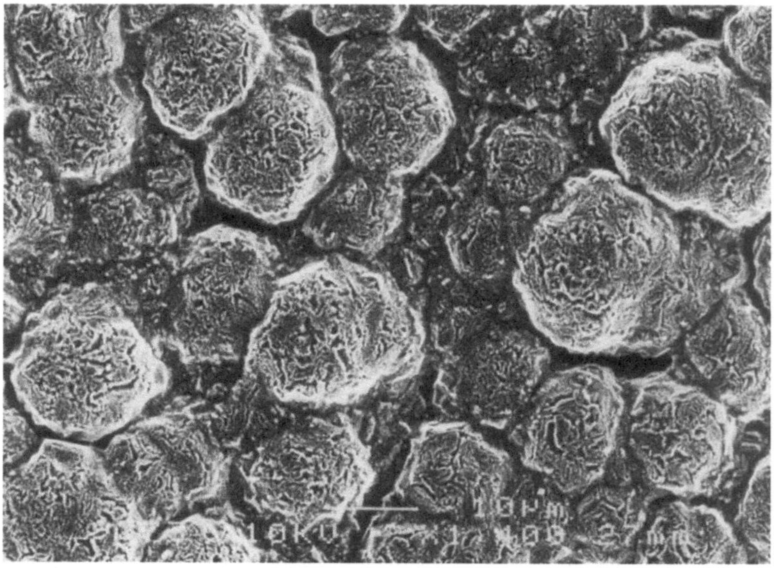

2.18 High magnification SEM of heat-treated electroless copper deposit at solution side on dielectric substrate.

between the particles. The fine faceted grains in the as-deposited electroless copper have disappeared. However, this intensified growing might be the effect of different catalyst and substrate used in the plating process.

2.9 Conclusions

The scanning electron microscopy images of the deposit from different plating conditions show the different growing process of copper deposits and the effect of the conditions on the morphology of the final surface.

Due to the acidic nature of glyoxylic acid and its high sensitivity to temperature and concentration, it is less easy to control the plating solutions with glyoxylic acid than those with formaldehyde. This is one of the reasons that glyoxylic acid is not widely used in the electroless copper plating industry. Although the concentration of the copper sulphate is relatively low in the glyoxylic acid low concentration high temperature solution, with the increase of the concentration of the glyoxylic acid, the plating solution is easily decomposed at 60 °C, which makes the solution unable to sustain prolonged plating. When the temperature is lowered to 45 °C, although the concentration of glyoxylic acid is even higher, the plating solution remains relatively stable during the whole plating process.

By replacing formaldehyde with glyloxylid acid, we can obtain a smoother surface finish in the electroless copper deposit. The reason for this may be in the fine grains on the surface of the deposits from glyoxylic acid solutions in the early stage of the plating process. In the glyoxylic acid high concentration high temperature solution, although the 120-minute deposit has large grains on the surface, the surface morphology of the earlier stage deposit consists of small grains. Small grains help to fill the gaps between grains and hence form a smoother surface finish in prolonged plating.

The glyoxylic acid low temperature and formaldehyde high concentration low temperature solutions give the smoothest surface finish. The surface morphology of the substrate is fully covered by the copper deposit and can hardly be discerned on the surface of the deposit. By reducing the temperature, the copper deposit tends to form a smoother surface. The formaldehyde low concentration low temperature solution has a less smooth surface because of the large sized diamond pyramid structures forming at the very beginning of the plating process.

The diamond pyramid structures from the low temperature formaldehyde solutions consist of virtually pure copper and only a small amount of oxygen possibly from contamination during analysis. No other element is present. This, together with the X-ray diffraction analysis described in Chapter 4, proves that the diamond pyramid structures are formed with only copper atoms, and no other element.

The electroplated copper deposits on the substrate side and the solution side have very different surface microstructural features, likely caused by the substrates of the deposits. The deposit on the substrate side has small voids scattering across the surface of the deposit, and has relatively large cracks at some portions. In contrast, no voids are evident in the electroplated deposit on the solution side, which consists of a fine grain structure. Heat treatment of the electroplated copper deposit does not change the surface morphology of the deposit much. Copper is the only element present in the electroplated copper deposits both at as-deposited and heat-treated conditions. In the electroless copper deposits, the effect of the substrates on the deposits is strong. The number of large granular particles on the deposit plated on copper clad substrate is smaller than that in the deposit plated on the laminated substrate. After heat treatment at 140°C for one hour, the electroless copper deposit plated on a screen-printed dielectric substrate has its surfaces almost fully covered with the large granular particles. However, the effect might come from the use of a different catalyst and substrate during plating process, rather than from the heat treatment. Compared to the electroplated deposits, the electroless copper deposits are relatively rough and virtually void free.

3
Cross-section of electroless copper deposits and the void fraction

Abstract: Electroless copper deposits are plated on epoxy substrates in various plating solutions at either a high operating temperature (60 °C) or a low one (45 °C). The microscopy images of the over-etched cross-sections show voids in low temperature deposits and solid structure in high temperature ones. The surface morphology images also indicate such structures in low temperature deposits. The microscopy image of the cross-section of a stand-alone deposit shows similar voids observed on an etched cross-section. An image-processing program identifies the voids in the over-etched cross-sections of the deposits from low temperature solutions and thus the void fraction can be measured and compared with simulation data.

Key words: electroless plating, copper, over-etching, scanning electron microscopy (SEM), void.

3.1 Calculating the void fraction on an electron microscopy cross-section image

The objective of this chapter is to examine the inner structure of the electroless copper deposit from different solutions using the over-etching method. The etchant removes the surface layer of the polished cross-sections of the deposits and reveals the inner structures. The scanning electron microscopy (SEM) images of the over-etched cross-sections are then processed to extract the void area data. The void fraction of the cross-section can then be calculated.

The following procedures are used to calculate the void fraction on the SEM images of the over-etched cross-sections. The parameters are selected in the next section to ensure that the shape and the feature of the voids can be best captured.

1. For the SEM image of each deposit, three rectangular sub-images are randomly selected on the over-etched deposit section. These sub-images should cover only the deposit part in the SEM cross-section image and should be much bigger than the typical size of a void. The brightness of the sub-images should roughly be evenly distributed for the accuracy of the following procedures.
2. For each of the sub-images, the average grey scale value of all the pixels is calculated. A threshold of G_{bw} is thereafter set. For each pixel in the

sub-image, if its grey scale value is equal or below this threshold, its grey scale value is set to zero (black) or 255 (white).

3. For each of the black pixels (with a grey scale value of zero) in the sub-image, a neighbour black pixel search is performed in the 5×5 pixel in the area. If the percentage of the black pixel (including itself) surrounding it is lower than a value P, this black pixel will be changed to a white pixel by switching its grey scale value from 0 to 255. After this procedure, most of the black pixels due to the noises of the brightness are removed.
4. For each of the white pixels (with a grey scale value of 255), the procedure in step 3 is performed to remove the white pixels introduced by the brightness noises.
5. The percentage of the black pixels within each sub-image is calculated by dividing the number of the black pixels by the number of all the pixels. This value is then used as the void fraction of this sub-image.
6. For each SEM image, the average of the void fraction of its three sub-images is calculated and used as the overall void fraction of the deposit.
7. The standard deviation of the void fractions from the three sub-images of each deposit is calculated and used as the error of the void fraction calculation of the deposit.

The threshold values G_{bw} and P involved in the above procedure are tested in several different values and the processed images are compared with the original ones. The G_{bw} and P values that give the best-processed image are chosen for use in the image processing procedures.

3.2 Determination of the optimum threshold values of grey scale and noise

In order to determine the best G_{bw} value for the image processing procedure, the G_{bw} values of 0.7, 0.65, 0.6, 0.55 and 0.5 are used in the process and the processed images with these G_{bw} values after step 2 are shown in Fig. 3.1. Generally speaking, the higher the G_{bw} value is, the more likely a pixel is recognised as a black pixel, which means that more void information will be kept in the processed image but more noises will be introduced. In these images, when the G_{bw} value is 0.7 (Fig. 3.1(b)) or 0.65 (Fig. 3.1(c)), the voids on the original image are well captured but a large amount of the black noises remain on the processed images, which may not be removed by the subsequent image processing step. When the G_{bw} value is 0.55 (Fig. 3.1(e)) or 0.5 (Fig. 3.1(f)), much less noise remains on the processed images but some shallow voids are not completely captured. The best-processed image is Fig. 3.1(d), where the G_{bw} value is 0.6. This image extracts most of the

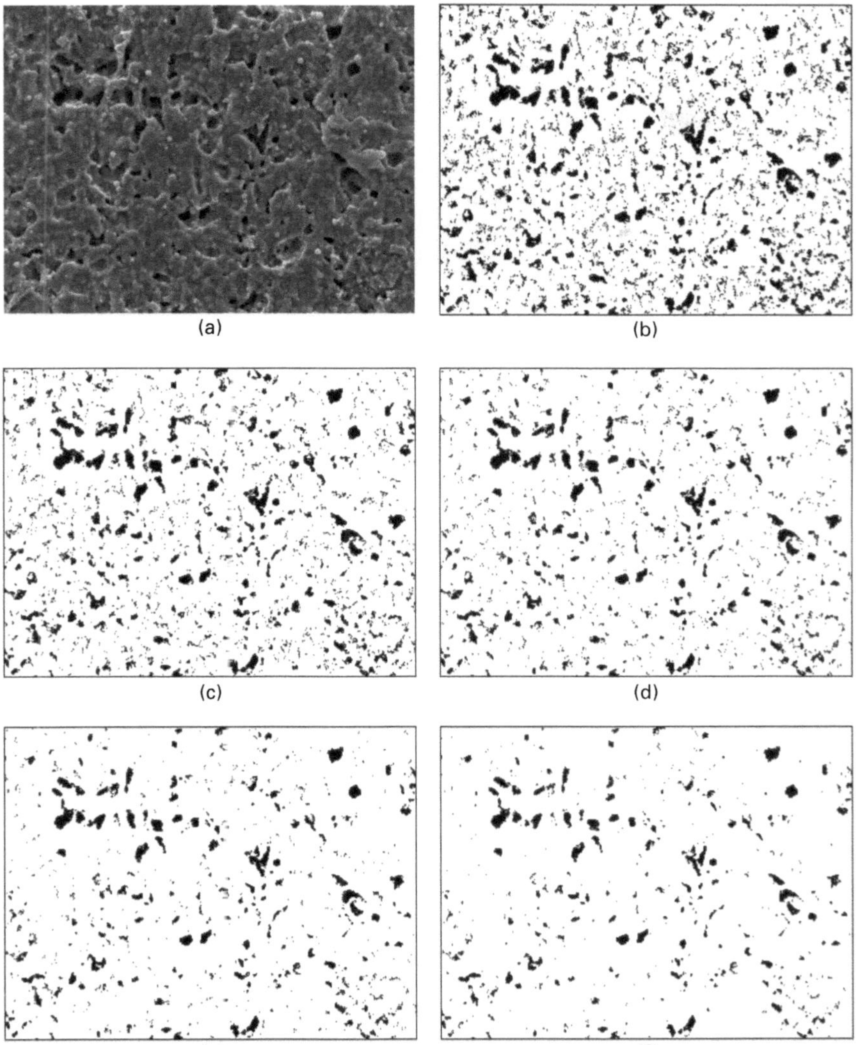

(a) (b) (c) (d) (e) (f)

3.1 The original SEM image (a) and the processed images after step 2 in the image processing procedure with a G_{bw} value of (b) 0.7, (c) 0.65, (d) 0.6, (e) 0.55 and (f) 0.5.

void information from the original image and at the same time keeps the noise at a controllable level for the subsequent procedures. Therefore, the best G_{bw} value for the image processing procedure is 0.6.

In order to determine the best *P* value for the image processing procedure, the *P* values of 0.25, 0.3, 0.35, 0.4 and 0.45 are used in the process with the G_{bw} value of 0.6. The processed images after step 4 are shown in Fig. 3.2. The higher the *P* value is, the bigger the black and white particles on the

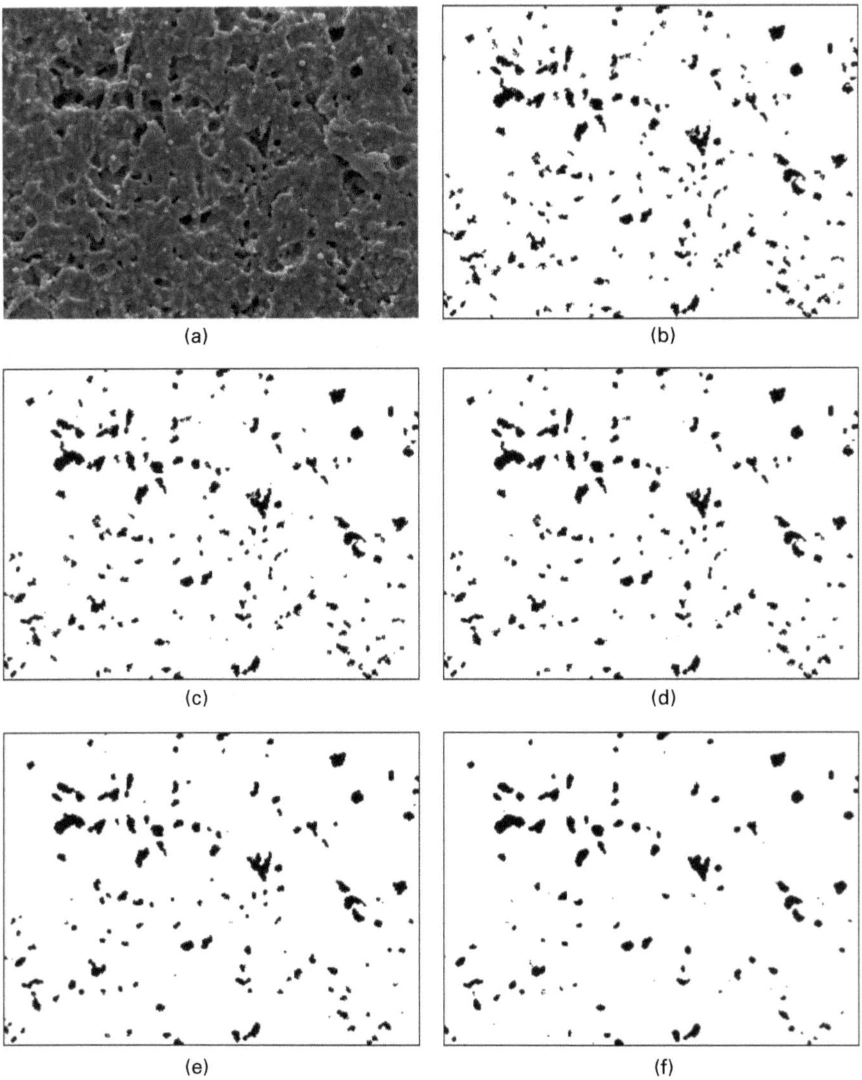

3.2 The original SEM image (a) and the processed images after step 4 in the image processing procedure with a G_{bw} value of 0.6 and the P value of (b) 0.25, (c) 0.3, (d) 0.35, (e) 0.4 and (f) 0.45.

image after step 2 are recognised as noise and removed. The best *P* value will remove the black and white particles that are actually noise and at the same time keep the small voids on the image. In Figs 3.2(b) and 3.2(c), where the *P* value is set to 0.25 and 0.3, respectively, some black noise and some white noise in and at the edge of the voids remain on the processed images. In Figs 3.2(e) and 3.2(f), where the *P* value is set to 0.35 and 0.4, respectively, some small voids are recognised as noise and removed. The best image where most of the noise is removed and the small voids are well kept is in Fig. 3.2(d). Therefore, the best P value for the imaging processing procedure is 0.35.

For the remaining image processing procedure, the G_{bw} value is set to 0.6 and the P value is set to 0.35 according to the above explanation.

3.3 The voids

In Fig. 3.3, the lighter shade area in the middle level of the image is the copper deposit. The upper dark area is the epoxy substrate with the fibres in the epoxy board. The lower dark area is the epoxy resin used to hold the deposit upward together. The copper area is brighter in the SEM image because copper atoms emit more secondary electrons than epoxy molecules do when bombarded by high-energy electrons, so that more electrons are collected by the detector, which gives a brighter image for the copper area. In this image, we can see that the deposit has a relatively flat cross-section

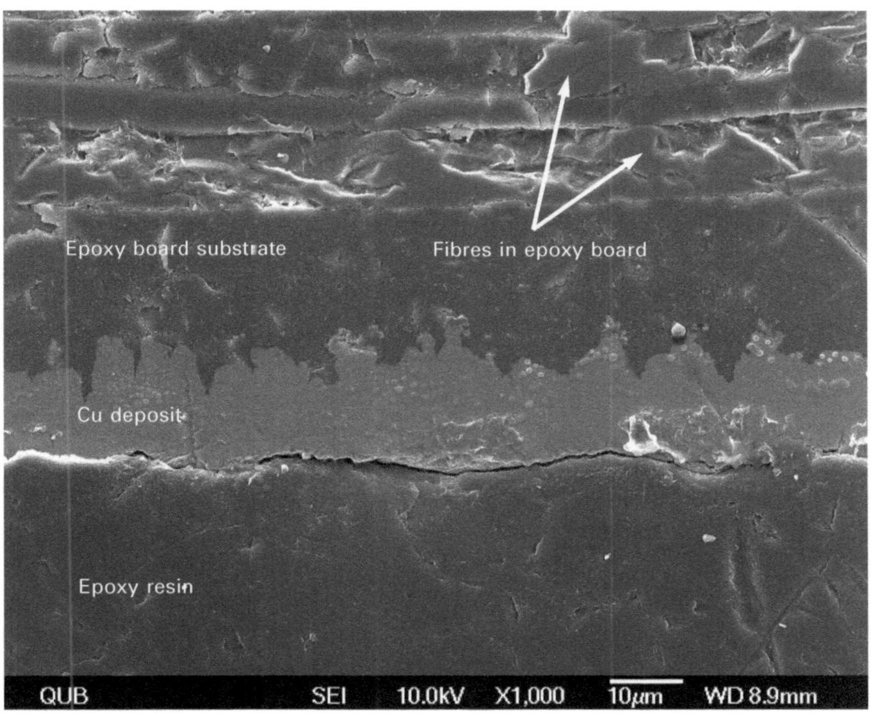

3.3 The SEM image of the polished cross-section of the electroless copper deposit from the glyoxylic acid low concentration low temperature solution on epoxy board substrate.

after being polished although some dirt and debris remain. The copper deposit has a tight bond with the epoxy substrate, but there is a gap between the deposit and the epoxy resin. The copper deposit and the substrate are level after polishing with no significant gap between them. The image also shows the relatively flat surface of the deposit compared to the rough surface of the substrate, which means that the thickness of the deposit is uneven across the deposit.

In the polishing process, soft copper smears may be generated on the polishing surface and cover the sub-micron voids. These smears can be removed by the 1:1 1% nitride acid and 0.2% silver nitride solution etchant and, after etching for 10–30 seconds, the voids can thus be revealed with SEM. Compared to Fig. 3.3, Fig. 3.4 shows great difference in the level of depth between the copper deposits and the substrates, which means that the deposits are strongly over-etched. The smooth surface layer of copper deposit on the cross-section of the polished deposits is removed by the etchant and the inner structures are exposed. In the SEM over-etched and the void images in Fig. 3.5, the shape and the distribution of the voids are clearly demonstrated. The void fractions can then be calculated from the void images (Table 3.1).

After over-etching, the copper deposit still has a relatively smooth cross-section although some very small crystal structure patterns are apparent (Fig. 3.4(a)). Due to the strong attacking effect of the etchant on both the grain boundaries and the grains in the over-etching process, the contrast between them is not strong enough in Fig. 3.4(a) to show the grain structures clearly. No significant void is on the over-etched cross-section indicating the compact structure within the copper deposit from the formaldehyde high concentration high temperature solution. The void image in Fig. 3.5(a) confirms only a small amount of void present and the void fraction is 0.05% (Table 3.1).

The deposit from the formaldehyde low concentration high temperature solution (Fig. 3.4(b)) has similar features to the formaldehyde high concentration high temperature solution deposit shown in Fig. 3.4(a) despite some debris at the near-resin side. The over-etched deposit cross-section is still relatively smooth with some traces of the crystal structure. Some small and shallow voids are at the near-substrate side, which means that the deposit may have a slightly porous structure in the beginning of the plating process. This could be because of the rough surface of the substrate. When copper starts to grow perpendicularly to the surface at very sharp spots on the surface of the substrate, it is hard for the deposit to connect with each other, leaving voids within. However, the overall structure of deposit from the formaldehyde low concentration high temperature solution as we can see from Fig. 3.4(b) is still compact with no significant voids, which is substantiated by the void image shown in Fig. 3.5(b). Its void fraction is 0.33% (Table 3.1).

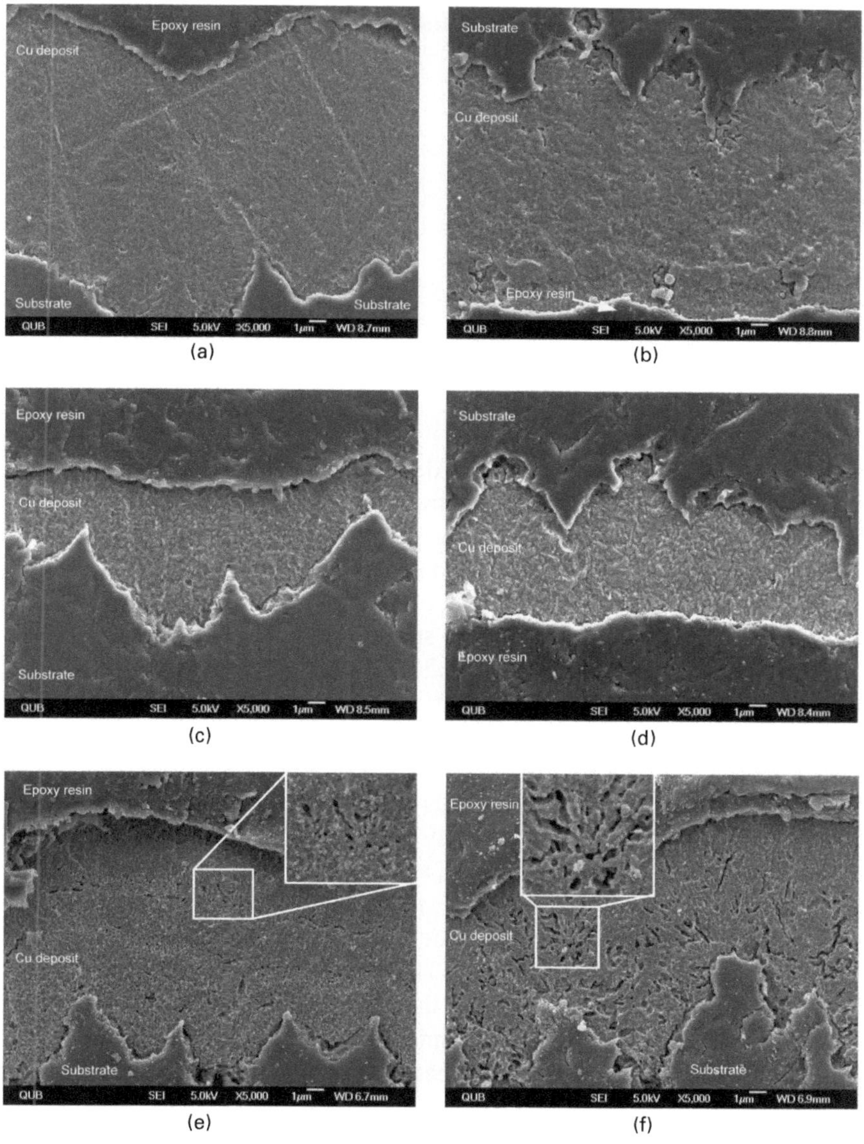

3.4 The SEM images of the over-etched polished cross-sections of electroless copper deposits on epoxy board from different solutions. (a) Formaldehyde high concentration high temperature; (b) formaldehyde low concentration high temperature; (c) glyoxylic acid high concentration high temperature; (d) glyoxylic acid low concentration high temperature; (e) formaldehyde high concentration low temperature; (f) formaldehyde low concentration low temperature; (g) glyoxylic acid high concentration low temperature; (h) glyoxylic acid low concentration low temperature.

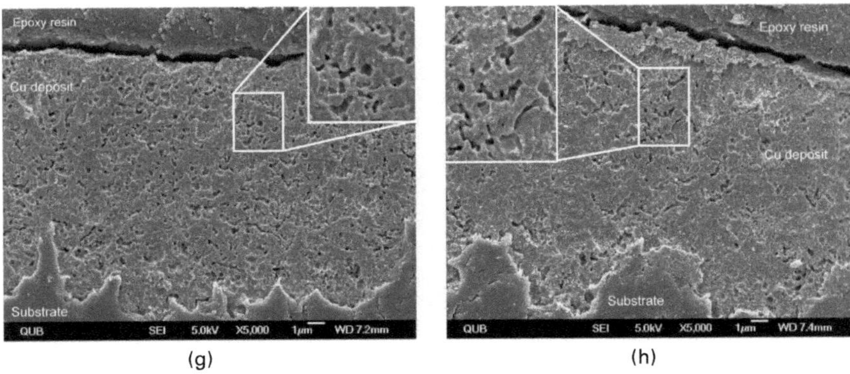

3.4 Continued

In the over-etched cross-section of deposits from glyoxylic acid high (Fig. 3.4(c)) and low (Fig. 3.4(d)) concentration high temperature solutions, the thickness of the deposits is smaller than that of the formaldehyde deposits shown in Figs 3.4(a) and 3.4(b), due to the properties of the plating solutions. Slightly rougher grain structures are on the over-etched cross-sections but with few voids, indicating compact structures within these two deposits. The void images in Figs 3.5(c) and 3.5(d) show that only a small amount of voids is present and the void fractions are 0.32% and 0.17%, respectively.

In the over-etched cross-section of the deposit from the formaldehyde high concentration low temperature solution (Fig. 3.4(e)), the magnified image at top right shows some details of the cross-section. Different from all the images above, there are clearly many small voids. The contrast and appearance of the voids in the image, and their absence in high temperature deposits with the same etching process indicate that they are unlikely to be etching pits. They are more likely the voids left in the deposit during the growing process of copper deposit. The void image in Fig. 3.5(e) clearly shows the size and the distribution of the voids. These voids measure around 0.1–0.3 μm and some of them are connected with each other forming a void network as shown in the magnified image. Most voids are round in shape and separated, but when connected, the length of the void network can reach up to more than one micron, as in the magnified image within Fig. 3.4(e) and the void image in Fig. 3.5(e). The void fraction of the formaldehyde high concentration low temperature deposit reaches 4.3%, significantly larger than those of the high temperature deposits (Table 3.1).

In the over-etched cross-section of the deposit from the formaldehyde low concentration low temperature solution (Fig. 3.4(f)), more and much bigger voids are present. As also shown in the void image in Fig. 3.5(f), a clear network is formed by these voids. The size of the big voids measures around 0.2 μm by more than 1 μm. These voids, together with the deposit,

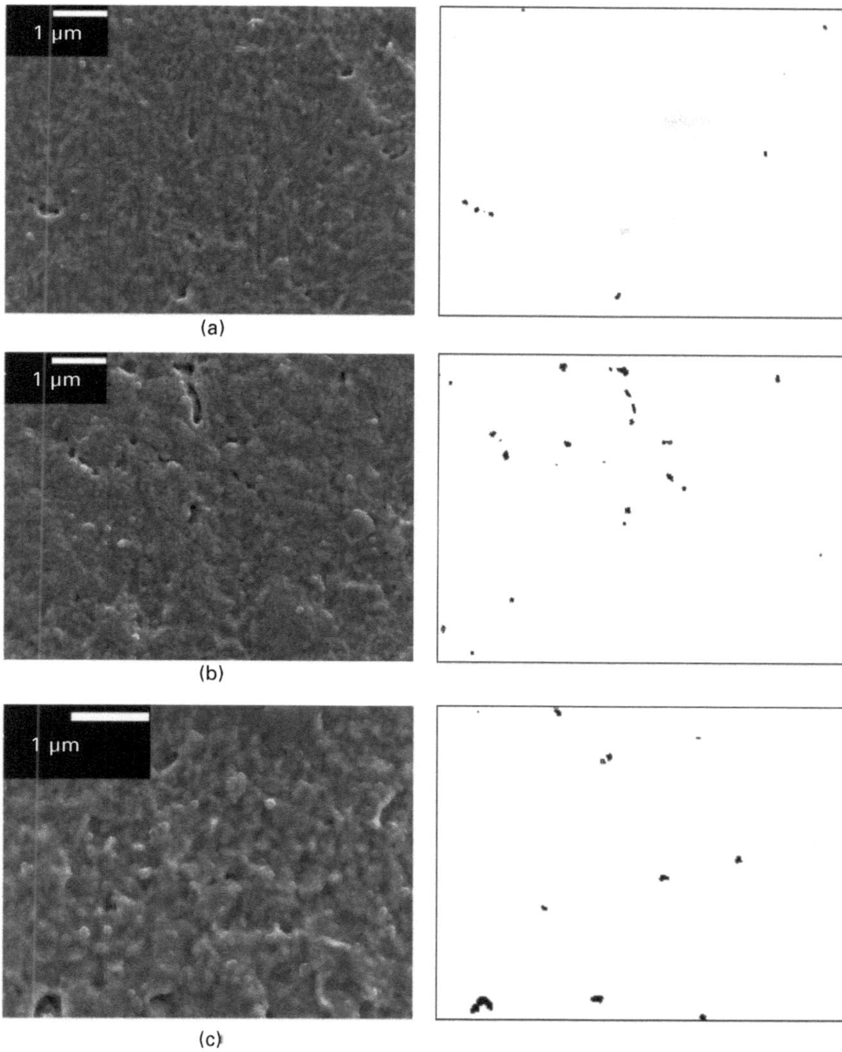

3.5 The sub-images of the electroless copper deposits from different solutions before (left) and after (right) image processing. The black areas in the images on the right-hand side show the voids on the cross-sections. (a) Formaldehyde high concentration high temperature; (b) formaldehyde low concentration high temperature; (c) glyoxylic acid high concentration high temperature; (d) glyoxylic acid low concentration high temperature; (e) formaldehyde high concentration low temperature; (f) formaldehyde low concentration low temperature; (g) glyoxylic acid high concentration low temperature; (h) glyoxylic acid low concentration low temperature.

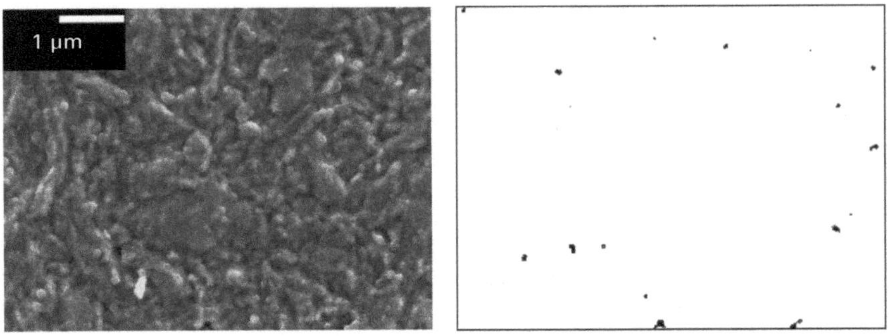

(d)

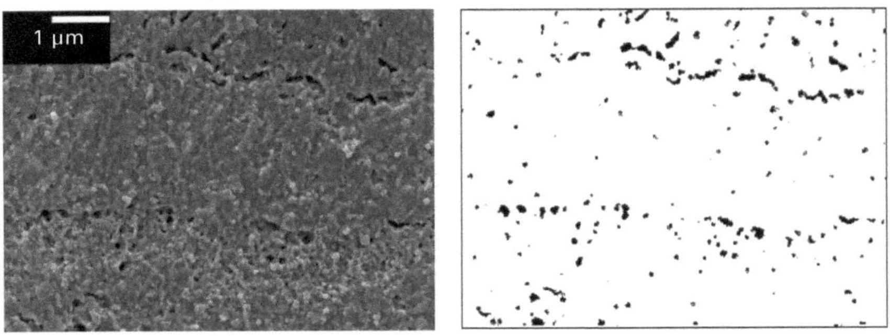

(e)

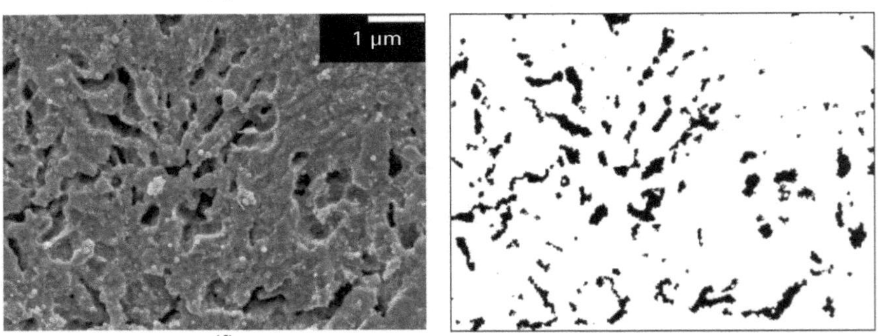

(f)

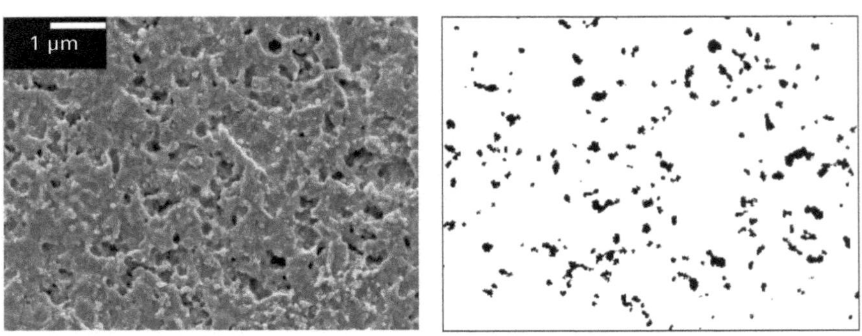

(g)

3.5 Continued

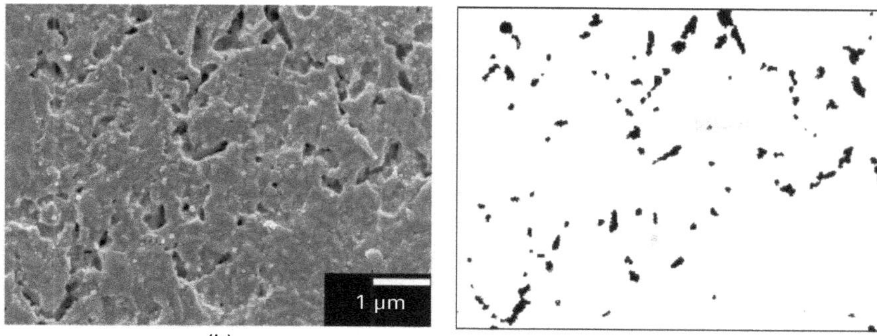

(h)

3.5 Continued

Table 3.1 The void fractions

Solution	Void fraction (%)	Standard deviation (%)
Formaldehyde high concentration high temperature (FHCHT)	0.05	0.04
Formaldehyde low concentration high temperature (FLCHT)	0.33	0.08
Glyoxylic acid high concentration high temperature (GHCHT)	0.32	0.14
Glyoxylic acid low concentration high temperature (GLCHT)	0.17	0.06
Formaldehyde high concentration low temperature (FHCLT)	4.3	0.7
Formaldehyde low concentration low temperature (FLCLT)	9.0	1.6
Glyoxylic acid high concentration low temperature (GHCLT)	5.5	0.4
Glyoxylic acid low concentration low temperature (GLCLT)	4.4	0.5

have strong directionality pointing from the substrate to the resin where the surface of the deposit is, which means that the growth of the copper deposit in this solution is more out of surface than laterally within the surface, leaving large percentages of porosity in the deposit. Some smaller round shaped voids scatter on the cross-section of the deposit. The void fraction of the formaldehyde low concentration low temperature deposit reaches 9.0%, which is the highest among all the deposits.

A two dimensional Monte Carlo model can simulate the deposition process of electroless copper and calculate the cross-section and the void fraction of the deposits from plating solutions with the ethylenediaminetetraacetic acid (EDTA) or other complexing agents. Column shaped voids are discovered on the cross-section of the simulated deposit from the EDTA plating solution, and the void fraction is 6.0%. The shape of the simulated voids closely matches that of the voids we observe here in the formaldehyde low concentration low temperature deposit (Fig. 3.5(f)). The simulated void fraction is also in agreement with the observed ones listed in Table 3.1.

In the over-etched cross-section of the deposit from the glyoxylic acid high (Fig. 3.4(g)) and low (Fig. 3.4(h)) concentration low temperature solutions, there are also clear voids, as also shown in the void images in Figs 3.5(g) and 3.5(h), respectively. In the high concentration deposit, voids are mainly spherical and are separated and evenly distributed with low directionality though with very high number and volume densities. These voids measure from around 0.1 to 0.3 μm in diameter, and the overall void fraction is 5.5%. In the low concentration deposit, as demonstrated in Fig. 3.5(h), there is a lower density of voids, mainly connected and measuring around 0.2 × 1 μm. The void fraction is 4.4%.

Due to the flat surface of the stainless steel substrate, the deposit on both the substrate side and the surface side is flat and even. When cut by scissors, without polishing and etching, the cross-section shows fracture patterns of the deposit (Fig. 3.6). Despite this, there are some very dark holes or connected voids, as marked with the white circles in the image, which means that no

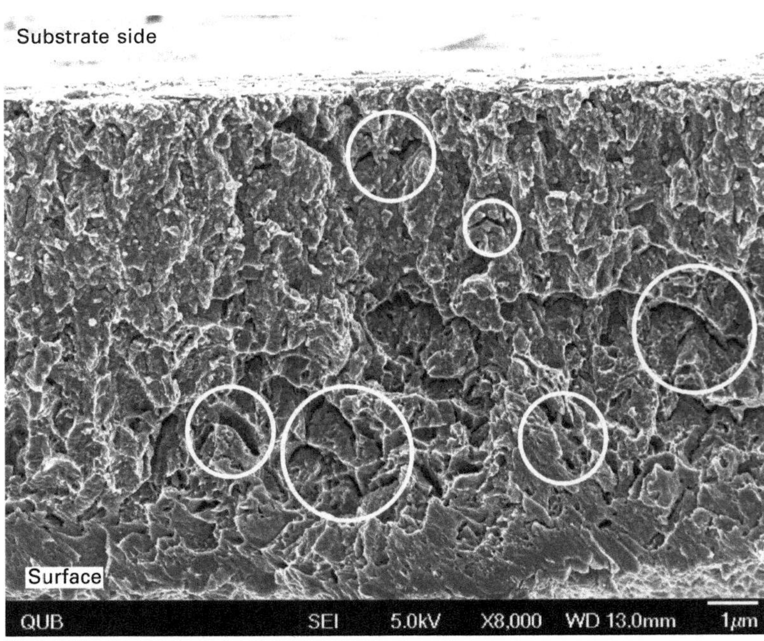

3.6 The SEM image of the cross-section of the stand-alone deposit from the formaldehyde low concentration low temperature solution. The white circles mark some of the voids in the cross-section.

copper occupied there at the near top of the cross-sections. They are not likely induced by pulling out during the cutting process but more likely, the voids left during the plating process.

Another proof of the porosity in the electroless copper deposits from the low temperature solutions is shown in Fig. 3.7 where the surface morphology of the deposit from the formaldehyde low concentration low temperature solution on epoxy substrate is demonstrated. In this image, deep gaps are at the joints of copper deposit clusters as marked with white circles. These gaps could be the extension of the porosity within the deposit. The magnified image on the top left of this figure shows a large amount of layered diamond pyramid structures on the surface of the deposit. Similar structures can be found on the surface of the deposit from the formaldehyde high concentration and glyoxylic acid low temperature solutions and the stand-alone deposit from the formaldehyde low concentration low temperature solution. The last deposit has the densest and finest diamond pyramid structures. These diamond pyramid structures measure around 0.2×0.5 μm in size and could be bigger below the surface. On the surface of other deposits, only grains smaller than 0.2 μm are present.

3.7 The SEM image of the surface morphology of electroless copper deposit on epoxy board substrate from the formaldehyde low concentration low temperature solution. The white circles mark some of the voids on the surface. The high magnification image on the top left shows the diamond pyramid structures on the surface.

As we have already noticed, the voids are only present in a large amount in the deposit from solutions operated at low temperature (45 °C). It can be concluded that temperature plays an important role in forming such structures. Temperature affects the diffusion of the molecules in the plating solutions and the movement of the copper atoms within the deposits. During the plating process, fine diamond pyramid structures form and grow on the surface of the deposits in low temperature solutions. Due to the special shape of these diamond pyramid structures, voids may be left in between them during the growing process and chemicals in the bulk area of the plating solution have more restricted access to these voids in lower temperature solutions due to slower diffusion. Once the concentration of the plating solution in and around the vacant areas reaches a certain low level, the reaction is stopped and the voids cannot be filled with copper and hence are left within the deposits. Since temperature can affect the plating rate dramatically (Hanna *et al.*, 2004; Lantasov *et al.*, 2000), the reduction of the chemical concentration, especially that of the hydroxyl, which affects the pH of the plating solution, can more strongly lower the plating rate in low temperature than in a higher one. On the other hand, lower temperature leads to less active copper atoms, so the copper atoms have less chance to move around on the deposit to fill the voids in the deposits. Another effect that low temperature may cause on the void in the electroless copper deposit is that the hydrogen bubbles that are generated during the plating process are less active to move away from the surface of the deposits during the plating. Some hydrogen bubbles may be trapped on the surface and copper is later plated surrounding them, forming round-shaped porosities. These hydrogen-containing voids may greatly affect the mechanical properties of the copper deposit.

3.4 Conclusions

In summary, this chapter reveals the inner structures of the electroless copper deposits. There is a critical effect of the temperature on the structure of the deposits. Voids of 4–9% are present in the low temperature (45 °C) deposits, which closely matches the void fraction data calculated in simulation.

3.5 References

Hanna F, Hamid Z A and Aal A A (2004), 'Controlling factors affecting the stability and rate of electroless copper plating', *Mater Lett*, **58**, 104–9.

Lantasov Y, Palmans R and Maex K (2000), 'New plating bath for electroless copper deposition on sputtered barrier layers', *Microelectron Eng*, **50**, 441–7.

4 Crystal structure and surface residual stress of electroless copper deposits

Abstract: Crystal structure and surface residual stress are measured using the X-ray diffraction (XRD) method. There is a typical fine copper crystal structure on the deposits and no significant amount of residual stress is present in the deposits, either as plated or heat-treated. No significant signal of copper oxide is in the X-ray patterns, which means that little or no copper oxide is presented on the surface of the deposits. The atomic model of the diamond pyramid structure is based on this key fact. The grains on most of the deposits are randomly oriented.

Key words: crystal structure, surface residual stress, X-ray diffraction (XRD), heat treatment, grains.

4.1 X-ray normal scan patterns and the crystal structures of the deposits

The objective of this chapter is to examine the crystal structure and the surface residual stress of the electroless copper deposits from different plating conditions. The X-ray pattern of a normal scan can give information on the crystal structures of the deposit. A detailed tilted scan focusing on one peak of the sample can provide information on the surface residual stress of the deposit.

For copper crystals, strong peaks can be observed at 2θ of around 43°, 50°, 74°, 90° and 95°, and for copper oxide, peaks should be at 2θ of around 37°, 43°, 62°, 74° and 78°, according to the relative intensities in the Cu and CuO powder diffraction files (Fig. 4.1 and Table 4.1).

Experimentally, five strong and sharp peaks are in the pattern at the scanning angle 2θ positions of around 43°, 50°, 74°, 90° and 95°, respectively (Fig. 4.2). These peaks correspond to the copper {111}, {200}, {220}, {311} and {222} lattice plane, respectively. The positions of these peaks closely match the copper powder diffraction file (PDF) and the shape of the peaks is narrow and sharp, which means that the electroless plated copper deposits have strong crystallised copper structure. In the X-ray pattern, only peaks close to the copper PDF are present, with no significant CuO peaks. This means that no significant amount of CuO exists on the electroless copper deposits. The X-ray pattern of the copper deposits before and after heat treatment shows little difference, which means that the heat treatment at

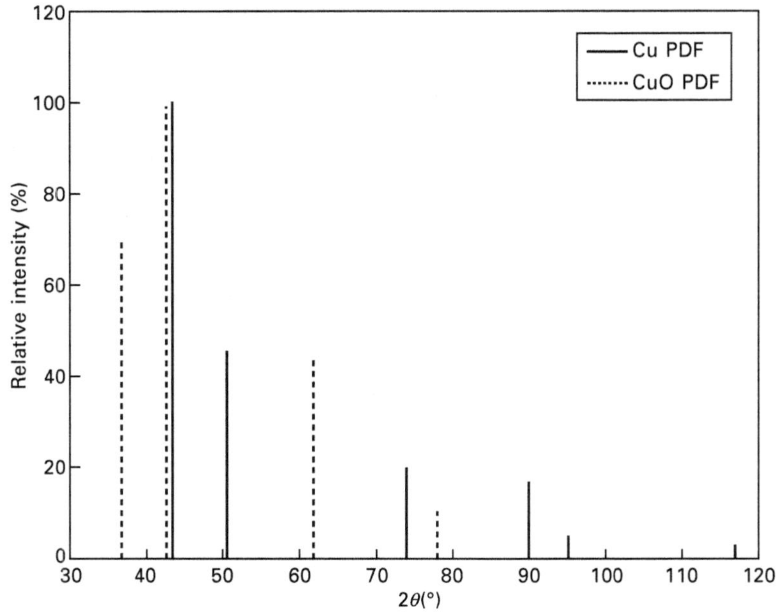

4.1 The powder diffraction files (PDF) of copper (Cu) and copper oxide (CuO).

Table 4.1 List of the positions of the peaks and their relative intensities of Cu PDF and CuO PDF

Cu			CuO		
Peak index	2θ position (°)	Relative intensity (%)	Peak index	2θ position (°)	Relative intensity (%)
{111}	43.298	100	{111}	36.637	77.0
{200}	50.434	46	{200}	42.559	100.0
{220}	74.133	20	{220}	61.761	45.0
{311}	89.934	17	{311}	74.003	17.1
{222}	95.143	5	{222}	77.893	10.9
{400}	116.923	3			

140 °C for one hour has little effect on the crystal structure of the copper deposits.

Where the X-ray intensities of the peaks at the same positions are slightly different between as plated and heat-treated deposits, it can be related with the machine setup and the positioning of the X-ray samples. The crystallography needs to be characterised by using relative intensities.

The traditional relative intensities of the electroless copper deposits (Table 4.2) show big variance with those of the copper PDF, indicating that the grains in electroless copper deposit are not randomly oriented. With the integrated

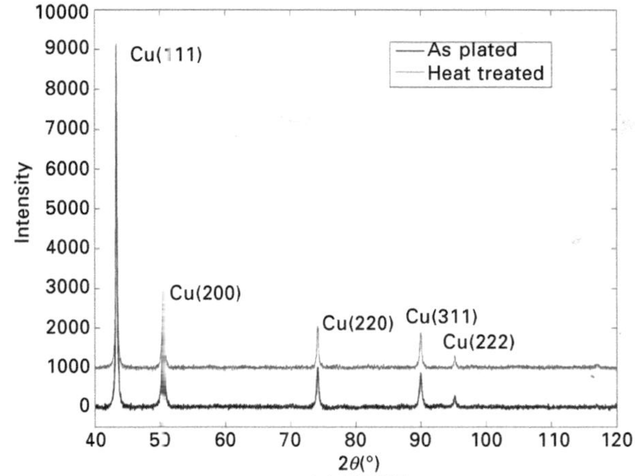

(a) FHCHT

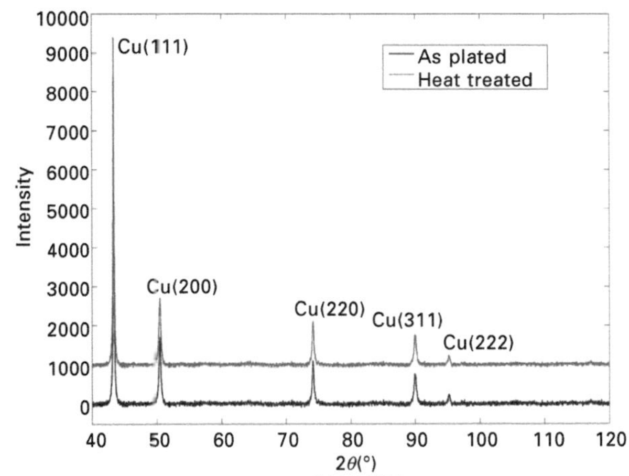

(b) FLCHT

4.2 Normal scan X-ray diffraction patterns of electroless copper deposits on epoxy board both before and after heat treatment from the plating solutions of (a) formaldehyde high concentration high temperature; (b) formaldehyde low concentration high temperature; (c) glyoxylic acid high concentration high temperature; (d) glyoxylic acid low concentration high temperature; (e) formaldehyde high concentration low temperature; (f) formaldehyde low concentration low temperature; (g) glyoxylic acid high concentration low temperature; (h) glyoxylic acid low concentration low temperature. For the convenience of demonstration, the X-ray pattern of the heat-treated deposit is shifted upward.

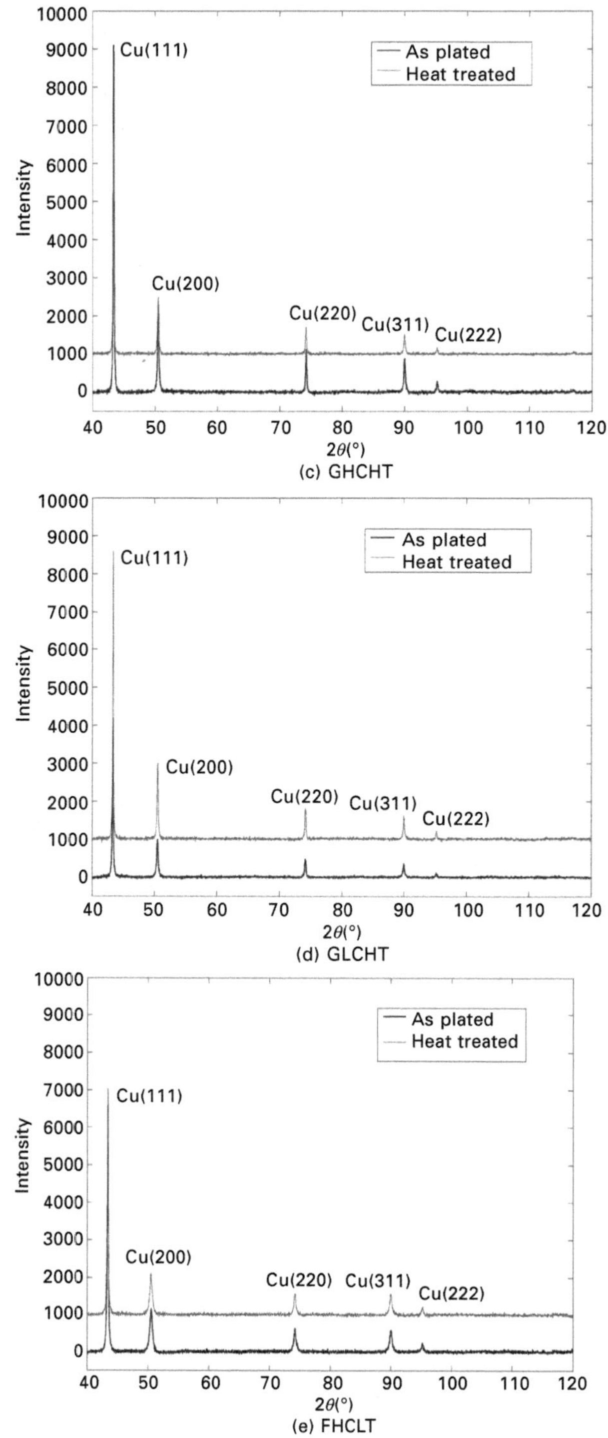

4.2 Continued

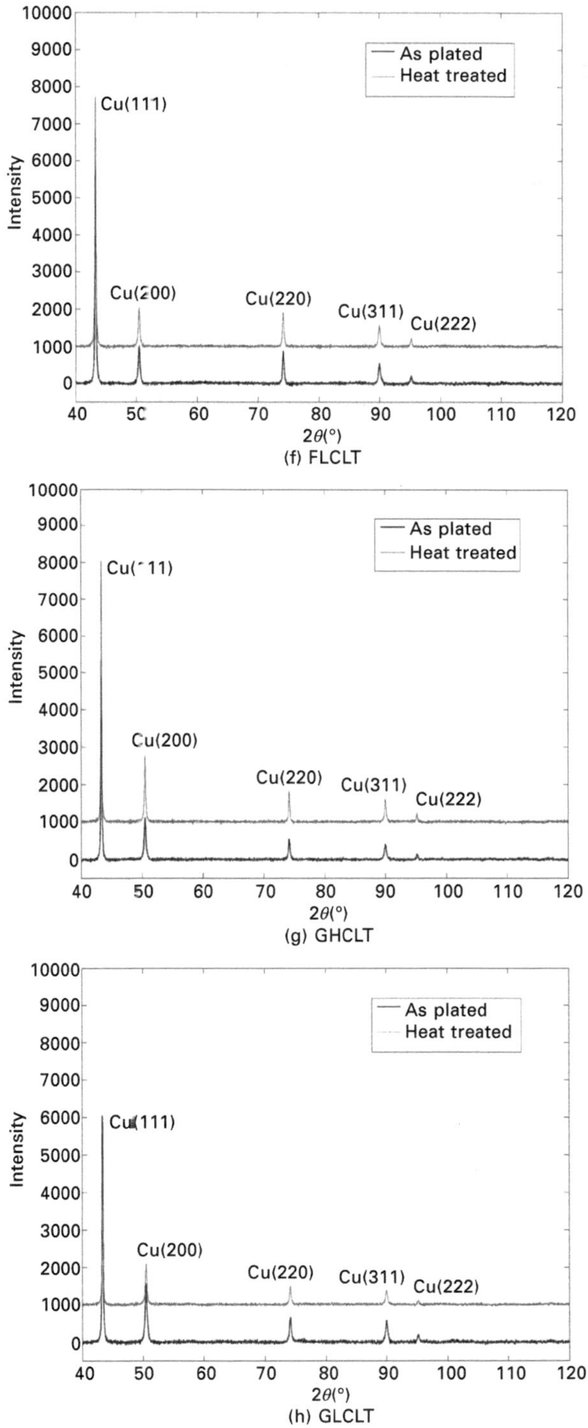

4.2 Continued

Table 4.2 Traditional relative intensities I_{rel} (%) of the first five peaks in the X-ray patterns of as plated and heat-treated deposits from all solutions, compared with the Cu PDF data

		FHCHT		FLCHT		GHCHT		GLCHT	
Index	Cu PDF	As plated	Heat treated	As plated	Heat treated	As plated	Heat treated	As plated	Heat treated
{111}	100	100.0	100.0	100.0	100.0	100.0	100.0	100.0	100.0
{200}	46	40.1	36.1	36.7	36.5	44.7	38.2	36.5	41.8
{220}	20	16.2	18.7	22.2	22.3	15.3	18.8	16.5	16.8
{311}	17	18.3	17.8	18.8	19.0	18.1	18.7	18.6	15.6
{222}	5	6.9	5.8	6.2	6.4	5.7	6.6	6.0	3.9

Table 4.2 continued

		FHCLT		FLCLT		GHCLT		GLCLT	
Index	Cu PDF	As plated	Heat treated	As plated	Heat treated	As plated	Heat treated	As plated	Heat treated
{111}	100	100.0	100.0	100.0	100.0	100.0	100.0	100.0	100.0
{200}	46	37.9	43.6	24.0	25.9	45.0	43.7	46.8	52.9
{220}	20	22.2	20.1	17.4	21.3	19.1	17.8	23.4	23.7
{311}	17	22.3	21.9	15.3	16.9	18.6	20.7	17.8	23.4
{222}	5	8.6	6.8	5.1	4.9	4.9	7.6	7.9	4.9

intensity of the first peak set as the standard 100%, the second peak of the formaldehyde low concentration low temperature (FLCLT) deposit has the smallest integrated intensity compared to all the other deposits and the copper PDF, with only 24% in as plated and 25.9% in heat-treated samples. These values are only just over half of that of the copper PDF, 46%. The glyoxylic acid low concentration low temperature (GLCLT) deposit has the strongest second peak compared to the rest of the deposits and the copper PDF, with 46.8% in as plated deposit and 52.9% in heat-treated deposit, close to that of the copper PDF. The second peaks of the rest of the deposits are all above 36% relative intensities. The relative intensities of the third, fourth and fifth peaks of all the deposits have quite low fluctuation compared to the copper PDF.

The fractional relative intensities show the percentage of the integrated intensity of each peak within the overall intensity. Higher fractional relative intensity of a peak means higher percentage of this type of grain orientation in the deposit. From Table 4.3, the FLCLT deposit has the highest percentage of copper {111} peak, with 61.8% in as plated deposit and 59.2% in heat-treated deposit. At the same time, it has the lowest percentage of copper {200} peak, with 14.8% in as plated deposit and 15.3% in heated treated deposit. Compared with the data of the copper PDF, where the percentage of the first and the second peaks are 53.2% and 24.5% respectively, the FLCLT

Table 4.3 Fractional relative intensities I/I_{total} (%) of the first five peaks in the X-ray patterns of as plated and heat-treated deposits from all solutions, compared with the Cu PDF data

		FHCHT		FLCHT		GHCHT		GLCHT	
Index	Cu PDF	As plated	Heat treated	As plated	Heat treated	As plated	Heat treated	As plated	Heat treated
{111}	53.2	55.1	56.0	54.4	54.3	54.4	54.8	56.3	56.1
{200}	24.5	22.1	20.2	20.0	19.8	24.3	21.0	20.6	23.5
{220}	10.6	8.9	10.5	12.1	12.1	8.3	10.3	9.3	9.5
{311}	9.0	10.1	10.0	10.2	10.3	9.9	10.3	10.5	8.8
{222}	2.7	3.8	3.2	3.4	3.5	3.1	3.6	3.4	2.2

Table 4.3 continued

		FHCLT		FLCLT		GHCLT		GLCLT	
Index	Cu PDF	As plated	Heat treated	As plated	Heat treated	As plated	Heat treated	As plated	Heat treated
{111}	53.2	52.4	52.0	61.8	59.2	53.3	52.7	51.1	48.8
{200}	24.5	19.9	22.7	14.8	15.3	24.0	23.0	23.9	25.8
{220}	10.6	11.6	10.4	10.8	12.6	10.2	9.4	11.9	11.6
{311}	9.0	11.7	11.4	9.5	10.0	9.9	10.9	9.1	11.4
{222}	2.7	4.5	3.6	3.2	2.9	2.6	4.0	4.0	2.4

deposit has a significantly high percentage of the first peak intensity but very low second peak intensity. The GLCLT deposit has the lowest percentage of copper {111} peak, 51.1% in as plated deposit and 48.8% in heat-treated deposit, and its percentage of copper {200} peak is almost the highest among all the deposits. For the rest of the deposits, the fractional relative intensities of each peak are quite close to those of the copper PDF. These show that in the FLCLT deposit, a much larger amount of the copper {111} planes and a much smaller amount of copper {200} planes are oriented parallel to the surface of the deposit, while in the GLCLT deposit, less copper {111} planes are parallel to the surface of the deposit. For the rest of the deposits, a small fluctuation is present on the orientation of the copper lattice planes but overall the grains are roughly randomly distributed.

4.2 Tilted scan patterns and the surface residual stress of the electroless copper

The copper {311} peak shown in Fig. 4.3 becomes weaker and flatter with the increase of the tilted angle. This is due to reduced reception of the X-ray signal with the increase of the tilted angle. There is no significant change of the position of the peaks with different tilted angles in Fig. 4.3, which

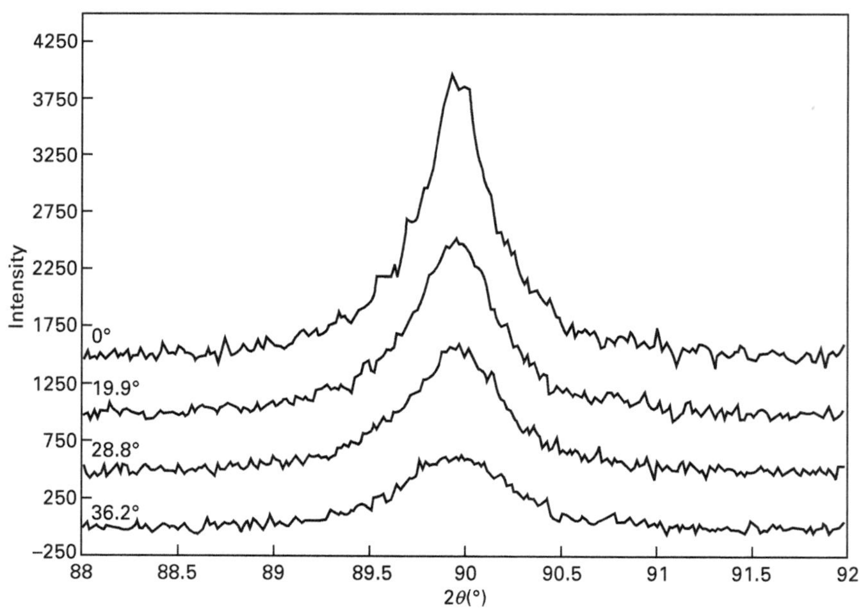

4.3 X-ray patterns of the tilted scan around the copper {311} peak of as plated deposit from the formaldehyde high concentration high temperature solution. The X-ray patterns have been shifted for the convenience of comparison and demonstration.

means that there is no significant amount of surface residual stress on the electroless deposited copper. The same is the case for deposits from other electroless solutions, both before and after heat treatment.

The basic cause of residual stresses in a coating is the remains of non-uniform plastic stress flow in the coating. These internal stresses can be either tensile or compressive. More than 95% of the reflected X-ray beam for the Cu {311} diffraction peak is from within the depth of penetration at 24 µm. Since the thicknesses of the electroless copper deposits are smaller, the entire coatings are subjected to the penetrated X-ray beams.

The electroless copper deposits plated on the different kinds of substrates, namely copper cladded substrate and laminated substrate, both have an insignificant amount of residual surface stress.

4.3 The error in calculating the position and the relative intensities of the peaks

Although before any calculation is carried out the X-ray patterns are background removed and Kα_2 stripped, some background noise is still present and visible in the X-ray patterns shown in Fig. 4.2 in the form of the small amount of vibration of the base line. These noises, varying around zero, can affect the precision of the calculation of the peak positions and the relative intensities.

The background noise affects the calculation precision in two ways. One is from the magnitude of the noise. The bigger the magnitude of the noise is, the higher the error level may be acquired during the calculation. For weak X-ray peaks, high magnitude noise signals can be considered as nearby smaller peaks, which, when included in the calculation, may greatly affect the results. However, for stronger X-ray peaks, the effect of the background noise can be tiny. Another way that the background noise affects the calculation is related with the data range selected for calculation. For the same peak, different ranges of data chosen for calculating the position of the peak may affect the results. Again, this effect might be greater for weaker X-ray peaks but smaller for stronger X-ray peaks.

In order to estimate the error involved in the calculation of the positions and the relative intensities, different data ranges are selected. Same as for the method used in calculating the positions of the peaks of the tilted X-ray scans, the data ranges used include: the full data range (that is 150 data on both sides of the selected peak point), the full data range less 10, 20, 30 and 40 data each on both left and right ends. The average and the drift in percentage of the peak positions and the relative intensities for each peak of each deposit condition are then calculated.

As an example, the formaldehyde high concentration high temperature (FHCHT) deposit is used to perform the error analysis (Table 4.4). For peak

Table 4.4 Error analysis in percentages of the average value of the peak positions and the fractional relative intensities of the formaldehyde high concentration high temperature (FHCHT) deposit

Index	Peak positions error		Fractional relative intensity error	
	As plated	Heat treated	As plated	Heat treated
{111}	±0.002%	±0.002%	±0.434%	±0.496%
{200}	±0.037%	±0.030%	±0.401%	±0.255%
{220}	±0.016%	±0.014%	±0.570%	±0.393%
{311}	±0.012%	±0.015%	±0.570%	±0.418%
{222}	±0.107%	±0.115%	±6.374%	±6.924%

positions, all the errors are in very low levels, with the highest percentages below 0.1% except for the error of the fifth, the copper {222} peak, which exceeds 0.1%. A similar situation arises with the error of the fractional relative intensities, where the error of the fifth peak exceeds 6% but that of the rest of the peaks remains well below 1%. As analysed above, this is because the fifth peak is very weak in the X-ray normal scan pattern. The small vibration of the background introduces greater error to the calculation of its position and relative intensities compared to the rest of the stronger peaks. The errors of calculating the position and the relative intensities of the peaks in X-ray patterns remain in very low levels, and the result is highly reliable.

4.4 The error in linear regression for surface residual stress analysis

In the calculation of the surface residual stress, the linear regression of $d(\theta, \Psi)$ to $\sin^2\Psi$ is required to acquire the slope of the linear fit, $\partial d(\theta, \Psi)/\partial \sin^2\Psi$. Here, Ψ is the tilting angle, the angle between the diffracting lattice planes and the specimen surface, and $d(\theta, \Psi)$ is the lattice spacing at scanning angle 2θ with tilting angle $\Psi>0$. The reliability and the error of the linear regression will affect the calculated surface residual stress. In order to estimate the reliability and the error of the linear regression, the coefficient of determination of the linear regression and the error of the slope of the linear fit are calculated. The error of the slope is then used to replace the slope term to calculate the error of the surface residual stress due to the linear regression.

If the t test of the null hypothesis $H_0 : \beta_1 = 0$ passes, it gives the statistical result that the surface residual stress is zero.

4.5 Conclusions

In this chapter, the crystal structure of both the as plated and heat-treated electroless copper deposits is revealed with the X-ray diffraction method

(XRD). The positions of the peaks in their X-ray patterns are calculated. The relative intensities of the peaks are also calculated. The surface residual stresses of the deposits are analysed with the X-ray tilted scan method.

The electroless copper deposits from all the plating solutions, in both as plated and heat-treated conditions, have a strong fine copper crystal structure. No significant peaks of the copper oxide are in the X-ray patterns, which means that there is no or only little copper oxide present on the surface of the deposits.

The grains on most of the deposits are more or less randomly distributed, as indicated by the relative intensities of the peaks in X-ray patterns. However, a significant amount of the surplus of the copper {111} planes is on the FLCLT deposit compared with the copper PDF data. This means that on this deposit, more grains are oriented with {111} plane parallel to the surface.

There is no significant residual surface stress present. The precision of the data of the position of the peaks and the relative intensities is in a very high level, which indicates the high reliability.

5
The atomic model of the diamond pyramid structure in electroless copper deposits

Abstract: Based on the diamond pyramid structures discovered on some of the deposits, and the chemical analysis and X-ray diffraction (XRD) results, an atomic model of the structure is proposed and developed. The deduction of the atomic model of the diamond pyramid structure starts from a unit diamond pyramid model found in a unit face centred cubic lattice. The multi-layer diamond pyramid model and that with twinning are then developed. Theoretical calculation based on the surface stress is conducted. The results prove this copper {111} stacking model with twinning for the diamond pyramid structures.

Key words: electroless copper deposit, low temperature, diamond pyramid structure, atom model, surface stress.

5.1 The unit diamond pyramid structure in a face centred cubic crystal lattice

In Chapter 2, a large amount of the diamond pyramid structures are observed on the surface of the electroless copper deposit from the formaldehyde low concentration low temperature solution. In Chapter 4, the relative intensity of the X-ray diffraction (XRD) pattern of this deposit shows an excess amount of the copper {111} plane parallel to the surface of the deposit. These evidences indicate that the diamond pyramid structures on the deposit are likely stacking of the copper {111} planes. In this chapter, an atomic model is brought forward to explain the mechanism of forming such diamond pyramid structures. Further analysis based on the surface stress of copper {100} and {111} planes is conducted.

The size of the structures varies with the number of layers that they have, but the common aspect is that the four angles of the diamond base measure around 60° and 120°, respectively, which means that the diamond shape can be divided into two equilateral triangles (Fig. 5.1).

When we examine a unit face centred cubic (fcc) lattice of copper, we can actually find a basic diamond shape that consists of four atoms on each of the {111} planes of the copper fcc lattice. The four atoms of the diamond shape, together with the one atom on the corner closest to the diamond shape in the cubic lattice, form a unit diamond pyramid structure, which is marked with the thick lines in Fig. 5.2. In this diamond pyramid structure which consists

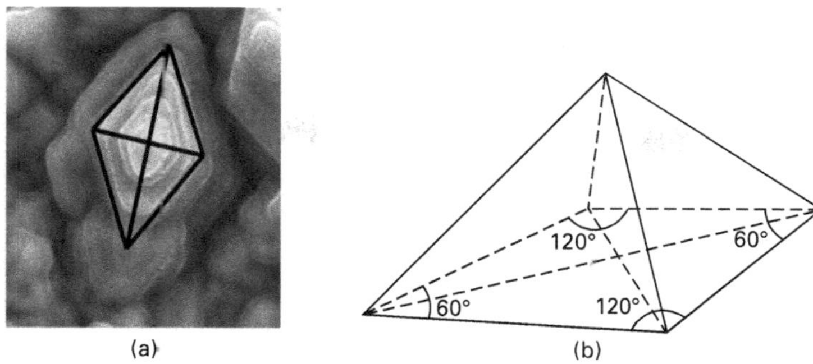

5.1 (a) The outline masked on a single diamond pyramid structure in a SEM image of electroless copper deposit and (b) the basic pyramid-like feature of the diamond pyramid structure with diamond-shaped base.

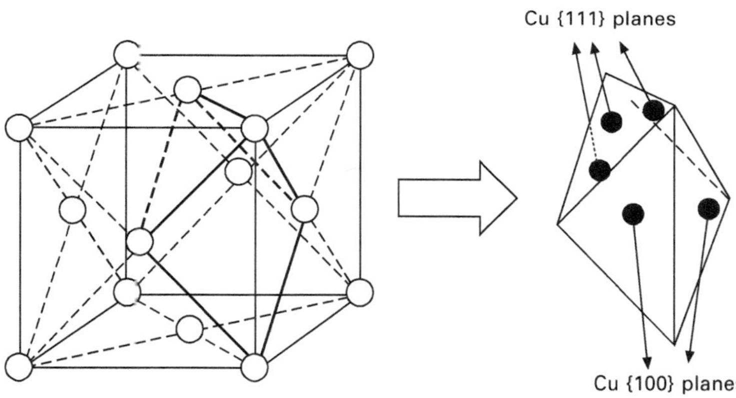

5.2 The unit diamond pyramid structure in a single fcc crystal lattice and the five faces of the structure.

of five faces, the base surface and two side faces are Cu {111} planes, while the other two side planes are Cu {100} planes, as illustrated in Fig. 5.2. If we lay this unit diamond pyramid structure down with its diamond-shaped {111} plane horizontally, it will look similar to the structure illustrated in Fig. 5.1(b). This is the interpretation of the unit diamond pyramid structure. This unit diamond pyramid structure can grow into a much bigger multi-layer diamond pyramid structure that we can see in the scanning electron microscopy (SEM) image shown in Fig. 5.1(a).

To establish the coordinates of the atoms in a unit diamond pyramid model, the five atoms are marked with A, B, C, D and E, respectively, in Fig. 5.3. The x-axis is in the AD direction. The y-axis is in the BC direction and the z-axis is in the FE direction, as shown in Fig. 5.3. Given the position of the atom A as (0, 0, 0) and the distance between the nearest

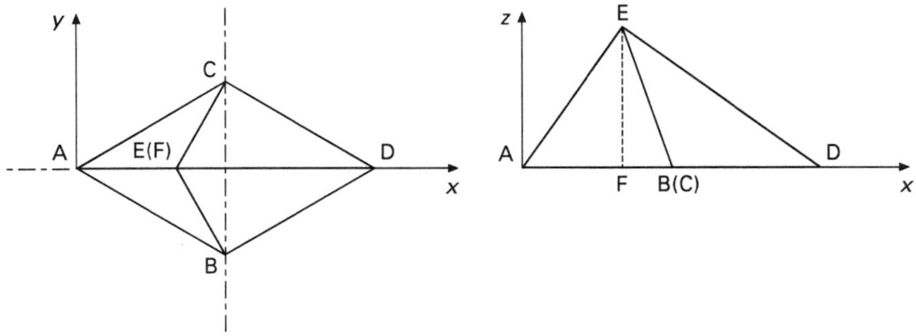

5.3 Sketch of the position of the five atoms in a unit diamond structure. Left: plan view and right: side view.

Table 5.1 Position of the five atoms in a unit diamond pyramid structure

Atom	x	y	z
A	0	0	0
B	$\sqrt{3}/2$	−1/2	0
C	$\sqrt{3}/2$	1/2	0
D	$\sqrt{3}$	0	0
E	$\sqrt{3}/3$	0	$\sqrt{6}/3$

atoms is 1, the position of the rest of the atoms can be calculated, as listed in Table 5.1.

5.2 Multi-layer atomic model

The first part of this section will set up coordinates of the atoms in an n-layer diamond pyramid model. From the unit diamond pyramid structure, we can build up a diamond pyramid structure with more atoms. In Fig. 5.4, each atom in the n-layer model is marked with three notations as A (I J K). The first two notations, I and J, mark the in plan position of the atoms, as the row number and the column number, respectively, in a matrix. The third notation, K, marks the layer in which the atom is positioned. The first layer, with K equal to 1, starts from the top of the diamond pyramid structure, where there is only one atom. The second layer, with K equal to 2, has 4 atoms making a 2 × 2 matrix in plan. Therefore, the ith layer has an $i \times i$ atom matrix making the number of the atoms within the layer i^2.

For an n-layer model, if we let the coordinate of the first atom in the bottom (nth) layer, A (1 1 n), be (0, 0, 0) and the bottom layer sits on the z = 0 plane, following the same system and the deduction in Section 5.1,

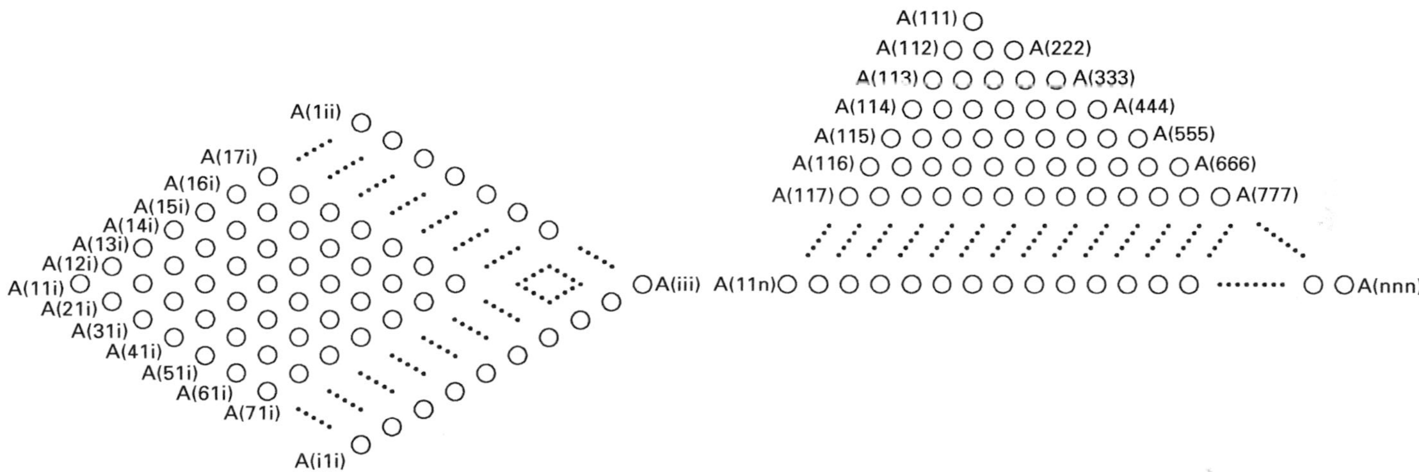

5.4 Sketch of the positions of the atoms in an *n*-layered diamond pyramid structure model. Left: plan view of the *i*th layer and right: side view.

the coordinates of the atom A (i 1 n) will be $\left(\frac{\sqrt{3}}{2}(i-1), -\frac{1}{2}(i-1), 0\right)$. For the atom of A (1 i n), the x and the z coordinates are the same with those of A (i 1 n) and the y coordinate is the negative of that of A (i 1 n). Therefore, the coordinates of atom A (1 i n) is $\left(\frac{\sqrt{3}}{2}(i-1), \frac{1}{2}(i-1), 0\right)$.

Atom A (i j n) in the nth layer can be considered as the atom A (i 1 n) offset by the coordinates of atom A (1 j n). As deduced above, the coordinates of atom A (i 1 n) are $\left(\frac{\sqrt{3}}{2}(i-1), -\frac{1}{2}(i-1), 0\right)$ and those of atom A (1 j n) are $\left(\frac{\sqrt{3}}{2}(j-1), \frac{1}{2}(j-1), 0\right)$. So the coordinates of atom A (i j n) are $\left(\frac{\sqrt{3}}{2}(i+j-2), -\frac{1}{2}(i-j), 0\right)$.

According to the deduction in Section 5.1, atom A (1 1 $n-1$), the first atom in the layer above the bottom layer, has the coordinates of $\left(\frac{\sqrt{3}}{3}, 0, \frac{\sqrt{6}}{3}\right)$, the same as atom E in Fig. 5.3. Therefore, atom A (1 1 k), the first atom in the kth layer, has the coordinates of $\left(\frac{\sqrt{3}}{3}(n-k), 0, \frac{\sqrt{6}}{3}(n-k)\right)$. Atom A ($i$ j k) can be considered as atom A (i j n) offset by the coordinates of atom A (1 1 k). Therefore, the coordinates of any atom in the n-layer diamond pyramid model, A (i j k), are $\left(\frac{\sqrt{3}}{2}(i+j-2) + \frac{\sqrt{3}}{3}(n-k), -\frac{1}{2}(i-j), \frac{\sqrt{6}}{3}(n-k)\right)$.

For the n-layer diamond pyramid model oriented at (x_0, y_0, z_0), the coordinates of atom A (1 1 1) will be (x_0, y_0, z_0), and the coordinates of any atom in the structure A (i j k) will simply be $\left(\frac{\sqrt{3}}{2}(i+j-2) + \frac{\sqrt{3}}{3}(n-k) + x_0, -\frac{1}{2}(i-j) + y_0, \frac{\sqrt{6}}{3}(n-k) + z_0\right)$.

In order to interpret the n-layer diamond pyramid model, Fig. 5.5 illustrates the front-top 45° view and the top view of a 10-layer diamond pyramid structure that is built up according to the unit diamond pyramid structure shown in Fig. 5.2 and the deduction in this section so far, compared with a SEM image. The bottom face and the two side faces (one in the front and another at the back) on the left side of the structure are copper {111} planes and the two side faces on the right side of the structure are copper {100} planes. When compared with the SEM image (Fig. 5.5(c)), the multi-layered

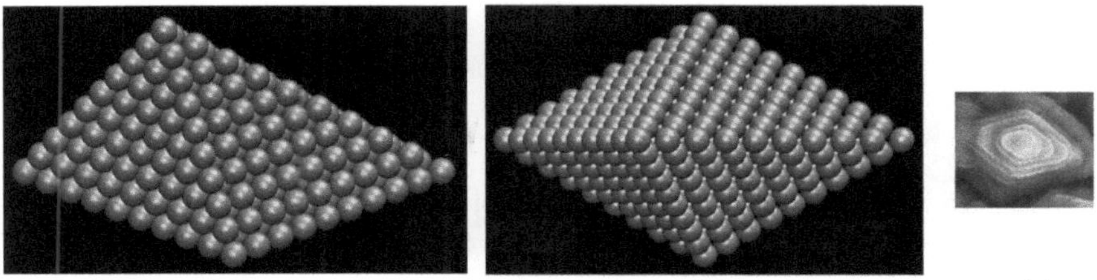

(a) (b) (c)

5.5 (a) Side view and (b) top view of a ten-layer diamond-shaped copper {111} plane stacking, compared with (c) the SEM image of a diamond pyramid structure.

atom model shows a prime feature close to what we can see in the image, but we cannot see on the side faces of the model the steps shown in the image. In addition, the structure in the image looks symmetric in both left–right and up–down directions but the top view of the model does not show a left–right symmetric structure because the left two side-faces of the model are copper {111} planes and the right two side-faces are copper {100} planes.

5.3 Twinning in the diamond pyramid model

Note that from the point of view of the copper {111} plane, the copper lattice has a close packed 'ABCABC…' sequence. According to the atom model given above, the diamond pyramid structure has two copper {100} planes on two side faces. However, it is preferable to form more copper {111} planes and avoid forming a large integrated area of copper {100} plane because the surface energy of the copper {100} plane is higher than that of the {111} plane. Nonetheless, the copper {100} plane is inevitable in the above diamond pyramid structure model. During the growth of the diamond pyramid structure, the whole structure should tend to avoid forming a big area of the copper {100} plane in order to lower the local energy. In this case, the original stacking chain such as 'ABCABCABC…' is broken and it may become like 'ABCABACBA…', whereas if 'ABCABC' forms copper {111} planes on the left side of the structure, the 'CBACBA' stacking forms copper {111} planes on the right side of the structure, illustrated in Fig. 5.6. If we mark the continual layers that have copper {111} planes on their left sides as section 'L' and those having copper {111} planes on their right sides as section 'R', the twinning diamond pyramid model can be described as the repeating stacking of 'L' and 'R' sections. The layers standing in between an 'L' section and an 'R' section are twinning planes. In a top view, for certain layers of atom stacking, the {111} plane sides look narrower and denser than the {100} plane sides, and thus, the step feature can be observed from the top.

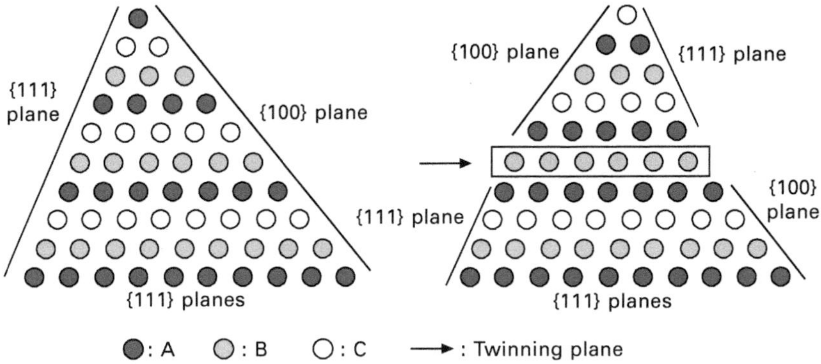

5.6 Illustration of a 10-layer diamond pyramid structure (left) and a twinned 10-layer diamond pyramid structure.

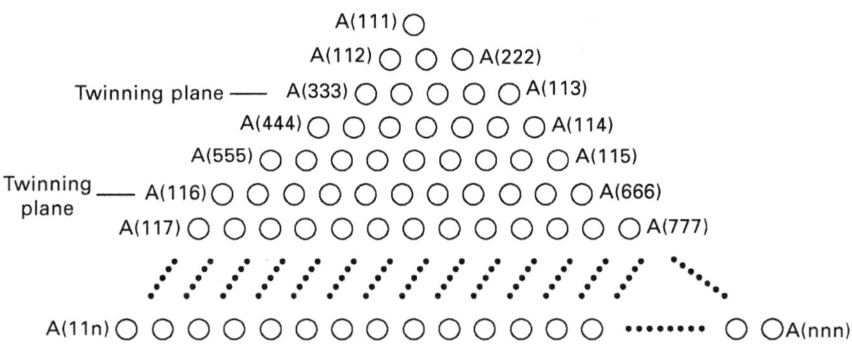

5.7 The notation arrangement of the atoms in the multi-layer diamond pyramid model with twinning.

Figure 5.7 builds a diamond pyramid structure with twinning. In each 'L' section, the notation starts from the atom in the left end, and in each 'R' section, the notation starts from the atom in the right end.

For the 'L' sections, the coordinates of the left end atom in its bottom layer is calculated according to Section 5.1 based on the coordinates of the left end atom in one layer below. Following the deduction in Section 5.2, the coordinates of all the atoms in the 'L' section can be calculated. For example, if the coordinates of the left end atom in one layer below the 'L' section are (x_0, y_0, z_0), the coordinates of the left end atom in the bottom layer of the 'L' section A (0 0 0) are $\left(x_0 + \frac{\sqrt{3}}{3}, y_0, z_0 + \frac{\sqrt{6}}{3}\right)$, and the atom with notation $(i\ j\ k)$ in this section (not in the whole model) is $\left(\frac{\sqrt{3}}{2}(i + j - 2) + \frac{\sqrt{3}}{3}(n - k) + x_0 + \frac{\sqrt{3}}{3}, -\frac{1}{2}(i - j) + y_0, \frac{\sqrt{6}}{3}(n - k) + z_0 + \frac{\sqrt{6}}{3}\right)$.

The bottom layer of the 'L' section is an $n \times n$ layer and the atom A $(i\ j\ k)$ is in a $k \times k$ layer.

For the 'R' sections, the coordinates of the right end atom in its bottom layer are calculated based on the right end atom in one layer below the section. The coordinates of the rest of the atoms can then be calculated by offsetting the atoms from right to left. Similar to the 'L' sections, if the coordinates of the right end atom in one layer below the 'R' section are (x_0, y_0, z_0), the coordinates of the right end atom in the bottom layer of the 'R' section A (0 0 0) are $\left(x_0 - \frac{\sqrt{3}}{3}, y_0, z_0 + \frac{\sqrt{6}}{3}\right)$. The atom with notation $(i\ j\ k)$ in this section (not in the whole model) is $\left(-\frac{\sqrt{3}}{2}(i+j-2) + \frac{\sqrt{3}}{3}(n-k) + x_0 - \frac{\sqrt{3}}{3}, -\frac{1}{2}(i-j) + y_0, \frac{\sqrt{6}}{3}(n-k) + z_0 + \frac{\sqrt{6}}{3}\right)$. The bottom layer of the 'R' section is an $n \times n$ layer and the atom A $(i\ j\ k)$ is in a $k \times k$ layer.

Following the deduction above, a program can be written to write a text file with the coordinates of the atoms in a diamond pyramid structure of any layer with or without twinning. The MatLab program used to generate the coordinates of the atoms in a pyramid structure is shown below. The program produces a diamond.xyz file containing the coordinates of all the atoms in a diamond pyramid structure with chosen number of layers and twinning.

```
% This program generates the coordinates of the atoms
for a diamond-shaped structure.
fprintf ('\n ######## Diamond Generator
########\n');

Nlayer = 0; % Total number of the layers.
while (Nlayer == 0) | (Nlayer > 99)
  Nlayer = input ('\n Please input the total number
of the layers of the diamond (1-99): ');
  if isempty (Nlayer)
    fprintf ('\n No input detected.');
    Nlayer = 0;
  end
end

fprintf ('\n The total number of the layers is set to
%d. Please press any key to continue.\n', Nlayer);
pause;
```

```
fprintf ('\n Please input the number of the layers in
each dislocated section (>0 and <%d) [%d]:', Nlayer,
Nlayer);
d = input (' ');
if (isempty (d)) | (d > Nlayer) | (d < 0)
  d = Nlayer;
end

fprintf ('\n The number of the layers in each section is
set to %d. Please press any key to continue.\n', d);
pause;

% Calculate the number of the dislocated sections.
Nsection = ceil (Nlayer/d);

scale = input ('\n Please input the scale [3]:');
if isempty (scale)
  scale = 3;
end

Natom= 0; % Total number of atoms.
for i = 1 : Nlayer
  Natom = Natom + (Nlayer-i+1)^2;
end

foutput = fopen ('diamond.xyz','w');
fprintf (foutput, '%d \n', Natom);
fprintf (foutput, 'This file is generated by Matlab
program
''Diamond.m.''');

% Start to write coordinate file.

layer = 0; %Initialize the layer counter.
a0 = 0; b0 = 0; c0 = 0; %Initialize the starting
point.
a = sqrt(3)/2; b = 1/sqrt(3); c = sin(acos(b));
%Initialize some constants.

for s = 1 : Nsection
  ssign = (-1)^(s-1);

  for n = 1 : d
```

```
    layer = layer + 1;

    for m = 1 : (Nlayer-layer+1)    %line

      for l = 1 : (Nlayer-layer+1) %atom

        x  =  a0  +  (ssign*(a*(l+m-2)  +  b*(n-
1)))*scale;
        y = b0 - (ssign*(l-m)/2)*scale;
        z = c0 + (c*(n-1))*scale;

        fprintf (foutput, '\nCu \t %8.5f \t%8.5f
\t%8.5f', x, y, z);

      end

    end

    if layer == Nlayer

      break;

    end

  end

  a0 = x - ssign*b*scale; % Reset starting point.
  b0 = 0;
  c0 = z + c*scale;

end

fprintf ('\n###### Coordination file prepared. Please
check file ''diamond.xyz''. ######\n\n');

fclose (foutput);
clear;
```

In Fig. 5.8, there are four-layer atoms in each ‘L’ and ‘R’ section. The base diamond layer measures around 60 × 35 nm in diagonals, which is only about 1/10 of a real diamond pyramid structure on the electroless copper deposit. The step feature can be clearly observed on the four side faces of the structure, and the model looks symmetric in both left–right and up–down

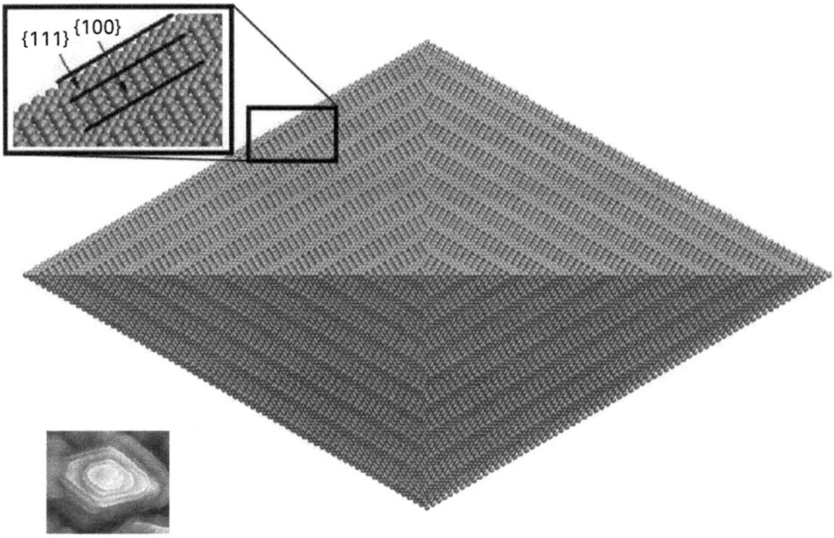

5.8 The top view of a 70-layer diamond pyramid structure with twinning, compared with a SEM image of the diamond pyramid structure.

directions, which is already very close to the SEM image. The only significant difference between the appearance of the model and that of the image is that the edges of the structure in the image are smooth and round, while those in the model are sharp. This is because the model is an ideal model, in which each atom is placed exactly on the ideal copper fcc lattice. If the model structure is left relaxed for only a very short time, the sharp edges in the structure, where the surface energy is relatively high, will be rounded. Then, the model structure will look more similar to structure shown in the image. Furthermore, notice that the model shown in Fig. 5.8 is only about 1/10 of the real structure. By increasing the number of the layer in the model, the appearance of the model will also be closer to the real structure. The model itself further supports the assumption stated above that the layered diamond pyramid structure observed on the surface of electroless copper deposit is copper {111} plane stacking.

5.4 The role of surface stress in the formation of twinning in a diamond pyramid

5.4.1 Surface stress on copper {111} and {100} planes

As stated above, the possible reason for forming twinning in a diamond pyramid structure is the surface energy. The copper {100} surface plane has a higher level of surface energy than the copper {111} surface plane. Below is a detailed discussion.

The surface stress $\tau_{\alpha\beta}$ can be expressed as:

$$\tau_{\alpha\beta} = A^{-1}\frac{\partial(\gamma A)}{\partial \varepsilon_{\alpha\beta}} \quad 5.1$$

where A is the area of the surface unit cell, γ is the surface energy, and $\varepsilon_{\alpha\beta}$ is the strain tensor. Assume that the surface lies in the x–y plane and its normal is in the z direction. For a free surface, the stress in the z direction τ_{zz} is zero. For the {100} and the {111} planes in fcc lattice, the surface stress is isotropic, which means $\tau_{xx} = \tau_{yy} = \tau_0$ and A can be considered as a constant. In this case, the surface energy γ can be expressed as:

$$\gamma = \gamma_0 + \tau_0(\varepsilon_{xx} + \varepsilon_{yy}) \quad 5.2$$

According to calculation, for a relaxed copper lattice, the γ_0 value of the {100} plane and the {111} plane are 0.1031 and 0.0879 eVÅ^{-2}, respectively, and the τ_0 value of the {100} plane and the {111} plane are 0.1312 and 0.1203 eVÅ^{-2}, respectively. Therefore, the surface energy of the relaxed copper {100} plane and the copper {111} plane can be written as:

$$\gamma(100) = 0.1031 + 0.1312(\varepsilon_{xx} + \varepsilon_{yy}) \quad 5.3$$

$$\gamma(111) = 0.0879 + 0.1203(\varepsilon_{xx} + \varepsilon_{yy}) \quad 5.4$$

The positive values of the surface stress of the copper {100} and {111} shown above indicate the compressive strain within these two surfaces (Gill, 2007). These stresses will induce an uneven match at the edge of the copper {100} and {111} face.

5.4.2 Calculation of the normal direction of the edges

The normal direction of each edge of the copper {111} plane and the copper {100} plane (Fig. 5.9) is calculated according to the coordinates of the points listed in Table 5.1.

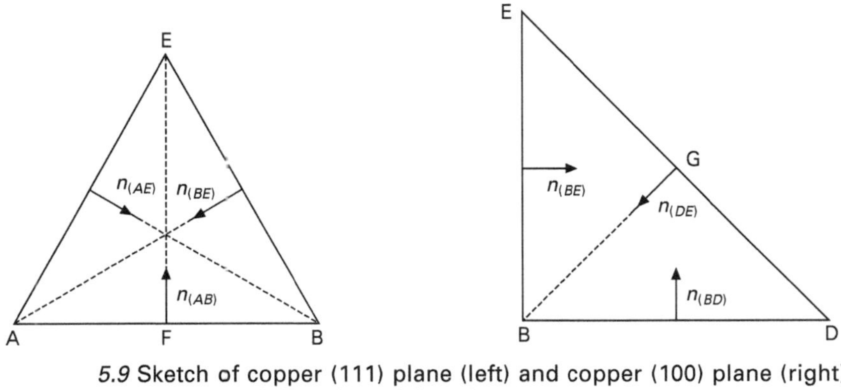

5.9 Sketch of copper (111) plane (left) and copper (100) plane (right) and the normal direction of the edges.

As an example, the calculation of the normal direction of line AB is demonstrated below. The normal direction of line AB, $\vec{n}(AB)$, is the same as the direction from point F, the mid point of line AB, to point E. Let the length of AB equal 1; we have:

$$\begin{aligned}\vec{n}_{(AB)} &= \frac{\overrightarrow{FE}}{|\overrightarrow{FE}|} \\ &= \frac{E-(A+B)/2}{|\overrightarrow{FE}|} \\ &= \frac{(\sqrt{3}/3, 0, \sqrt{6}/3)-[(0,0,0)+(\sqrt{3}/2, -1/2, 0)]/2}{\sqrt{3}/2} \\ &= (1/6, \sqrt{3}/6, 2\sqrt{2}/3)\end{aligned} \tag{5.5}$$

Following the same way, the normal direction of all the rest of the edges can be calculated. The results are listed in Table 5.2.

5.4.3 The surface stress induced external force on the tip of the diamond pyramid

Let us consider a small diamond pyramid structure, outlined with ABCDE in Fig. 5.10, at the tip of a big diamond pyramid structure. As analysed in Section 5.4.1, the small diamond pyramid structure ABCDE will get surface stress induced external force from the rest of the big diamond pyramid structure within the copper {111} plane and the {100} plane. The net external force on the small structure ABCDE will attempt to push it to do a translation movement to a stable position. Due to the symmetric nature of the structure, only the forces in the $-y$ half space need to be calculated and

Table 5.2 Normal direction of the edges on copper {111} and {100} planes in a diamond pyramid structure

Face	Edge	n_x	n_y	n_z
{111}	AB	1/6	$\sqrt{3}/6$	$-2\sqrt{2}/3$
	BE	–5/6	$\sqrt{3}/6$	$-\sqrt{2}/3$
	AE	2/3	$-\sqrt{3}/3$	$-\sqrt{2}/3$
{100}	BD	$-\sqrt{3}/6$	1/2	$\sqrt{6}/3$
	BE	$\sqrt{3}/2$	1/2	0
	DE	$-\sqrt{6}/6$	$-\sqrt{2}/2$	$-\sqrt{3}/3$

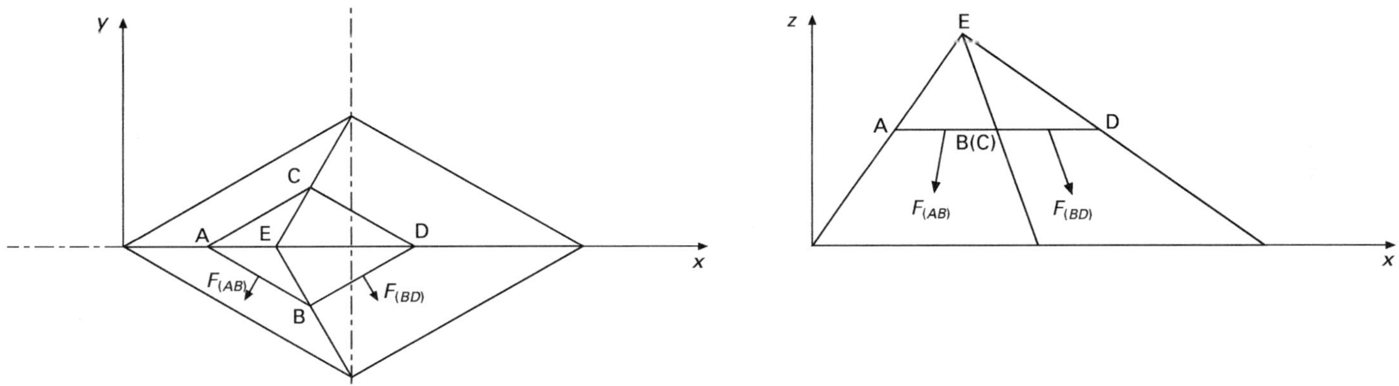

5.10 The small diamond pyramid structure cut off from a big one and the external force on it induced by the surface stress of the structure below it in $-y$.

can be doubled to get the overall force on the whole structure ABCDE in the x and z directions. The overall force on the y direction will be zero.

As shown in Fig. 5.10, the direction of the surface stress induced forces is perpendicular to the edge within the planes, the same as the normal direction of the corresponding edges. The magnitude of the forces equals the surface stress τ_0 times the length of the edges. The direction of the forces can be acquired from Table 5.2. For edge AB on the copper {111} plane, the surface stress induced external force will be:

$$\begin{aligned}\vec{F}_{(AB)} &= \tau_{0(111)} \times (-\vec{n}_{(AB)})l_{(AB)} \\ &= 0.1203 \times \left(-\frac{1}{6}, -\frac{\sqrt{3}}{6}, \frac{2\sqrt{2}}{3}\right) \times l_{(AB)}\end{aligned} \quad 5.6$$

For edge BD in the copper (100) plane, similarly we have:

$$\begin{aligned}\vec{F}_{(BD)} &= \tau_{0(111)} \times (-\vec{n}_{(BD)})l_{(BD)} \\ &= 0.1312 \times \left(\frac{\sqrt{3}}{6}, -\frac{1}{2}, -\frac{\sqrt{6}}{3}\right) \times l_{(BD)}\end{aligned} \quad 5.7$$

Let $l_{(AB)} = l_{(BD)} = l_0$. The overall surface stress induced external force on structure ABCDE in x direction will be:

$$\begin{aligned}F_x &= 2 \times (F_{x(AB)} + F_{x(BD)}) \\ &= 2 \times \left(-0.1203 \times \frac{1}{6}l_{(AB)} + 0.1312 \times \frac{\sqrt{3}}{6}l_{(BD)}\right) \\ &= 0.0178 l_0\end{aligned} \quad 5.8$$

This indicates that for the top part of a diamond pyramid structure, the surface stress induced external force from the rest of the structure is positive in the x direction, which will push it to do a translation movement in positive x direction.

The above calculation does not change when the structure ABCDE becomes infinitely small. This means that a fine diamond pyramid is not a stable structure. The surface stress in the copper {111} and {100} planes will push the tip of the structure to move from the copper {111} side to the {100} side.

5.4.4 The surface stress induced forces on the side edges of a diamond pyramid

The surface stress induces forces on the side edge EB within the copper {100} plane and the {111} plane in a fine diamond pyramid structure (Fig.

5.11). The net force on EB in the x direction will push the edge to rotate around point B if the bottom surface ABDC is position fixed. Therefore, the position of the atoms in the structure will be shifted within the x–y plane.

From Fig. 5.11 and Table 5.2, following a procedure similar to that in Section 5.4.3, the surface stress induced forces can be calculated as below:

$$\vec{F}_{(BE)(111)} = \tau_{0(111)} \times \vec{n}_{(BE)(111)} l_{(BE)}$$
$$= 0.1203 \times \left(-\frac{5}{6}, \frac{\sqrt{3}}{6}, -\frac{\sqrt{2}}{3}\right) \times l_{(BE)} \quad 5.9$$

$$\vec{F}_{(BE)(100)} = \tau_{0(100)} \times \vec{n}_{(BE)(100)} l_{(BE)}$$
$$= 0.1312 \times \left(\frac{\sqrt{3}}{2}, \frac{1}{2}, 0\right) \times l_{(BE)} \quad 5.10$$

The overall force on edge BE in the x direction will be:

$$F_x = \vec{F}_{x(BE)(111)} + \vec{F}_{x(BE)(100)}$$
$$= -0.1203 \times \frac{5}{6} l_{(BE)} + 0.1312 \times \frac{\sqrt{3}}{2} l_{(BE)} \quad 5.11$$
$$= 0.0134 l_0$$

The same result will be acquired for edge CE.

Similarly, for edge AE and DE, the overall surface stress induced force can be calculated as below:

$$\vec{F}_{(AE)} = 2 \times \tau_{0(111)} \times \vec{n}_{(AE)} l_{(AE)}$$
$$= 2 \times 0.1203 \times \left(\frac{2}{3}, -\frac{\sqrt{3}}{3}, -\frac{\sqrt{2}}{3}\right) \times l_{(AE)} \quad 5.12$$

$$\vec{F}_{(DE)} = 2 \times \tau_{0(111)} \times \vec{n}_{(DE)} l_{(DE)}$$
$$= 2 \times 0.1312 \times \left(-\frac{\sqrt{6}}{6}, -\frac{\sqrt{2}}{2}, -\frac{\sqrt{3}}{3}\right) \times l_{(DE)} \quad 5.13$$

The overall force on these two edges in the x direction will be:

$$\vec{F}_{(AE)x} = 2 \times 0.1203 \times \frac{2}{3} \times l_{(AE)} = 0.1604 l_0 \quad 5.14$$

$$\vec{F}_{(DE)x} = 2 \times 0.1312 \times \left(-\frac{\sqrt{6}}{6}\right) \times l_{(DE)} = -0.1515 l_0 \quad 5.15$$

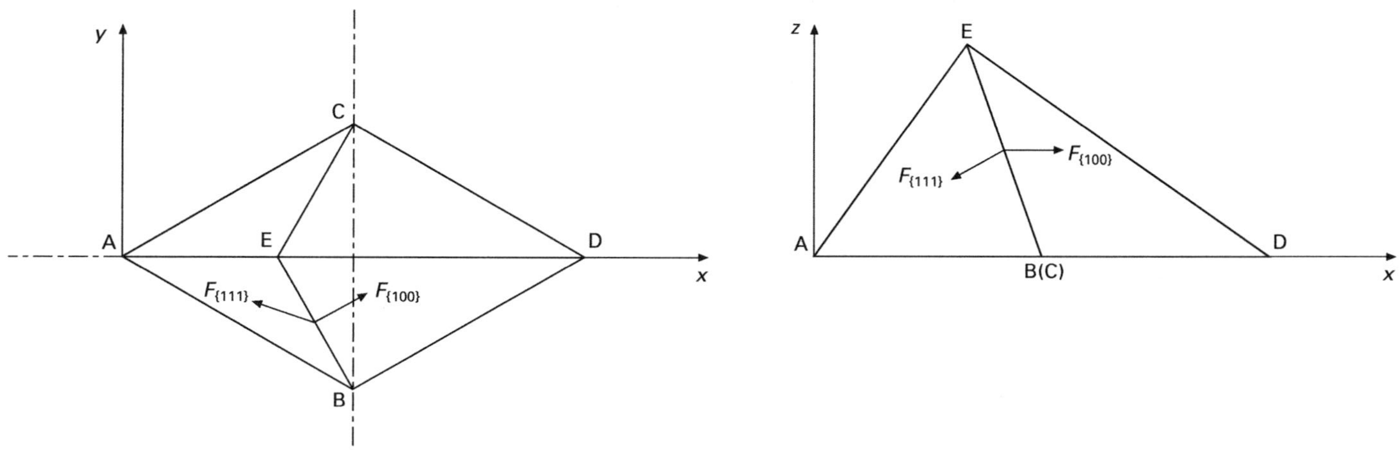

5.11 Sketch of the surface stress and the tensile forces on the edge of the copper {100} and {111} planes in a fine diamond pyramid structure.

The overall force on these two edges in the z direction will be:

$$\vec{F}_{(AE)z} = 2 \times 0.1203 \times \left(-\frac{\sqrt{2}}{3}\right) \times l_{(AE)} = -0.1134 l_0 \tag{5.16}$$

$$\vec{F}_{(DE)z} = 2 \times 0.1312 \times \left(-\frac{\sqrt{3}}{3}\right) \times l_{(DE)} = -0.2142 l_0 \tag{5.17}$$

This result indicates that for a fine diamond pyramid structure, a net internal force due to the surface stress of the copper {100} and {111} planes is present on the side edge of the structure, pointing from the {111} side to the {100} side. For the copper {111} side edge, the surface stress induced force is in $+x$ and $-z$ direction, and for the copper {100} side edge, it is in $-x$ and $-z$ direction.

The analysis results in both Section 5.4.3 and this section show that, for a fine diamond pyramid structure, the unbalanced surface stress in the copper {111} plane and the copper {100} plane will change the shape of the structure to a more stable appearance. The tip of the structure will move from the {111} side toward the {100} side due to the surface stress induced external force, and a net internal force in the x direction is present in the structure. The overall effect is to shape the structure in a symmetric way, not only the mirror symmetry in the y direction, which the structure has already had, but also in the x direction.

For a diamond pyramid structure with an even number of twins of the same thickness, the surface induced forces on the side edges in the x direction within adjacent twins are the same in magnitude but opposite in direction according to the calculation above (Fig. 5.12). Therefore, the overall internal force due to the surface stress in the x direction in the entire twinned structure is zero. In this case, the formation of twins in the diamond pyramid structure helps to reduce the surface stress induced force in the structure and hence the twinned diamond pyramid structure is more stable than the fine one. However, twinning itself may slightly increase the energy of the structure (Kibey *et al.*, 2006). The barrier energy that the atoms have to overcome to form one layer of twinning is also very high in a big fine diamond pyramid structure. Therefore, it is not possible to form twinning by shearing under the surface stress induced forces in a large-scale diamond pyramid structure. For a small sized diamond pyramid structure, e.g., of five layers, the surface stress induced forces are large enough to shift the atoms into a twinning position (see Chapter 6 for details). So, during the growth of the diamond pyramid structure, twins tend to form to acquire symmetric structure due to the surface stress within in the very early stage. Once twins form, they will be able to keep growing. This analysis further proves the reliability of the twinned diamond pyramid model.

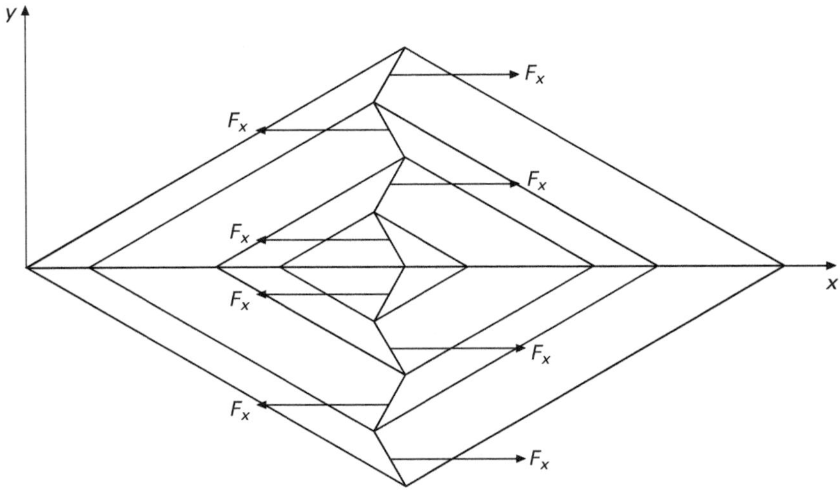

5.12 Sketch of the surface stress induced forces in a twinned diamond pyramid structure in the *x* direction.

5.5 Conclusions

In this chapter, an atom model is brought forward to explain the diamond pyramid structures observed in the surface morphology SEM images of some electroless copper deposits. The deduction starts from a unit diamond pyramid model found in a unit face centred cubic lattice. The multi-layer diamond pyramid model and that with twinning are then developed. The coordinates of the atoms in the model are deduced analytically in a general form, with a program developed in MatLab to generate text files containing the coordinates of the atom in an atomic model with a given number of layers and twins. With the program, the diamond pyramid model can be visualised and compared directly with the diamond pyramid structure observed in the SEM images. The appearance of the twinned diamond pyramid model matches that of the diamond pyramid structure observed in the microscopy morphology images, which indicates that the model is reliable. A further analysis based on the surface stress of copper {100} and {111} planes proves that in a fine diamond pyramid model, a net surface stress induced force is present on the side edges and the tip of the structure pointing from the {111} side to the {100} side. In a twinned model with an even number of twins of the same thickness, this force is balanced in the whole structure. The twinned diamond pyramid structure is more stable than the fine one and tends to form during the growing process.

5.6 References

Gill S P A (2007), 'The effect of surface-stress on the concentration of stress at nanoscale surface flaws', *Int J Solids Struct*, **44**, 7500–9.

Kibey S, Liu J B, Johnson D D and Sehitoglu H (2006), 'Generalized planar fault energies and twinning in Cu–Al alloys', *Appl Phys Lett*, **89**, 191911.

6
Molecular dynamics (MD) simulation of the diamond pyramid structure in electroless copper deposits

Abstract: Molecular dynamics (MD) simulation is shown based on the atomic model to investigate the relaxation and the growing processes of the diamond pyramid structures. The simulation of the effect of the surface stress on the shape of the diamond pyramid structure shows the movement of atoms under the surface stress induced forces. The final structures of the simulation are compared with the microscopy images and the atomic model. The mechanism of forming such structure is analysed. The radial distribution function from the final structures of the simulation and from the X-ray diffraction (XRD) pattern of the electroless copper deposit is compared.

Key words: copper, plating, molecular dynamics simulation, radial distribution function, relaxation.

6.1 Simulation setup

In Chapter 2, we observed a large amount of diamond pyramid structures on the deposit from the formaldehyde low concentration low temperature solution. In Chapter 5, an atomic model is brought forward and further analysis based on surface stress is shown to explain the mechanism of forming such a structure in an electroless copper deposit. The analysis shows that the diamond pyramid structure is highly likely to be copper {111} stacking with twinning. In this chapter, molecular dynamics (MD) simulation procedures are established to simulate the relaxation and the growing process of the diamond pyramid structure. The software package used in this chapter is XMD, which was originally designed for the simulation of metals and alloys.

6.1.1 Relaxation of the diamond pyramid structure

The purpose of the relaxation calculation is to examine if a fine diamond pyramid structure created according to the atomic model given in Chapter 5 can be relaxed into a structure that looks more similar to the structures we see in the scanning electron microscopy images. In this calculation, diamond pyramid structures are created on copper {111} oriented rectangular substrates. All substrates contain five layers of copper atoms and the bottom three layers of atoms are fixed on all *x*, *y* and *z* directions. The differences among the three models are the number of layers of the diamond pyramid structure and the number of twins in it (Table 6.1). In models 1 and 2, we have fine diamond pyramid structures of 5 and 11 layers, and in model 3, we have a 15-layer diamond pyramid structure with three twins, i.e., two twinning planes. The simulation temperature is set to 273 K isothermally for all three models and the pressure is set to zero for convenience. The time step is 1×10^{-15} s and all the models are relaxed for 50,000 steps.

In addition, three 5-layer diamond pyramid structure models (numbers 4 to 6 in Table 6.1) with temperature set to 0 K, 50 K and 293 K, respectively, are also set up for simulation. These models have the same configuration with the above 5-layer model except for the temperature. Among them, the total simulation time of the 0 K model is set to 2,000,000 steps in order to view significant movement of atoms. The results of these simulations are used to show the temperature effect on the relaxation of the structure and the movement of the atoms in the diamond pyramid structure under the surface stress induced forces.

6.1.2 The temperature effect on voids and the formation and growth of the pyramid

One other type of calculation is to show the effect of the plating temperature on the formation of voids in the electroless copper deposit (Chapter 3). In this calculation, a 5-layer diamond pyramid structure is created centred on a 5-layer base the same as in the relaxation calculation. The original model is relaxed for 10,000 steps in isothermal condition and then extra atoms are added above the structure. The extra atoms are added in the following process:

1. Randomly assigning the coordinates of 100 atoms above the original diamond pyramid structure and the base.
2. Assigning an external force of 5×10^{-5} N on each of these 100 atoms. The direction of the external forces points to the centre of the top layer of the base. The reason for using central directed forces is that during the electroless copper plating, reactions initiate on the spots with negative charges. Normally, MD simulation will not be able to simulate the

Table 6.1 Setup of the relaxation models

Model	1	2	3	4	5	6
Number of the layers of the diamond pyramid structure	5	11	15	5	5	5
Number of twins in the structure	0	0	3	0	0	0
Relaxation temperature (K)	273	273	273	0	50	293

chemical reaction process but by using directed forces, the feature of the reaction can be partly described.

3. Assigning the temperature to the whole system including the extra atoms just added. This will automatically assign a random velocity to each of the extra atoms according to the Maxwell–Boltzmann distribution.
4. Performing a MD calculation for 30,000 steps, after which all of the extra atoms that are added will be placed on the substrate.
5. Removing all the atoms below the bottom layer of the base. Due to the random velocity assigned on each atom in step 3, not all the extra atoms can successfully land on the substrate. The failed atoms are removed to reduce the load of calculation.
6. Steps 1 to 5 are repeated 20 cycles and the morphology of the deposit is examined.

The temperature of the simulation is set to 50 K and 273 K, in order to show the trend of the temperature effect on the structure of the deposit.

A further type of calculation is to examine the growth of the diamond pyramid structure to see if the twin structure proposed in the atomic model in Chapter 5 can self-form. In this calculation, a 1-layer diamond pyramid structure (with only 1 atom) is created centred on a 5-layer base the same as in the relaxation calculation. The rest of the calculation procedure is the same as above in this section, and steps 1 to 5 are repeated until diamond pyramid structures can be observed.

6.1.3 Radial distribution function and its calculation from X-ray diffraction (XRD) profile

The radial distribution function (RDF) describes how the density of surrounding matter varies as a function of the distance from a point. It gives the information concerning the frequency with which certain distances occur. Similar to the X-ray diffraction (XRD) profile, for a crystalline structure, strong peaks can be observed on its RDF graph, indicating certain atomic intervals frequently present in the structure. In MD simulation, the RDF is calculated by counting the number of the atom pairs between given ranges of separation. The results can be transferred to the average density of atoms as a function of distance. The general expression to calculate the RDF $G(r)$ is:

$$rG(r) = 4\pi r^2(\rho(r) - \rho(0)) \qquad 6.1$$

where $\rho(r)$ is the density of the atoms at distance r and $\rho(0)$ is the bulk density of the atoms.

The appropriate Fourier transform of the XRD intensity data yields the RDF. While the XRD pattern gives the intensity counts versus scanning angle 2θ (and the interatomic distance d by Bragg's law), its Fourier transform gives the frequency with which certain interatomic distance may occur in the sample. The transform can be described with the following function:

$$rG(r) = \frac{2r}{\pi}\int_0^{\infty} si(s)\sin(sr)ds \qquad 6.2$$

where $s = (4\pi \sin\theta)/\lambda$ and $i(s)$ is defined as the reduced intensity, described as follows:

$$i(s) = \frac{I_t(s) - I_b(s)}{[f(s)]^2} \qquad 6.3$$

where $I_t(s)$ and $I_b(s)$ are, respectively, the total and the background intensity of the XRD and $f(s)$ is the atomic scattering factor. For a system with more than one type of atom, the denominator of the above function should be the sum of the weighted scattering factor of each type of atom.

After the deposition calculation, the RDF of the simulated model (with base atoms removed) is calculated and compared to the results that are calculated from the experimental XRD according to the above equations.

6.2 Preparing models for molecular dynamics calculation

In XMD, the simulation model is built up by filling the boundary box with repeating crystalline lattice cells oriented in a given direction. In order to build an n-layer diamond pyramid structure, the steps below are followed:

1. Determine the geometry parameters by the given number of layers of the diamond pyramid structure to be built.
2. Fill a cubic boundary box with copper atoms with the orientation of {100}, {010} and {001} lattice direction in x, y and z direction of the boundary box, respectively. The size of the box can be appointed or pre-determined by the size of the diamond pyramid structure required in step 1.
3. According to the size of the diamond pyramid structure required, the diamond pyramid structure is shaped by removing excess atoms above a copper {111} plane.
4. The rest of the structure is then rotated to make the diagonal directions of the diamond sit on the x and y directions, respectively. More atoms are then removed to create a rectangular base.
5. For the deposition calculation, the extra atoms are randomly positioned above the structure, all assigned with an external force pointing to the centre area of the structure. Atoms are added in groups for a longer deposition procedure.

The detailed procedure of creating the atomic model is described below.

Let the dimension of the cubic boundary box be a, with the unit of the copper lattice constant (which applies to all the length units used below). Let $a = 2^t$, which means that there will be even numbers of unit lattice in all three directions, and there will be an atom sitting at the centre of the boundary box.

Let the diamond pyramid structure have n layers of atoms, where n is an odd number, giving $n = 2k + 1$. This will make an atom sit on the centre point of the bottom layer of the diamond pyramid structure. Let the coordinates of this atom be the centre point of the boundary cube, which is F ($a/2$, $a/2$, $a/2$). In Fig. 6.1, ABDC is the bottom of the diamond pyramid structure, BDE and CDE are two copper {100} planes and ABE and ACE are two copper {111} planes.

The distance between two closest atoms in copper fcc lattice is $l_0 = \sqrt{2}/2$. For the n-layer diamond pyramid structure (Fig. 6.2), the length of the short diagonal of the bottom diamond BC (Fig. 6.1) will be $(n-1)l_0 = (n-1)\sqrt{2}/2 = \sqrt{2}k$ (Fig. 6.2). Note that the direction of CB line is (1, –1, 0), and the coordinates of points B and C can be calculated, giving B ($a/2 + k/2$, $a/2 - k/2$, $a/2$) and C ($a/2 - k/2$, $a/2 + k/2$, $a/2$) for an n-layer diamond pyramid structure with $n = 2k + 1$.

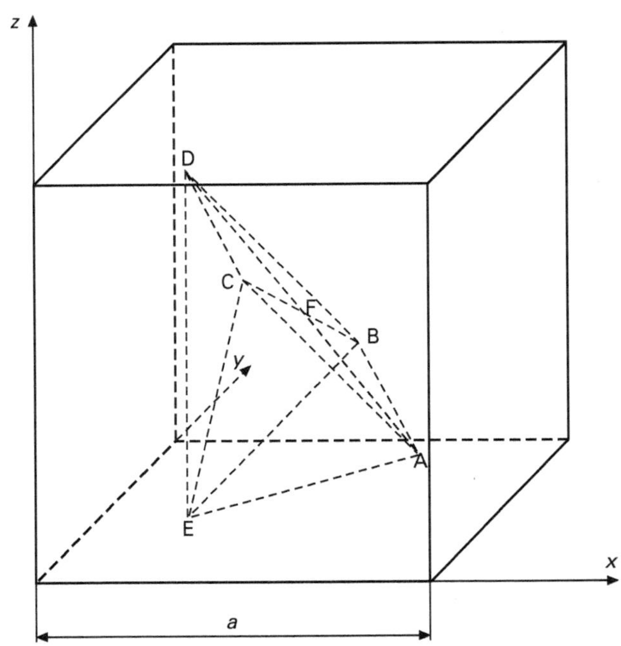

6.1 The orientation of the boundary box and the diamond pyramid structure.

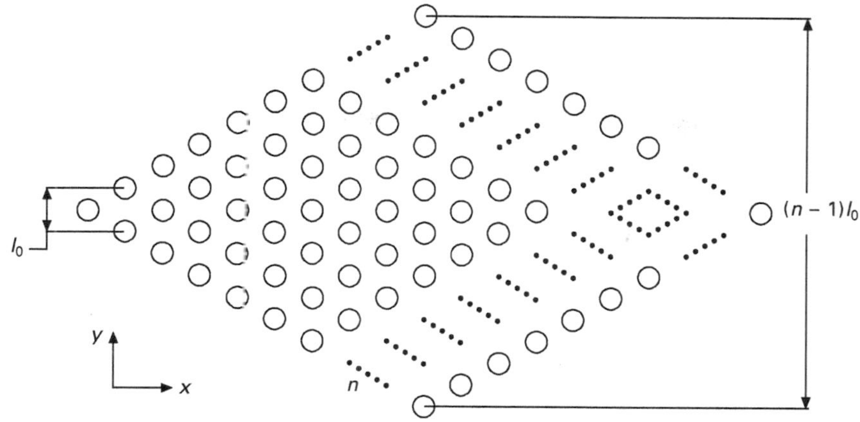

6.2 The dimension of an *n*-layer diamond.

Table 6.2 Positions of the five faces of the diamond pyramid structure shown in Fig. 6.1

Face	Key point within the face	Normal direction
ABE	B ($a/2 + k/2$, $a/2 - k/2$, $a/2$)	(1, −1, −1)
ACE	C ($a/2 - k/2$, $a/2 + k/2$, $a/2$)	(1, −1, 1)
BDE	B ($a/2 + k/2$, $a/2 - k/2$, $a/2$)	(0, 1, 0)
CDE	C ($a/2 - k/2$, $a/2 + k/2$, $a/2$)	(−1, 0, 0)
ABDC	F ($a/2$, $a/2$, $a/2$)	(1, 1, 1)

With the coordinates of points B, C and F, and the normal directions of faces ABE, ACE, BDE, CDE and ABDC, the positions of these five planes within the boundary box can be determined. Note that faces ABE, ACE and ABDC are copper {111} planes and faces BDE and CDE are copper {100} planes, and it is easy to get the normal directions of these faces. With the normal direction of the faces and the coordinates of a key point within the faces, the positions of the faces can be determined and are listed in Table 6.2.

With the positions of the five faces determined, the excess atoms within the boundary box can be selected and removed in XMD, and the diamond pyramid structure can be shaped.

The shaped diamond pyramid structure together with the base atoms is then rotated to make the long diagonal of the diamond layer sit on the x direction and the short one on the y direction. This is done by rotating the whole structure 125.26° (the supplementary angle of the angle between DA and EA, equal to $\tan^{-1}(\sqrt{2}/2) + 90°$) clockwise around line BC and then 45° around the z direction at F point clockwise.

The base is then shaped by removing the excess atoms on the x, y and z

directions, after which, a fine diamond pyramid structure with base is built up and ready for simulation (Fig. 6.3).

For a twinned diamond pyramid structure, the number of the layers in each twin, t, is required. The twinning in the diamond pyramid structure is built up by rotating t layers of atoms 180° around the z-axis through the centre of the bottom one in these layers. Twinning is created from the bottom of the diamond pyramid structure to its top. In Fig. 6.4, Twin 1 and the dashed atoms in Twin 2 (including the bottom layer in Twin 2 where the dashed circles are covered by the solid ones) have the parent lattice built

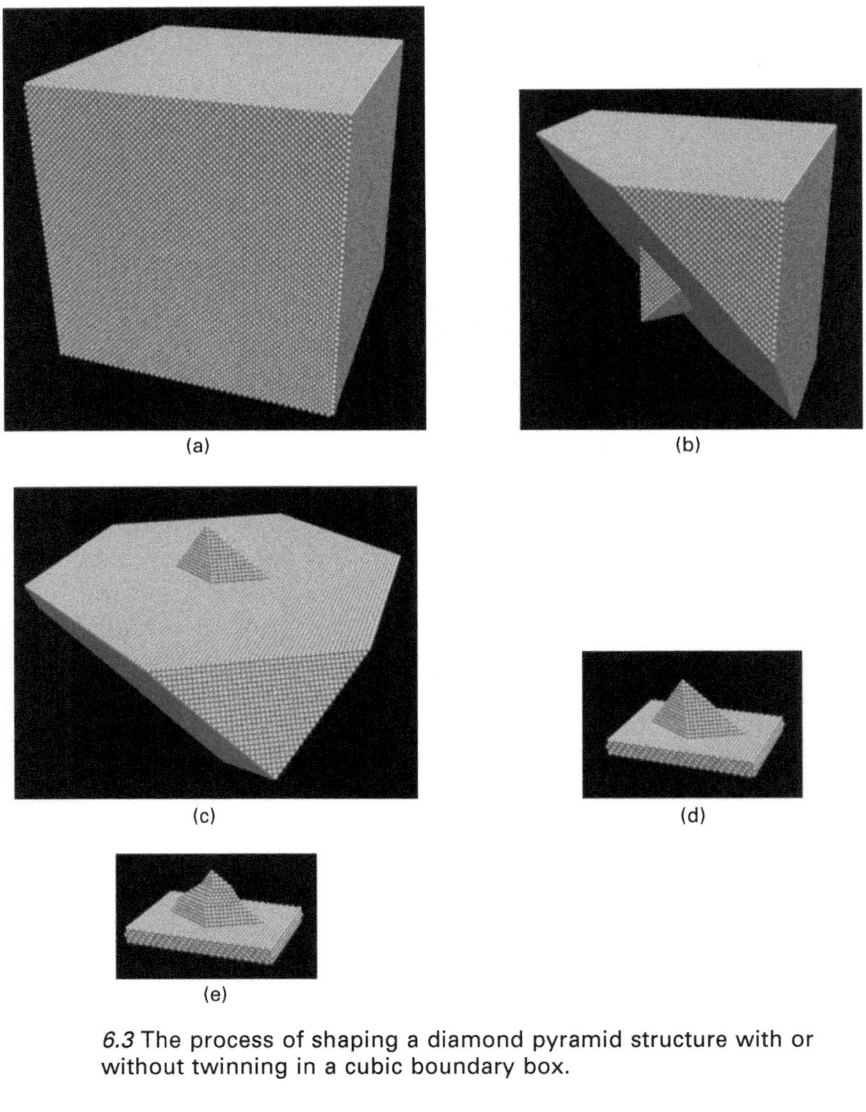

6.3 The process of shaping a diamond pyramid structure with or without twinning in a cubic boundary box.

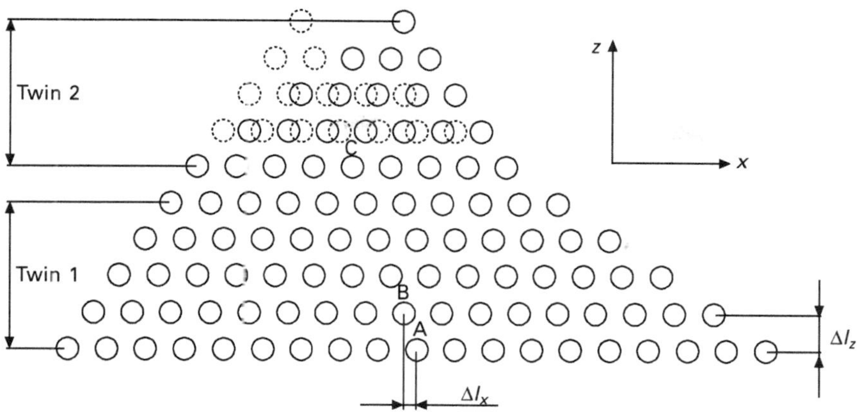

6.4 Creating twinning in a fine diamond pyramid structure by rotating. The dashed circles represent the original atoms in the fine diamond pyramid structure. These atoms (Twin 2) are rotated around the *z* direction through point C to be switched to the twinning position.

up by the procedure above. In order to create twinning in the structure, the original atoms in Twin 2 area are rotated 180° around z direction through the centre point of its bottom layer, at the C point in the figure. The centre points within each adjacent layer in Twin 1 shift $\Delta l_x = -\sqrt{6}/12$ in the x direction and $\Delta l_z = -\sqrt{3}/3$ in the z direction, and remain the same in the y direction. Therefore, it is easy to locate the centre point in the bottom layer of each twin. Figure 6.3(e) shows a diamond pyramid structure with three twins with five layers within each twin.

6.3 The effect of surface stress on the shape of the diamond pyramid structure

Significant change of the positions of the atoms in the structure can be observed (Fig. 6.5) The simulation temperature is set to 0 K. Therefore, only the surface stress affects the shape of the structure, and no temperature effect is included. The black dots indicate the original position of the perfect copper fcc structure before relaxation. After 2,000,000 steps relaxation, the atoms shift their positions under the surface stress induced force. As seen in Fig. 6.5(a), the atoms A–E on the left edge of the diamond pyramid structure (the edge between two {111} faces) move in the $+x$ and $-z$ direction and atoms K–M on the right edge of the structure (the edge between two {100} faces) move in the $-x$ and $-z$ direction. The whole diamond pyramid structure has the trend of moving rightwards, as can be indicated by the movement of the edge between the {111} and the {100} planes, outlined by atoms A and F–I. This result matches that in Section 5.4.4, where the surface stress

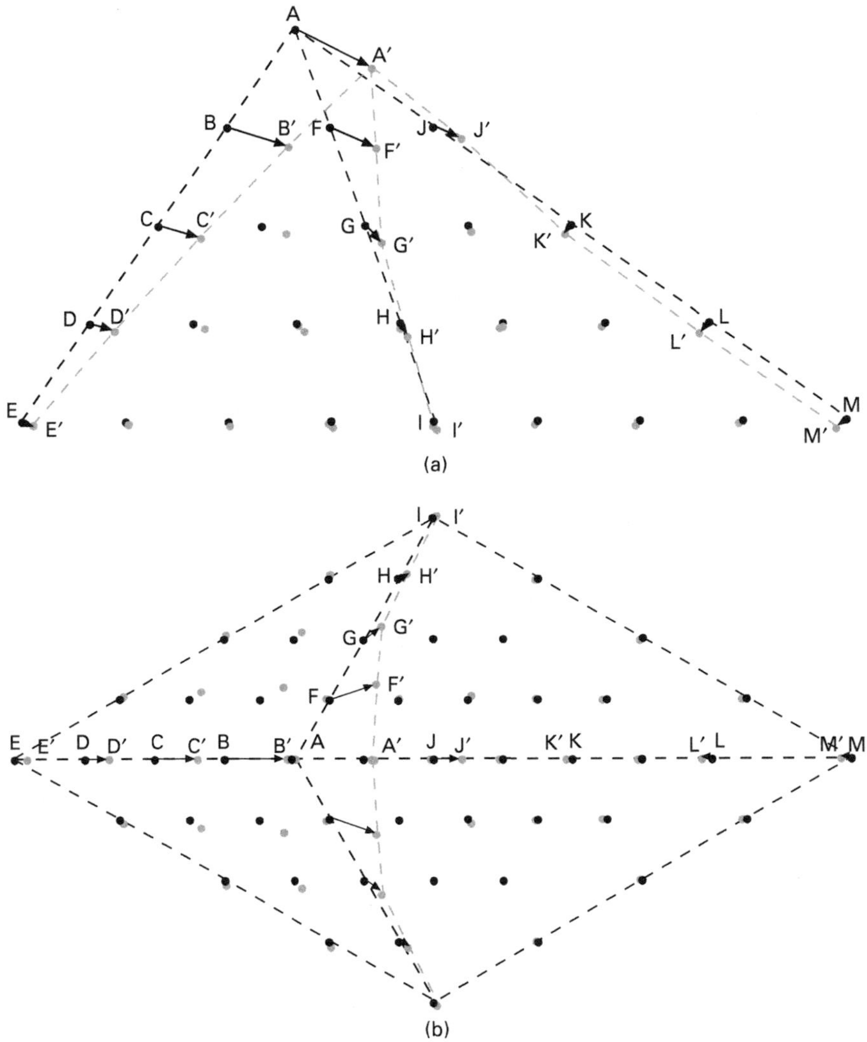

6.5 (a) Side view and (b) top view of the MD simulation result showing the effect of the surface stress induced forces on the shape of the 5-layer diamond pyramid structure. The black dots in the image are the atoms before relaxation and the grey ones are those after 2,000,000 steps relaxation. The arrows show the displacement of the atoms. The dashed lines outline the structures before and after relaxation. The base atoms are removed.

forces with similar directions to the movement of the atoms on the edges of the diamond pyramid structure, and the part structure, are predicted. The only different atom on the right edge of the structure is atom J, which moves in $+x$ and $-z$ direction. This is because the displacement of the atoms on the left side of atom J is so great that they push atom J to move in the $+x$ direction rather than let it move in the $-x$ direction under the surface stress induced forces the same as atoms K–M. As the structure increases in size, the short distance rejecting forces from the atoms in the inner part of the structure reduce the displacement of the atoms on the left edge (atoms C–E) and thus the atoms on the right edge (atoms K–M) can move in the direction of the surface stress induced force.

From Fig. 6.5, it can also be seen that the atoms at the edge and the tip of the structure move further than those atoms inside and closer to the base. This is because the exposed atoms have weaker bonds with the other atoms to resist the surface stress induced force on them. The more the atom is exposed, the further the atom moves during the relaxation. In Fig. 6.5, it is clear that the position of the topmost atom changes the most and the atoms in the bottom layer of the diamond pyramid structure barely move during the relaxation.

6.4 The relaxation of the diamond pyramid structure with different sizes

Figure 6.6 shows the three models before and after relaxation. In Fig. 6.6(a) and (b), we can see that the shape of the 5-layer diamond pyramid structure changes from non-symmetric to relatively symmetric. The atoms in the top layers of the structure move to positions with lower energy during the relaxation and thus the diamond pyramid structure is no longer a perfect face centred cubic (fcc) structure.

As the layer of the diamond pyramid structure is increased, the shift of the position of the atoms is only limited to the top five layers and the atoms in the rest of the layers remain almost in the original position (Fig. 6.6(c) and (d)). That is because the bonds within the atoms in the lower layers are much stronger than in the top five layers. Those atoms do not have enough energy to overcome these bonds to move in the structure during the relaxation.

A similar result is given by model 3 (Fig. 6.6(e) and (f)). Only the positions of the atoms in the top five layers are shifted during the relaxation and those of the atoms in the rest of the structure remain nearly the same.

Therefore, only the top five layers of a diamond pyramid structure that is built up according to the atomic model can be relaxed to a relatively symmetric shape due to the lack of the bond between atoms. Atoms in the rest of the layers in a diamond pyramid structure remain nearly on their original position during relaxation. Model 3 also proves that once twinning forms in the diamond pyramid structure in a big size, such as the twinning in the bottom ten layers in the diamond pyramid structure in Fig. 6.6(e) and (f), it can remain stable without being disturbed.

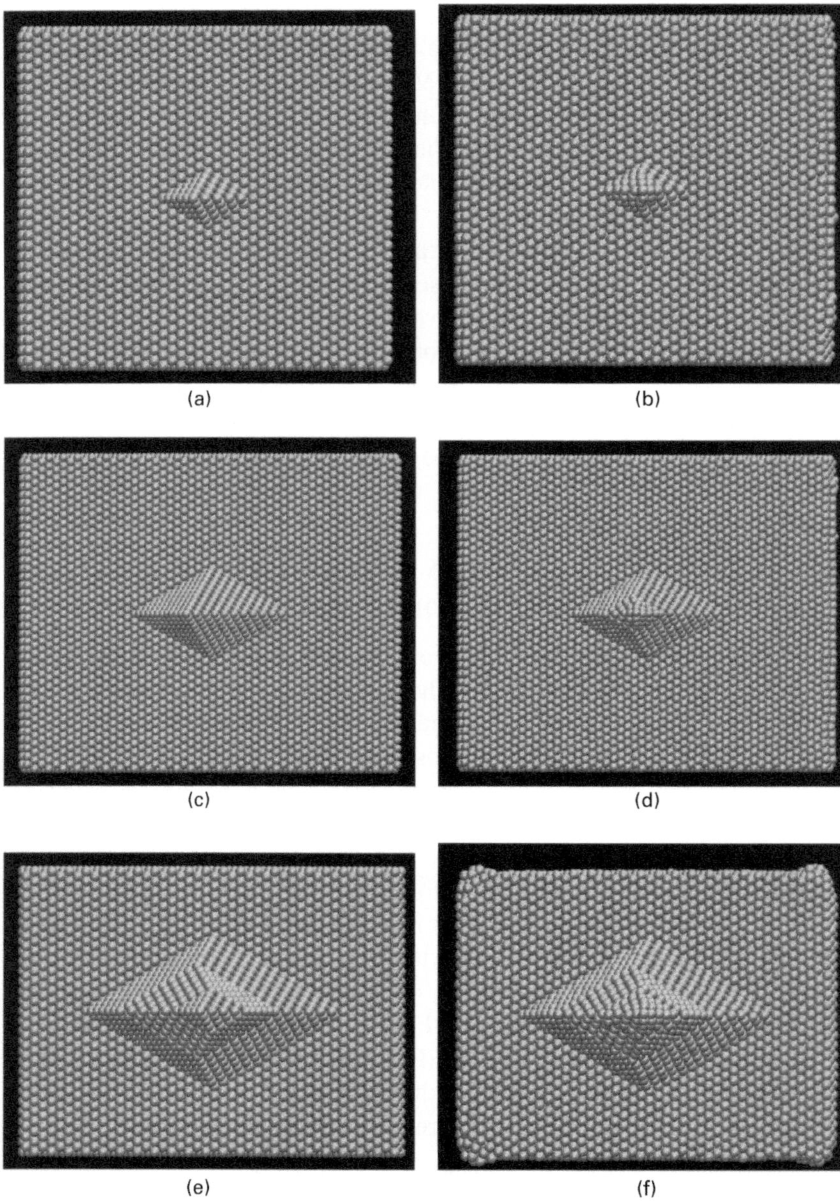

6.6 Top view of (a), (c), (e) the original model for relaxation calculation and (b), (d), (f) the calculation results of (a), (b) model 1, (c), (d) model 2 and (e), (f) model 3.

6.5 The effect of temperature on the relaxation of the diamond pyramid structure

The temperature has an effect on the shape of the relaxed diamond pyramid structure. In Fig. 6.7(a) and (b), there is no significant change in the appearance of the structure after 50,000 steps of relaxation, from both the top view image and the side view one. The reason for this phenomenon is that the relaxation time is too short for this 0 K model to show any difference from the original one. Only when relaxed for much longer time, as seen in Fig. 6.5, can the structure show significant difference from the original one. Since no temperature effect is involved in this model, this means that the effect of surface stress on the shape of the structure is relatively weak. It takes a much longer time for its effect to be visible, although in the macro scale the simulation time of Section 6.3 (Fig. 6.5), 2×10^{-9} s, is extremely small.

In Fig. 6.7(c) and (d), there are significant changes in the shape of the diamond pyramid structure after the 50,000-step relaxation. The tip of the structure moves to its {100} side and appears to be more symmetric in the *y* direction. Tendency towards twinning can be clearly observed in Fig. 6.7(d), which is shown down to smaller detail in Fig. 6.8. In Fig. 6.8, three types of atomic arrangements are in the structure, denoted by A, B and C in the figure. The original copper fcc atomic arrangement is A type. The type B arrangement is the mirror arrangement of type A and the mirror plane is the twinning plane. The appearance of type B arrangement indicates that atoms have overcome the energy barrier and they are positioned on the twinning points with relatively lower energy (compared with the barrier energy). The type C arrangement, with one angle of the triangle nearly 90°, is only discovered at the tip of the structure, where the bond to the atoms is weak and the barrier energy is relatively low. The same arrangement can also be seen in the 0 K model with longer relaxation time (Fig. 6.5). This means that the atoms there can easily stay at the barrier position in order to balance the forces from the surrounding atoms, including surface stress induced forces. When the structure grows, the bond from the surrounding atoms becomes stronger, and therefore the barrier energy increases. These atoms have to choose a position with relatively low energy to be stable, at either the original copper fcc position or the twinning position.

In Fig. 6.7(e) and (f), the increase of the temperature increases the vibration of the atoms and they move further away from their original lattice points. Further increase in temperature (Fig. 6.7(g) and (h)) makes it more severe. The shape of the structure is not very stable during the relaxation. The top few layers of the atoms vibrate in the *x* direction continuously and these images are just the snap shot at the end of the simulation. The twinning plane is not very clear in Fig. 6.7(f) and it is even less clear in Fig. 6.7(h) with the increase in temperature but the shape of the diamond pyramid structure

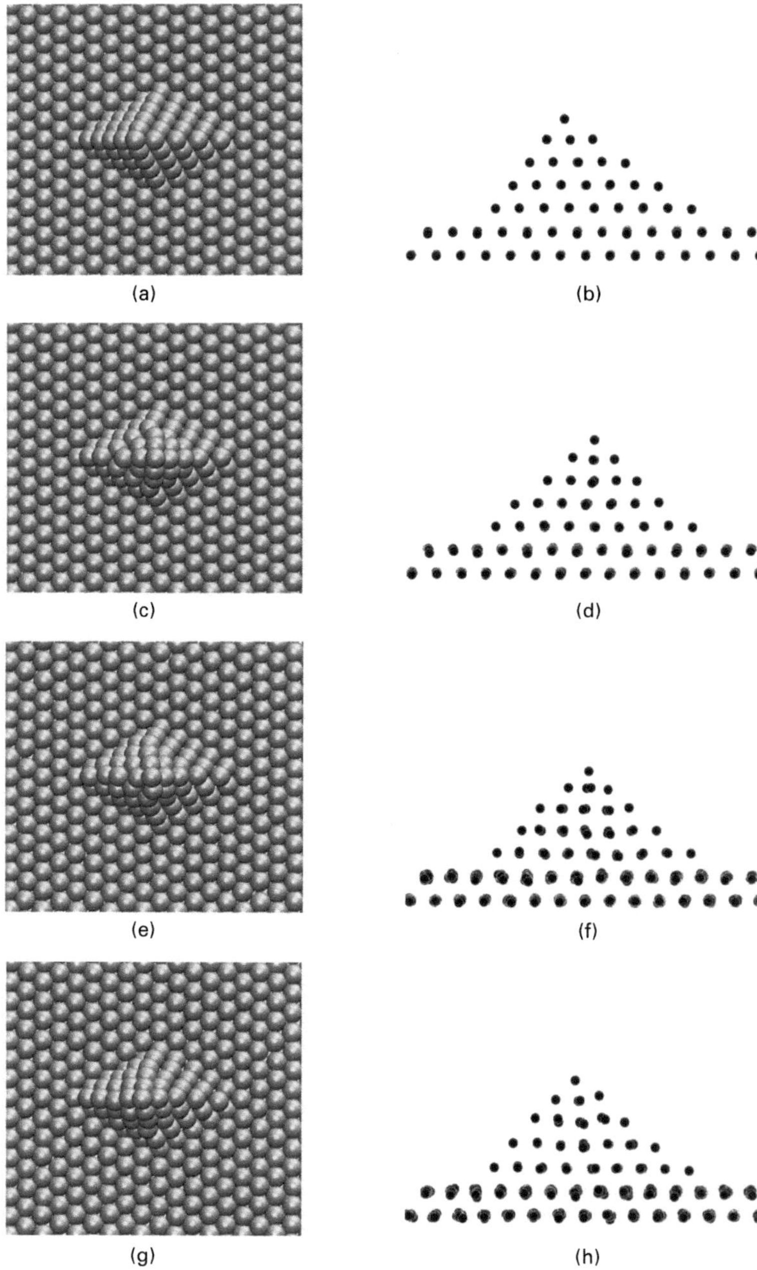

6.7 Top view (left) and side view (right) of the 5-layer diamond pyramid structure relaxed for 50,000 steps. The temperatures are (a) and (b) 0 K, (c) and (d) 50 K, (e) and (f) 273 K, and (g) and (h) 293 K.

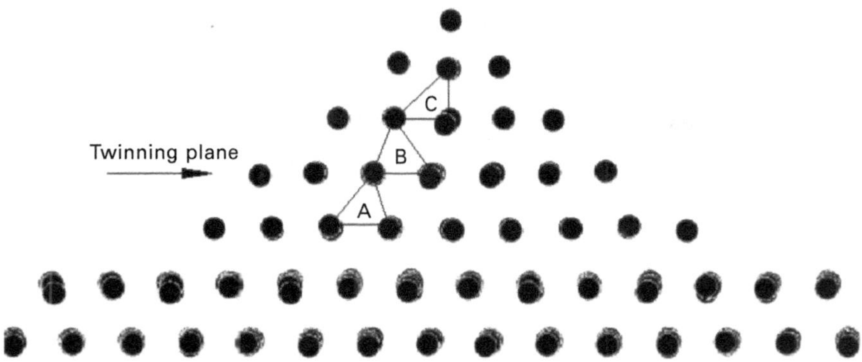

6.8 A detailed side view of the diamond pyramid structure relaxed in 50 K.

is overall symmetric. In Fig. 6.7(h), more type C arrangements are found, which means that the higher temperature makes the atoms more easily stay at the energy barrier point in order to balance the forces from the surrounding atoms. However, for the atoms at the bottom of the diamond pyramid structure, due to the strong bond from the surrounding atoms (including base atoms), the temperature does not strongly affect their positions.

There is a huge effect of the temperature on the shape of the diamond pyramid structure. Increase in temperature makes the atoms travel more easily across the barrier position or even stay around it. This increases the chance of forming twins in the structure. However, higher temperature also makes it easier for the atoms in twin positions to travel back to the copper fcc position. Figure 6.9 shows the extended relaxation. In Fig. 6.9(a), when the simulation step is 55,000, no clear twinning is observed at layer 3–5 from the top. Most of the atoms stay in type A arrangement and the atoms in the top 2 layers are in type C. In Fig. 6.9(b), when the relaxation step is 57,500, the atoms in the top three layers move across the barrier and a twinning plane is clearly in the 4th layer. However, with the relaxation continuing, twinning disappears at step 60,000 (Fig. 6.9(c)) and the arrangement of the 3rd layer atoms changes back to type A. This could be the reason that the diamond pyramid structure is only discovered in the lower temperature deposits.

6.6 The effect of temperature on the formation of voids in the electroless deposit

Figure 6.10 illustrates the effect of temperature on the formation of voids in an electroless copper deposit. The base atoms in the plan view are removed for the convenience of identifying the plated atoms. In Fig. 6.10(a), when the temperature is set to 50 K, almost all the atoms land on the original pyramid structure with only a few scattered elsewhere. The side view shows that the

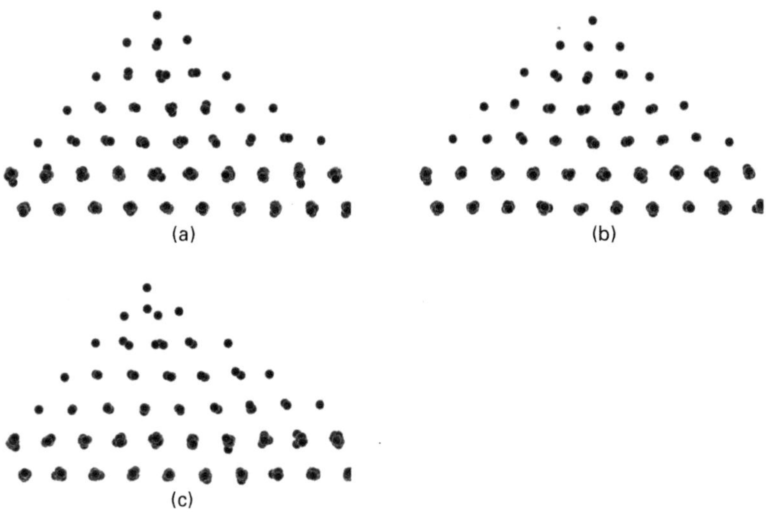

6.9 Side view of the 5-layer model structure in 293 K relaxation at step (a) 55,000, (b) 57,500 and (c) 60,000. The atoms in the bottom two layers are the base atoms.

extra atoms build up a mushroom shaped structure on the base with a cap slightly bigger than the neck. In Fig. 6.10(b), when the temperature is set to 273 K, a much more scattered distribution of atoms is illustrated. The side view of the structure also shows that a more even deposit is acquired.

The difference of the simulation results demonstrated in Fig. 6.10 shows the possible effects of the temperature on the structures of the deposits. Two main effects of the temperature, demonstrated by the MD simulations, include the effect on the vibration of the atoms before landing on the base and the effect on the possibility of mobility of the atoms that have landed on the base. Basically, the lower the temperature is, the weaker the vibration of the atoms is, and hence, the less chance the atoms have to deviate from the original direction that is placed on them by the external force, which means that more atoms will land on the initial diamond pyramid structure, forming the mushroom shaped structure as seen in Fig.6.10(a). In addition, for atoms that have landed on the base, the lower is the temperature, the less energy, and hence less chance they have to move around on the base. This means that at low temperature, atoms tend to stay where they land and it is not easy for them to move. As shown in Fig. 6.10(a), atoms that land on the mushroom shaped structures stay on them and are not able to move to the base due to lack of energy at low temperature. However, at higher temperature, as shown in Fig. 6.10(b), atoms have greater vibration when approaching the base and hence are more scattered on the base. In addition, the high energy induced by the high temperature gives atoms more chance

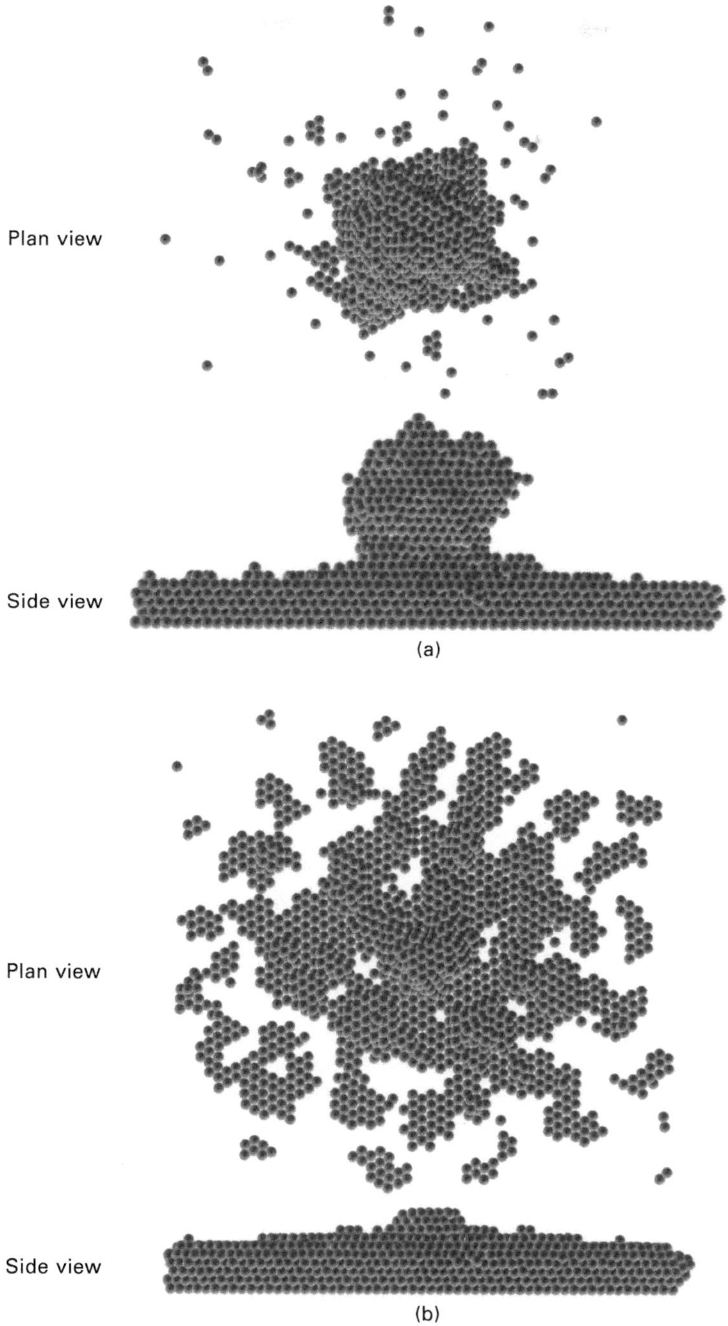

6.10 MD simulation result showing the plan view (with base atoms removed) of the model after simulation compared with its side view (with the bottom 5 layers of base atoms) at set temperature of (a) 50 K and (b) 273 K.

to reach peripheral positions, which makes the deposit flat and solid with less surface.

In the real plating environment, there will not be only one landing centre for copper on the substrate. The substrate goes through the catalyst step in the electroless copper plating process, which makes the surface of the substrate evenly covered with palladium–tin colloids as the initial location of the plating reaction. Correspondingly, for MD simulation, we can predict the following. If more initial pyramid structures as landing spots are placed on the base, to represent the many palladium–tin colloids on the substrate in the real plating environment, the mushroom structure that grows on each of the pyramid structures, as illustrated in the MD simulation result at 50 K, may join together and leave voids in between them. This will form similar voids in the deposit as observed in the scanning electron microscopy images of the over-etched cross-sections of the deposit from low temperature solutions.

6.7 The formation and growth of the diamond pyramid structure in deposit

From Fig. 6.11, we can see that although the original surface of the base is flat, the surface after deposition is not. During the simulation, atoms fallen onto the surface migrate and form into several clusters to reduce the surface energy. When atoms keep falling onto the surface, those clusters grow and join, covering most of the base area. In some places, enough atoms join together to form recognisable diamond pyramid structures, as marked with lines in Fig. 6.11. Some of the diamond pyramid structures actually join forming diamond chains, similar to those observed in the scanning electron microscopy images of the electroless copper deposit. The diamond pyramid structure can be self-assembled by atoms and this result shows the credibility of the atomic model brought forward in Chapter 5.

Notice that these diamond pyramid structures are flat-headed, which means that the top layer of these diamond pyramid structures usually consists of more than one atom, deviating from the atomic model and the relaxation calculation. This is reasonable because from the relaxation calculation results, the top five layers of diamond pyramid structures according to the atomic model are not stable. During plating, atoms tend to form stable bigger layers whenever possible.

Figure 6.12 is part of the projection of the atomic structures simulation. We can see that the atoms in the bottom three layers of the base remain exactly on their original position because their coordinates are fixed throughout the calculation. A few atoms seem to be out of their original positions in the bottom three layers of the base. These atoms are actually the extra atoms that fall onto the side of the base. The position of the atoms in the top two layers of the base is slightly shifted due to the temperature-induced vibration,

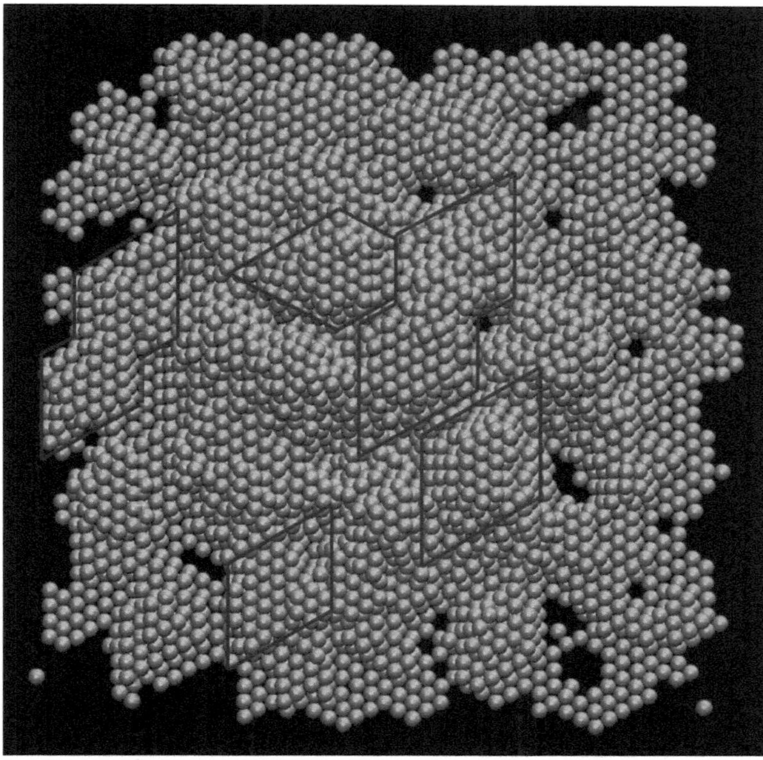

6.11 Top view of the surface morphology from deposition calculation. The base atoms are not shown for better demonstration.

but the stacking arrangement of the atoms is the same as that in the bottom three layers of the base, which is type 'A' in Fig. 6.12. In the deposit layers, twinning planes are found as the interface between type 'A' stacking structures and type 'B' ones. This means that the twinning structure can form by itself during the deposition process, and supports the model brought forward in Chapter 5.

In some of the positions in Fig. 6.12, atoms separate into two groups that are very close to each other, indicating that both type 'A' and type 'B' stacking structures occur in the projected view. This means that in some places along this projection line, atoms are stacked in type 'A', while in other places along the same projection line, atoms are stacked in type 'B'.

Notice that in the plated atoms, twinning only occurs above the first two layers of the deposited atoms. The reason could be that the atoms in the first two deposited layers are restricted by the bottom three layers of the atoms in the base, which are strictly fixed on all x, y and z directions. This makes the deposited atoms of the first two layers, which are within the cut-off distance of these fixed base atoms, less likely to move to the twinning positions,

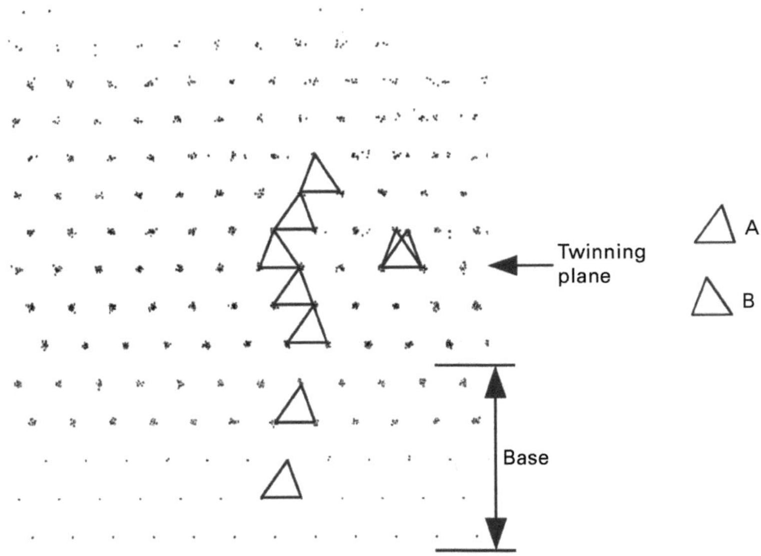

6.12 Partial projection along the *y*-axis on *x–z* plane of the deposition calculation result. Each atom is represented by a dot. The triangles show the atomic arrangement in the structure.

although the temperature-induced oscillation is inevitable. However, with the deposition continuing, the fixed base atoms have no more restriction to the newly deposited atoms when the distance between them is over the cut-off distance. This means that the only atoms that may affect the positions of the newly deposited atoms are the atoms that are previously deposited, and those not on the exact copper lattice positions due to the temperature-induced oscillation. The drift of the position from the oscillation of the atoms makes it more possible for the newly deposited atoms to form twins in the structures. As illustrated in Fig. 6.12, twins start to form from the third layer of the deposited atoms.

The MD simulation can be used to qualitatively investigate the possible mechanism of forming the diamond pyramid structure on the electroless copper deposit. It is exploratory rather than precise and quantitative in nature, due to the reasons given below. It is very hard indeed to directly or quantitatively compare the MD simulation with the experiment due to the assumptions and the scale in the simulation. The dimension of the biggest model used in the MD simulation in this chapter is about 5 nm, which is only about 1/50 of the dimension of the diamond pyramid structures observed in the scanning electron microscopy images. The total time in the MD simulation for relaxation model is 5×10^{-11} s and for the deposition model 1.5×10^{-9} s; these times are far shorter than the scale of the observation time in the experiments. The temperature is set to 0–293 K, which is lower than the operating temperature of 45 °C and 60 °C in the real plating condition. The equivalent plating rate in the deposition model of the MD simulation is around 43 kg h^{-1} mm^{-2}, which is indeed phenomenally high. However, the limit of the computing power and the nature of the MD simulation method itself make it impossible to set up a simulation model comparable with the real plating situation, i.e., with the dimension of the model around 200 nm and the plating rate in mg h^{-1} mm^{-2} level. So, the MD simulation results can only be used for qualitative interpretation of the mechanism and as a support of the reliability of the atom model in Chapter 5.

The calculation models of the MD simulation in this chapter are all based on a copper {111} plane oriented substrate. The reason for using this orientation is from the result of the XRD analysis of the deposit, which indicates that an excessive amount of copper {111} plane exists on the deposit parallel to the surface. The MD simulation can actually be used to interpret the plating process on all copper {111} planes, no matter which direction they are oriented.

Furthermore, in the simulation in this chapter, only the interaction between copper atoms is examined. Due to the lack of the potential information between copper and other atoms or molecules, it is very hard to establish a MD simulation system containing copper and other elements in the electroless copper plating system. Thus, the effect of the interaction between copper and other elements has not been addressed in the MD simulation. The role of the temperature in the MD simulation is to adjust the velocity of atoms and it will not be able to affect plating rate, which can be manually adjusted in the MD simulation, as dramatically as it does in the real system.

6.8 The radial distribution function (RDF) and the Fourier transform of X-ray diffraction (XRD)

The positions of the peaks in RDF indicate the interatomic distances that occur with high possibility. In Fig. 6.13, we can see that apart from the first two peaks (within 2 Å) and the peaks after 10Å, the positions of the peaks marked with 'a' to 'h' of both XRD and MD curves are well matched with each other. The nearest possible distance between atoms in a copper face-centred cubic (fcc) lattice is around 2.5 Å, which is marked with 'a' in Fig. 6.13. The appearance of the first two peaks in XRD curve is due to error in the Fourier transform of the XRD intensities and lack of the intensity data in higher scanning angles. There are no significant peaks for distance greater than 10 Å in MD results due to the limit of the volume of the atoms in the MD simulation that actually makes the counts, and hence the probability of occurrence, of atoms pairs with greater distance in the MD result much lower than in the bulk deposit. The match of the peaks between 2 Å and 10 Å shown in Fig. 6.13 indicates the comparability between the structure

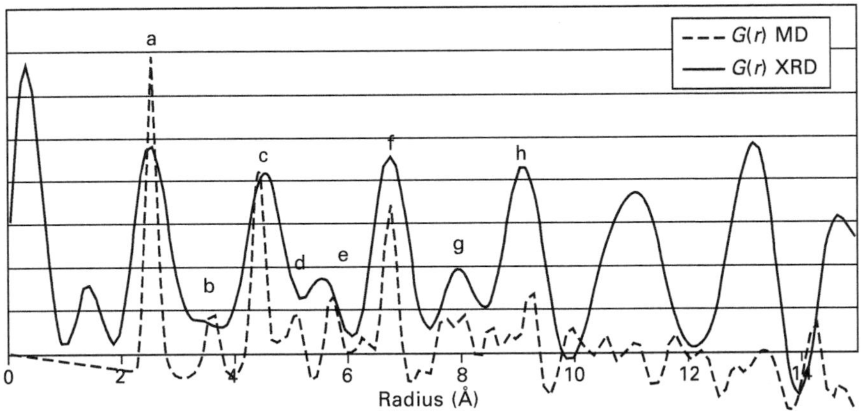

6.13 The radial distribution function $G(r)$ from experimental XRD using the Fourier transform (solid curve) and from the MD simulation (dashed curve). The horizontal axis represents the distance from an atom in the unit of Ångstrom.

of the MD simulation with that of the electroless copper deposit, and further supports the atomic model brought forward in Chapter 5.

The Fourier transform of the intensities of XRD requires an integration on s from 0 to infinity. As $s = (4\pi \sin \theta)/\lambda$, two parameters affect the integration interval of s, which are θ and λ. The maximum possible interval of θ is 0–90°, making $\sin \theta$ vary from 0 to 1. For a fixed wavelength (λ) X-ray scan, the smaller the wavelength the X-ray is, the wider the interval of s can be and so the more accurate the Fourier transform will be. The other solution is by varying the wavelength of the source beam in a wide interval, e.g., by using a synchrotron radiation source, to obtain an integration in a wide s interval.

The X-ray source that we used in this calculation for XRD is the commonly used copper source, providing the X-ray beam with typical wavelength of 1.54 Å (Kα radiation) which is not ideally suited for quantitative RDF analysis from XRD intensities. Moreover, the scanning θ is only between 20° and 50°, which may also produce greater error in the transform for quantitative analysis. However, for qualitative analysis, as demonstrated in this chapter, it is still reliable.

6.9 Conclusions

In this chapter, the molecular dynamics simulation is shown to investigate the effect of surface stress on the shape of the diamond pyramid structure, the size and temperature effects on the relaxation of the diamond pyramid structure, the temperature effect on forming of the voids, and forming and growth of the diamond pyramid structure in the electroless copper deposit.

The simulation of the effect of the surface stress on the shape of the diamond pyramid structure shows the movement of atoms under the surface stress induced forces. The result proves the reliability of the surface stress calculation in Chapter 5 and shows that surface stress does help to shape structure to a symmetric appearance.

The simulation of the relaxation of the diamond pyramid structure shows that, for a perfect or twinning diamond pyramid structure, only the positions of the atoms in the top five layers are significantly shifted due to the high energy within these atoms. Temperature can significantly speed up the relaxation process. Stable twinning is found in the model in 50 K, which means that twinning can self-form in the diamond pyramid structure during the relaxation process. In the simulation system, higher temperature makes both the formation of twins and the loss of twins easier. Therefore, a relatively low temperature is an essential condition for the forming of twins in the deposit, which could be the reason for the diamond pyramid structure to only exist in low temperature plating.

The simulation of the effect of temperature on forming of the voids in the electroless copper deposit shows that temperature plays an important role. The lower the temperature is, the less the atoms can be evenly plated on substrates and voids tend to form in the deposit. At higher temperature, the atoms have more chance to scatter and to be evenly plated on the substrate.

The calculation of the forming and growth of the diamond pyramid structure shows recognisable diamond pyramid structures on the surface of the deposit and twinning is observed in these structures. The calculation of the RDF shows comparable results between the XRD analysis in Chapter 4 and the MD simulation. These results support the atomic model brought forward in Chapter 5 and indicate that diamond pyramid structures with twinning can be self-established in atomic scale.

7 Adhesion strength of electroless copper deposit to epoxy board

Abstract: A direct pull testing kit is shown for the testing of the adhesion strength of the deposits to the FR4 substrate. A tensile machine is used for the adhesion testing. A program is written to calculate the pull-off fraction parameter of each testing specimen from its fracture surface photo. There is no strong relationship between this parameter and the adhesion strength. There is evidence, however, of greater adhesion strength between the low temperature deposits and the substrate. The load-displacement curve of the direct pull adhesion testing shows two load peaks. Two failure modes are proposed to explain this double peak feature.

Key words: adhesion strength, direct pull test, fracture surface, failure mode, load-displacement curve.

7.1 Adhesion testing

This chapter is concerned with the adhesion strength of electroless copper on the FR4 epoxy board substrate, tested using the direct pull method. In this test, a 12 mm diameter aluminium rod is glued on the deposit side of the plated epoxy board sample. The plating duration is two hours.

Copper plated epoxy samples for adhesion testing are cut into squares of 25 mm × 25 mm approximately and glued on aluminium rods with the dimensions of 12 mm in diameter on cross-sections and 40 mm in length. The glue used for this purpose is Araldite 2011 multipurpose two-component glue. The two components are equally mixed and applied on one end of the aluminium rods, which are attached firmly to the deposit side of the centre of the square samples. The samples are then left on a levelled heating plate in a 40 °C environment overnight so that the glue is fully cured. A 12 mm round saw is then used to remove the excessive glue at the edge of the interface between the aluminium rods and the copper deposit samples. The deposits on epoxy board are hence also cut into 12 mm diameter circular shapes attached with the rods (Fig. 7.1). This is to reduce the interference of the strength of the copper deposit itself with testing the adhesion strength between the deposit and the substrate.

The square shaped deposit sample together with the attached aluminium rod is inserted into the special aluminium sample holder cap (Fig. 7.2). Aluminium plates are used to make sure that the sample can be firmly

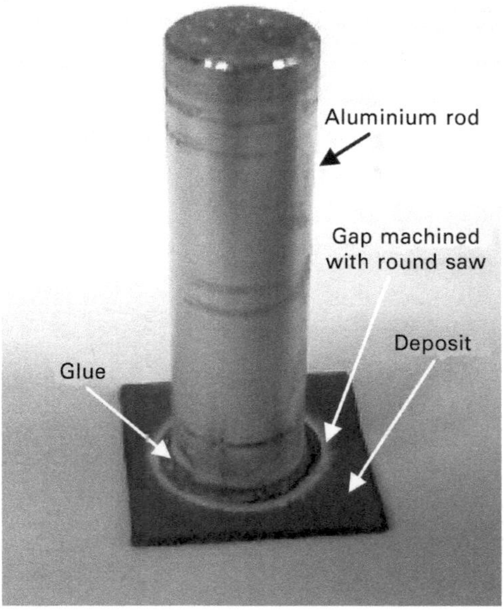

7.1 A direct-pull adhesion-testing specimen ready for testing. The light shade ring on the substrate shows the gap cut with the round saw.

attached to the holder. The core is then screwed into the cap and ready for testing.

During the test, the whole sample, including the holder, is fixed in the clamps on a tensile machine, and tensile force is applied to directly pull the rod off from the copper deposit on the epoxy board sample. The interface between the rod and the deposit is examined to see if the copper deposit is fully, partly or not pulled off. The peak force reached is recorded as the adhesion force between the deposit and the substrate for the fully pulled off situation and the adhesion strength is acquired by dividing the adhesion force by the contact area between deposit and the rod, which is the area of a 12 mm diameter circle.

For each plating condition, five samples were tested. The results were averaged for each condition as the average adhesion strength between the deposit from each plating condition and the epoxy board substrate.

7.2 Image processing for calculating pull-off fraction parameter of adhesion

For adhesion testing, a special imaging processing code is programmed. The purpose of this image processing program is to transfer the photos of the post-testing aluminium rods with the pulled-off electroless copper deposits

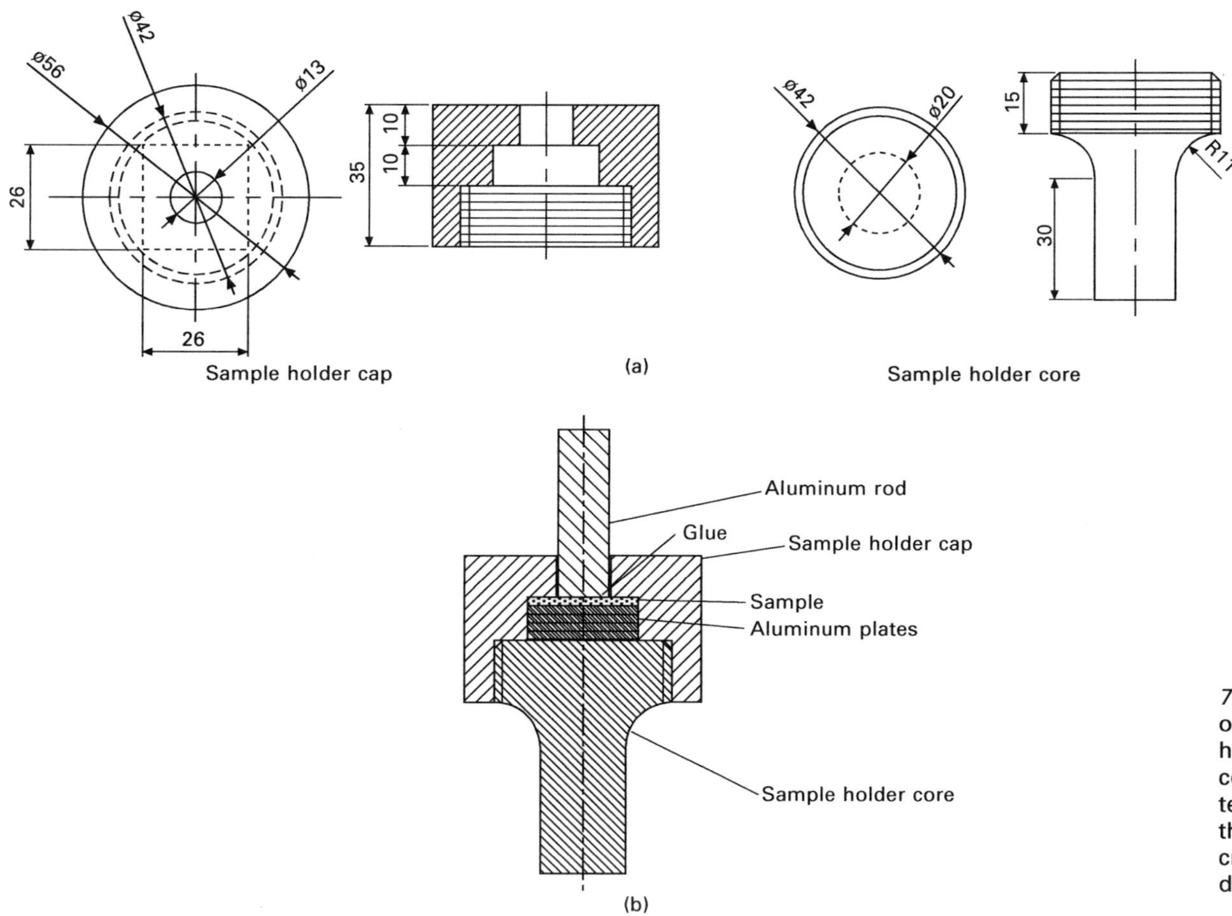

7.2 Illustration of (a) sample holder cap and core for adhesion testing and (b) the installation cross-section. All dimensions in mm.

into grey scale images so that the pull-off deposit areas can be clearly distinguished on the rods and thus the pull-off fraction parameter can be digitally calculated. The idea of the procedure is similar to that demonstrated in Chapter 3 but colour images are dealt with in this chapter. Background removal and edge treatment are needed. The detailed procedure is described including the following steps.

1. Photos are taken of the post-testing aluminium rods, with attached deposit, individually vertically on a monochromatic background with relatively high contrast to the colour of the aluminium rods and the pull-off deposits Each rod in the photo should be placed in the centre to avoid image distortion due to perspective, which is also the reason for shooting the rods individually rather than all together.
2. Each photo is then cropped into a square image consisting of mainly the rod in the centre with less background. This is to reduce unnecessary processing load and time.
3. Each square image is loaded into the software and becomes a three dimensional matrix. The matrix contains the scale value of red, green and blue (RGB value), each from zero to 255, of each pixel on the image.
4. The monochromatic background is removed by selecting the pixels whose RGB values fit the RGB properties of the background colour and setting them to medium grey. For example, in this chapter, the background colour used is yellow, for which the R and G values are both slightly higher than the B value. Therefore, all the pixels whose R and G values are greater than their B value by a threshold value $A_b = 15$ are selected. The RGB values of these selected pixels are all changed into 100 for the same medium grey colour.
5. A similar process to step 4 is carried out to select the pixels representing the pulled-off deposit. In this chapter, the deposit on the aluminium rod is mainly red, which means that the R values of these pixels are greater than their G and B values. A threshold $A_d = 1$ is used to select the deposit pixels and change their RGB values to 0 for black pixels. Both of the threshold values A_b and A_d are adjustable in order to get the best results in different conditions. The RGB values of the rest of the pixels on the image are then set to 255 for white pixels, representing the uncovered aluminium rod.
6. After step 5, the original colour images are changed into grey scale images in which the background, deposit and the uncovered aluminium rod can be easily distinguished and the areas of each can be digitally calculated. However, due to the imperfection of the photos, the round edge of the rod against the background sometimes is not perfectly recognised. Therefore, an edge recovery step is carried out, based on the image after step 5. The mean of the row numbers of the topmost and the bottommost white or black pixels (point V1 and V2, respectively) on

the central vertical line is used as the vertical coordinate of the centre of the round edge of the aluminium rod (point C). In the same way, the mean of the column numbers of the left-most and the right-most white or black pixels (point H1 and H2) on the central horizontal line is used as its horizontal coordinate. The radius of the circle r is determined by the average distance from point H1 (or H2) to C and V1 (or V2) to C. A circle is drawn with the centre on point C and the radius $k \times r$, where $0 \leq k \leq 1$ is a factor used to remove the extended white area due to colour variation at the edge of the rod, adjustable to get the best fit to the original photos. In this chapter, k is set to 0.95. The RGB values of all the pixels beyond this circle are set to 100 for the same scaled grey with the background. The RGB values of those pixels within the circle remain unchanged.

7. The percentage of the pulled-off deposit of each specimen tested can thus be calculated by dividing the number of the black pixels (area of deposit) by the number of both black and white pixels (area of aluminium rod).

7.3 Fracture surface

Electroless copper deposits are partly or fully pulled-off from the FR4 substrates. Ideally, the results of the fully pulled-off specimens are the best to describe the adhesion strength between the deposits and the substrates. However, due to the high surface roughness of the substrate and the deposit, it is possible that the adhesion between the deposit and the substrate, and that between the glue and the deposit are not uniform across the testing area. Therefore, it is possible to get a partly pulled-off result in an adhesion test in which only the weakest adhered part, between either the deposit and the substrate or the deposit and the glue, is pulled-off with the rest remaining on the substrate. As long as part of the deposit is pulled-off, the recorded highest applied load and the whole testing area are used to calculate the adhesion strength of the testing specimen. For example, the specimens from the formaldehyde high concentration high temperature solution (Fig. 7.3(a)) and the glyoxylic acid low concentration solutions are all partly pulled off. Some of the specimens from the formaldehyde low concentration high temperature solution (Fig. 7.3(b)), the glyoxylic acid high concentration high temperature solution, and the formaldehyde low temperature solutions have fully or nearly fully pulled-off the fracture surface.

For specimens from the glyoxylic acid high concentration low temperature solution (Fig. 7.3(c)), nearly no deposit is pulled-off from the substrate in all five tests. This means that the adhesion between the glue and the deposit is not strong enough to pull the deposit off from the substrate, and the testing results cannot be used as the reference to the adhesion strength between the

(a)

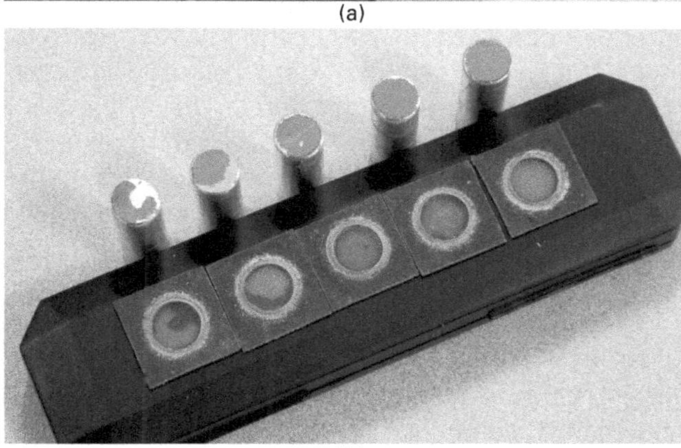

(b)

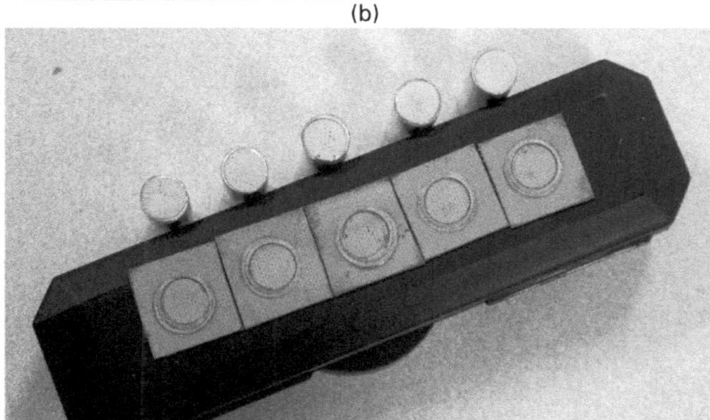

(c)

7.3 The fracture surface of the test specimens and the aluminium rods after adhesion testing. (a) Deposit from formaldehyde high concentration high temperature solution; (b) deposit from the formaldehyde low concentration high temperature solution; (c) deposit from the glyoxylic acid high concentration low temperature solution.

deposit and the substrate. The only conclusion we can get for this batch of specimens is that the adhesion strength between the deposit and the substrate is higher than the test result. The reason for poor adhesion between glue and deposit may lie in the surface morphology of the deposit. In Chapter 2, an extremely flat surface is shown on the 120-minute deposit from this solution in micro scale (Fig. 2.8(d)). A flat surface provides less bonding area with the glue and the contribution of the strong shear strength of glue is limited.

7.4 Image processing and the pull-off fraction parameter of adhesion testing

Figure 7.4 shows an example of the change of the image in the image processing procedure described in Section 7.2. Figure 7.4(a) is a photo of a fracture surface of the aluminium rod with pulled-off deposit on, cropped from the original photo. In Fig. 7.4(b), the background is replaced by a grey colour with uniform grey scale of 100. The pulled-off deposit on the rod is successfully extracted from the colour photo and marked with black colour (grey scale 0). The white colour in Fig. 7.4(b) shows the aluminium rod cross-section that is not covered by copper deposit, where the copper deposit is not pulled-off. However, the edge of the aluminium rod to the background is misrecognised as deposit and the shape of the rod cross-section in this figure is slightly distorted. In Fig. 7.4(c), after the edge restoration step is carried out, the black ring at the edge of the aluminium rod is removed and the shape of the rod fits the original photo well. This shows that the image processing procedure is capable of extracting the fracture surface information from the photo and can be used for pull-off fraction parameter analysis.

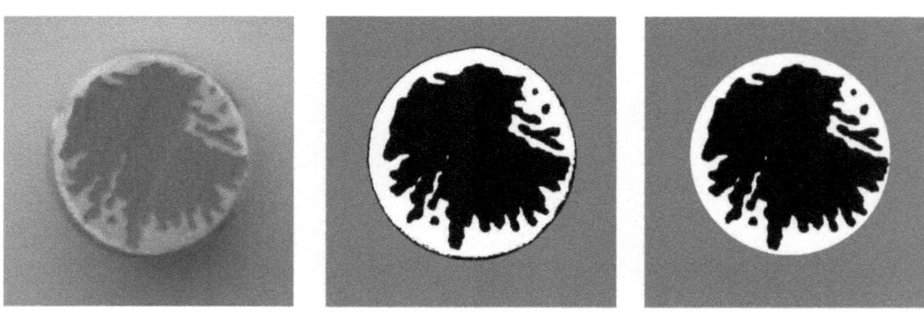

(a) (b) (c)

7.4 (a) The cropped photo of an aluminium rod (deposit side) with pulled-off deposit; (b) the image after step 5 of the processing procedure and (c) the final image. The grey area represents the background. The white area shows the testing area where the deposit remained on the substrate. The black area in (c) shows the pulled-off deposit.

The pull-off fraction parameter varies significantly for different specimens, even for the specimens from the same solution. In general, all the specimens from the formaldehyde low temperature solutions have a pull-off fraction parameter greater than 70%. The specimens from the glyoxylic acid high concentration low temperature solution have the lowest rate, all below 2%, which, as discussed above, is due to the extreme smoothness of the surface of the deposits. There is no specific relationship between the plating condition and the pull-off fraction parameter in the adhesion testing.

7.5 Adhesion strength of electroless copper deposits

The standard deviation is around 10% of the average adhesion strength, which means that the results are quite consistent (Fig. 7.5). The deposits from the formaldehyde high concentration and glyoxylic acid low concentration solutions, both at low temperature, give the highest adhesion strength of around 5000 kN m^{-2} (5 MPa), followed by those from the other two low temperature solutions at just over 4000 kN m^{-2} (4 MPa). The average adhesion strengths of the deposits from the high temperature solutions are below 4 MPa, with the lowest one of just above 3 MPa from the deposits of the glyoxylic acid low concentration high temperature solution. Note that for the deposits from the glyoxylic acid high concentration low temperature solution, due to the pull-off fraction parameter of nearly 0, the test results only indicate the

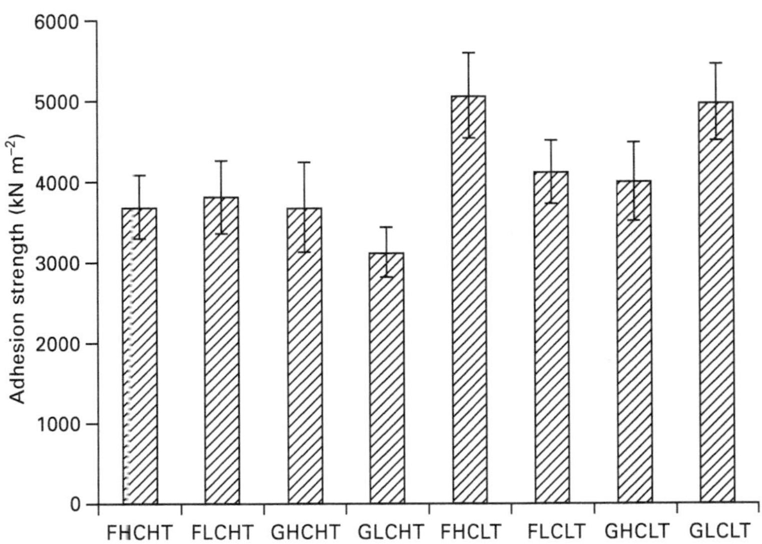

7.5 The average adhesion strength of the electroless copper deposit on FR4 substrate.

interface strength between the glue and the deposit surface and the actual adhesion strength can be higher. The adhesion strengths between the low temperature deposits and the FR4 substrates are higher than between the high temperature ones and the FR4 substrates.

A wide variation is seen in Fig. 7.6. For the same pull-off fraction parameter, a difference of the adhesion strength of more than 1.5 MPa is presented. For the deposits with the pull-off fraction parameter of lower than 70%, the tested adhesion strengths are within the range of 2.9–4.8 MPa. For the rest of the deposits, slightly higher adhesion strengths are acquired but the range, 3.5–5.6 MPa, is still wide. Focusing on particular solutions, we can say that there is no strong relationship between the tested adhesion strength and the pull-off fraction parameter. Actually, because the pull-off testing records the adhesion strength of the weakest part of the specimen, a lower pull-off fraction parameter, apart from that of nearly 0, means that the adhesion between the deposit and the substrate is not evenly distributed and the part with lowest adhesion strength affects the quality of the sample. Conversely, a higher pull-off fraction parameter means that the adhesion strength across the sample is generally evenly distributed, which indicates homogeneous pull-off strength.

7.6 Two failure modes for a partly pulled-off specimen

Two load peaks are present on the curve shown in Fig. 7.7. The gap between these two peaks means that there is a minor failure after peak 1 and the final

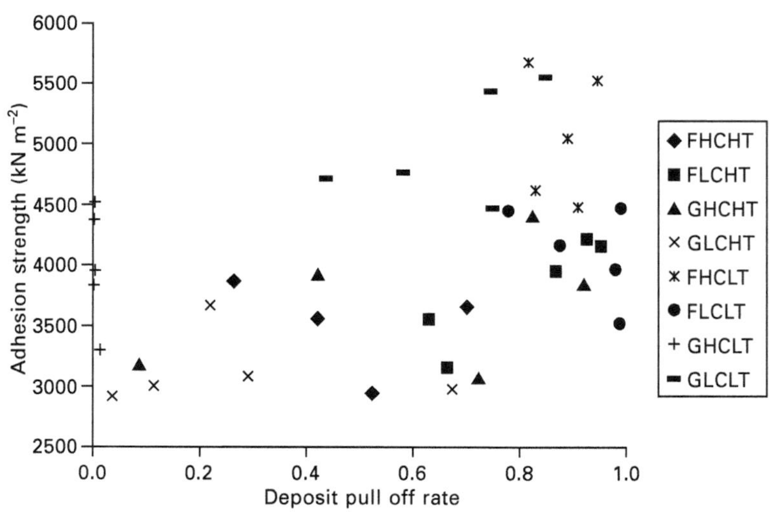

7.6 The distribution of adhesion strength against the pull-off fraction parameter of the deposit.

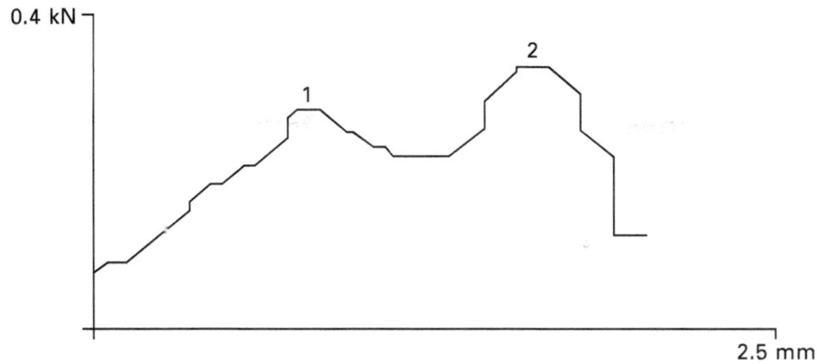

7.7 A typical load-displacement curve for partly pulled-off adhesion test. The numbers 1 and 2 mark the two peaks on the curve.

failure, when the test specimen is pulled apart, takes place after the load reaches peak 2. The possible reason for having a double peak on the load-displacement curve for adhesion testing is demonstrated in Fig. 7.8.

The original structure of the testing specimen is shown in Fig. 7.8(a). The aluminium rod is attached on the surface of the deposit via a thin layer of glue. Three interfaces are present in the system, which are between the aluminium rod and the glue, the glue and the deposit, and the deposit and the substrate. During the pull-off test, it is possible that any one of the three interfaces fails and the whole test assembly becomes separated. However, because the contact surface of the aluminium rods is treated with low grid number sandpaper, the interface between the aluminium rod and the glue is relatively strong. It is shown on the fracture surface that little failure on the aluminium rod and glue interface is present. Therefore, the failure of this interface is not concerned in the following failure modes.

When the applied load reaches the first peak, marked with 1 in Fig. 7.7 where the minor failure happens, it is possible that the failure happens either at the deposit/substrate interface (Fig. 7.8(b)) or the glue/deposit one (Fig. 7.8(d)), as marked with the thick lines in the figure. If the adhesion strength of the deposit on the substrate is not evenly distributed across the specimen, the weakest part of the deposit/substrate interface may not be able to withstand the tensile stress and fails first (Fig. 7.8(b)). Since the whole specimen is not fully pulled apart, the rest of the structure continues to stay integrated, but the load is reduced due to the reduction of the loaded area in an extension controlled test. The load continues to rise after the reduction until the whole specimen fails (Fig. 7.8(c)), and thus the second peak on the load-displacement curve forms.

The second failure mode is similar to the one discussed above but this time the glue/deposit interface fails first (Fig. 7.8(d)) at the first peak on the load-displacement curve. The rest of the glue that is attached to the deposit continues to withstand the load until the deposit/substrate interface fails at the second load peak.

Although the failure of the partly pulled-off specimen can be described with the two failure modes above, it is hard to say which peak corresponds to

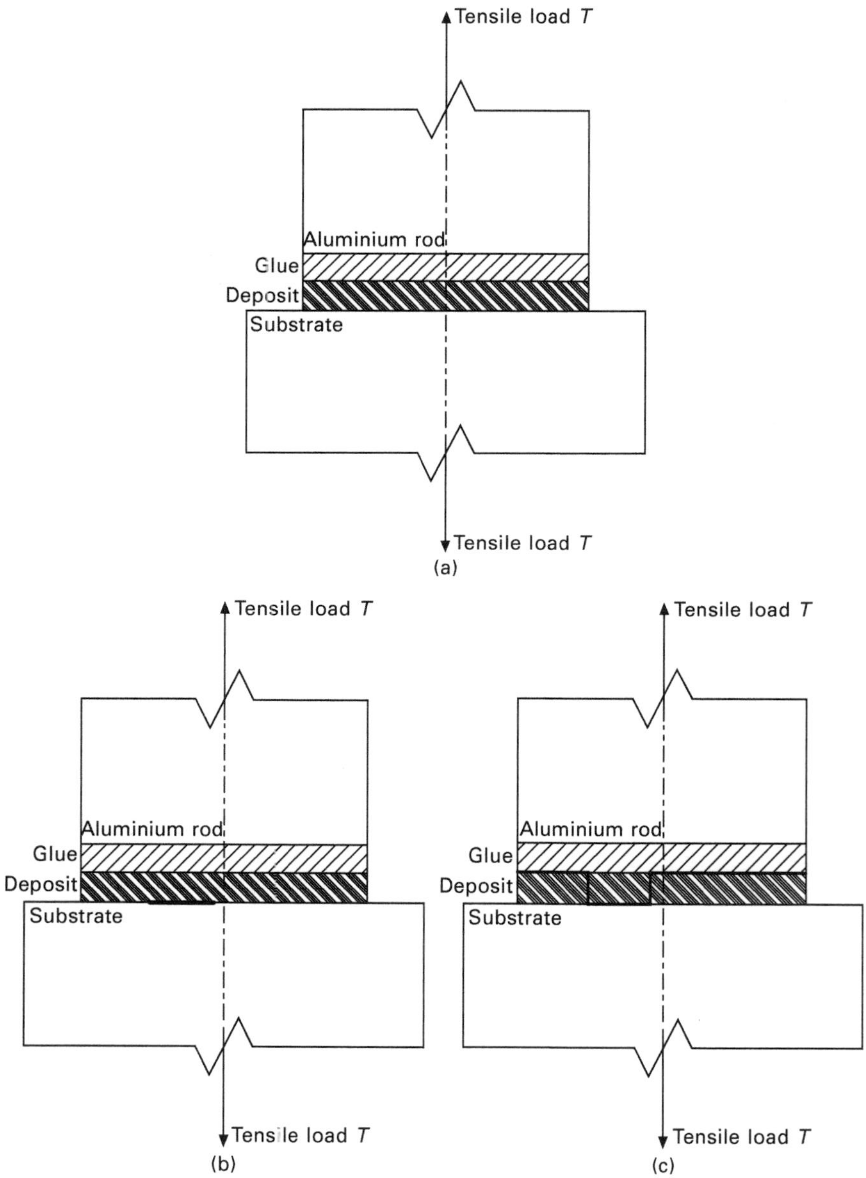

7.8 Sketch of the two possible failure modes (mode 1 and mode 2) of partly pulled-off specimen in the sequences of (a)–(b)–(c) and (a)–(d)–(e), respectively. The dark thick lines show the places that fail.

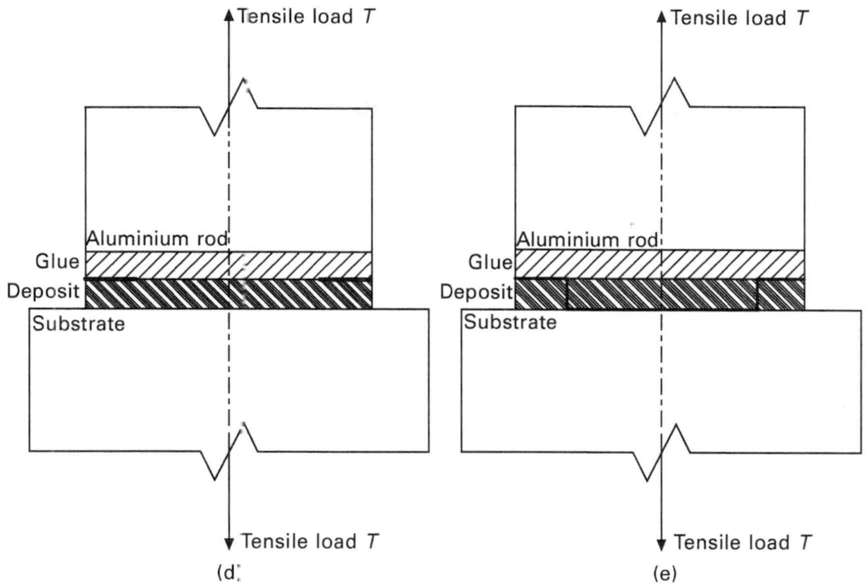

7.8 Continued

the failure of which interface. The height of the two peaks has no particular order, i.e., the height of the second peak can be lower than the first one. Therefore, it is hard to determine the real adhesion strength of a specimen if a double-peak load-displacement curve is obtained. However, as long as the multiple tests on the samples from the same condition show consistent results, the average can be used as the adhesion strength, or an indication of it, between the deposit and the substrate.

7.7 Conclusions

In this chapter, the direct pull method is used to test the adhesion strength of the electroless copper deposit to the FR4 substrate. For this purpose, an adhesion testing sample holder is shown. An image processing code is programmed to process the fracture surface images and calculate the pull-off fraction parameter. The adhesion strengths of the electroless copper deposits to the FR4 substrates are in the range of 3–5 MPa. The deposits from low temperature plating solutions have slightly higher adhesion strength to FR4 than those from high temperature solutions. Most of the specimens have partly pulled-off fracture surface and the pull-off fraction parameter has no significant relation with the adhesion strength acquired from the direct pull testing. The load-displacement curves of the partly pulled-off specimen have double peaks, possibly corresponding to the failure of the deposit/substrate interface and the glue/deposit interface, respectively. However, it is hard to determine which peak is from the failure of which interface. As long as the maximum load used to calculate the adhesion strength between the deposit and the substrate is reasonably consistent for all the specimens from the same plating solution, the result is acceptable.

8 Electrical resistivity of electroless copper deposit

Abstract: The electrical resistivity of the deposits is measured using the four-point probe method. A geometric correction method involving the length calculation of a digital curve is proposed to get a more accurate resistivity result. The results show relatively low resistivity in some low temperature deposits due to large grains and possible twinning. The resistivity of the formaldehyde low concentration high temperature (FLCHT) deposit is the highest among all the deposits. That of the formaldehyde high concentration low temperature (FHCLT) deposit is the lowest. The reason can be the twinned diamond pyramid structures found on the latter deposit.

Key words: electrical resistivity, four-point probe method, geometric correction, digital curve, diamond pyramid structure.

8.1 Four-point probe method for the electrical resistance analysis

The objectives of this chapter are to measure the sheet resistance of the electroless copper deposits from different plating solutions and investigate the influence of the different plating conditions on the electrical property of the deposits. The sheet resistance of the deposits is measured by the four-point probe method. Also in this chapter, a modified process, which can give a more precise result based on the real thickness of the deposit, to calculate the sheet resistance is proposed and applied. Both the results from the traditional method and the modified one are given and compared.

The four-point probe method (Fig. 8.1) is the most common tool used to measure sheet resistance. It comprises of four equally spaced probes, a voltmeter between the inner two probes and a direct current source between the outer two probes. To measure the sheet resistance of a sample, a constant current I goes through the outer two probes and the voltage drop between the inner two probes V is recorded. The differential resistance can be written as:

$$dR = \rho\left(\frac{dx}{A}\right) \qquad 8.1$$

where dR is the resistance between two points with the distance of dx, A is

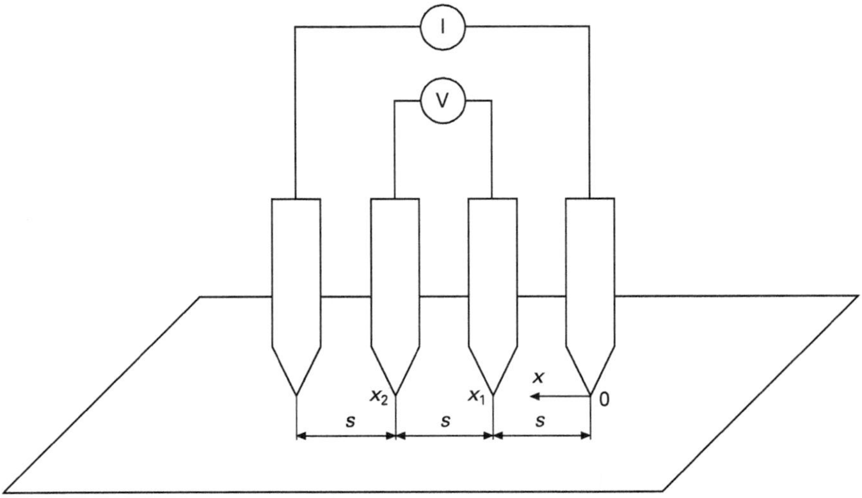

8.1 Sketch of a four-point probe system.

the area that the current goes through and ρ is the resistivity of the sample material.

For a bulk sample where the thickness $t >> s$, we assume a spherical protrusion of current emanating from the outer probes (Fig. 8.2(a)). Therefore, the integration between the two inner probes can be written as:

$$R = \int_{x_1}^{x_2} \rho\left(\frac{dx}{A}\right) = \int_{x_1}^{x_2} \rho\frac{dx}{2\pi x^2} = \frac{\rho}{2\pi}\left(-\frac{1}{x}\right)\Bigg|_{x_1}^{x_2} = \frac{\rho}{4\pi s} \qquad 8.2$$

where probe spacing is uniformly s. Due to the superposition of current at the outer two probes, the measured resistance can be calculated by:

$$R = \frac{V}{2I} \qquad 8.3$$

Therefore, the resistivity of a bulk sample is:

$$\rho = 2\pi s\left(\frac{V}{I}\right) \qquad 8.4$$

For a very thin sample where $t << s$, we get a current ring instead of a current sphere (Fig. 8.2(b)). Therefore, the area is the circumference of the circle times the thickness of the sample, which gives:

$$A = 2\pi xt \qquad 8.5$$

The integration is then expressed in a different way:

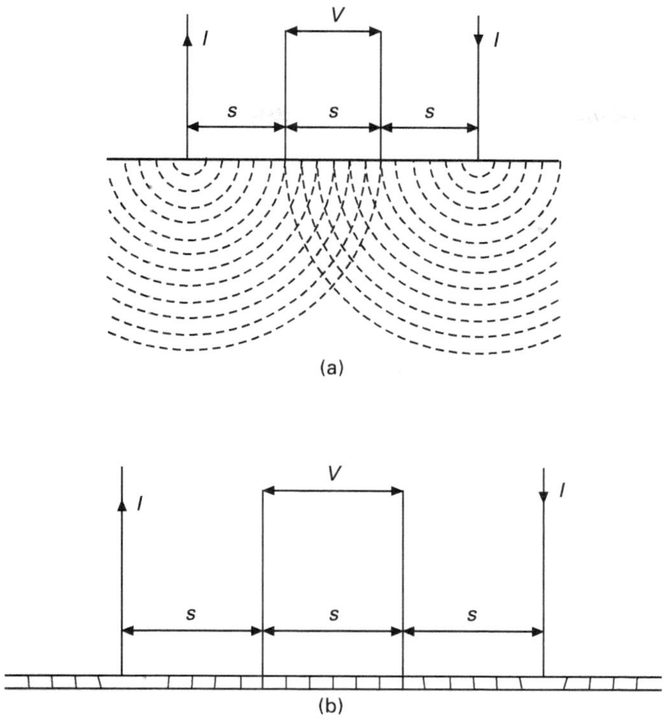

8.2 Illustrations of four point probe measurement on (a) a bulk sample and (b) a thin sample.

$$R = \int_{x_1}^{x_2} \rho\left(\frac{dx}{A}\right) = \int_{x_1}^{x_2} \rho\frac{dx}{2\pi xt} = \frac{\rho}{2\pi t}\ln(x)\Big|_{x_1}^{x_2} = \frac{\rho}{2\pi t}\ln 2 \qquad 8.6$$

Again, with the superposition of the current, the resistivity of the thin sample is:

$$\rho = \frac{\pi t}{\ln 2}\left(\frac{V}{I}\right) \qquad 8.7$$

Note that for the thin sample, the sheet resistivity is independent of the probe spacing. The sheet resistance of the sample can be expressed as:

$$R_s = \frac{\rho}{t} = \frac{\pi}{\ln 2}\left(\frac{V}{I}\right) = 4.53\left(\frac{V}{I}\right) \qquad 8.8$$

The above derivation is based on a semi-infinite bulk and thin sheet sample. For a non-ideal sample, a correction factor should be introduced. In the test, if the distance between the edge of the sample and the nearest probe is equal to or greater than five times the probe spacing, no correction is needed.

In this chapter, as examples showing this testing method, the samples for resistivity test are 5-minute, 10-minute, 15-minute and 20-minute deposits. The thickness of the samples is estimated to be less than 5 μm, which is far smaller than the probe spacing 1 mm of the four point probes used in the test. Therefore, the formula of the sheet resistance is used. The samples are around 70 mm × 40 mm in dimension and the contact areas between the sample and the four probes are in the centre of the sample, which guarantees that the distance between the probes and their nearest edge is greater than five times the probe distance. Therefore, no further correction to the formula is required. During the test, a 10 mA direct constant current source and a multi-functional voltmeter with the resolution of 10 μV are used.

8.2 Calculation of the thickness of the deposits using the weight gain method

8.2.1 The weight gain method

Traditionally, the thickness of the deposit on a fully covered epoxy board is calculated by the weight gain method. The method can be expressed using the following equation:

$$t = \frac{\Delta m}{A\rho} \tag{8.9}$$

where t is the thickness of the deposit, Δm is the weight gained on the substrate, A is the deposit area and ρ is the density of the deposit. The area A usually is calculated by the measured dimensions of the substrate. Note that for a double-sided substrate, the area A includes the area on both sides of the board.

This method is useful for a deposit on a smooth surface or thick deposit on a rough surface where the thickness of the deposit is far greater than the roughness of the surface of the substrate. However, for a thin deposit on a rough surface, the direct use of this method can introduce a large error. Therefore, a correction factor is needed for this specific situation. In Chapter 3, the cross-sections of the deposits are examined using a scanning electron microscope. For the electroless copper deposits with a plating time of 20 minutes or shorter, as the deposits used in this chapter, the thickness of the deposits is smaller than the surface roughness of the substrate. So, a correction factor is needed in order to get results that are more precise.

8.2.2 Method of calculating the length correction factor for FR4 substrate

In order to acquire the correction factor for the epoxy board substrates FR4 used in this book, the cross-section of a one-hour deposit on epoxy board substrate is shown by a scanning electron microscope under the material composition-mapping mode. In this mode, a higher acceleration voltage than the normal imaging voltage is used. In the image, areas of different materials (hence different atomic weight) are marked with high contrast so that the boundary between two materials can be easily distinguished. The image then goes through the following procedure to extract the surface information of the substrate enabling the calculation of the ratio of the actual length on the surface to the measured straight-line length.

1. The image is cropped so that the copper deposit is the main part of the image. Then, the contrast of the cropped image is increased until the deposit area is all white (255 in grey scale) to remove the scratches due to the grinding and polishing on the deposit.
2. For the pixels whose grey scale is less than 255, their grey scales are set to 0 (black), after which the interface between the substrate and the deposit becomes sharp.
3. For each of the black pixels in the image, a four-way search around it is performed (Fig. 8.3(a)). The black pixels with no white pixel in its four-way search are removed by changing their grey scale value to 255 (white). After this step, only the interface between the deposit and the substrate, and that between the deposit and the epoxy resin is shown as black pixels on the image.
4. For each black pixel on the image, a neighbouring black pixel search

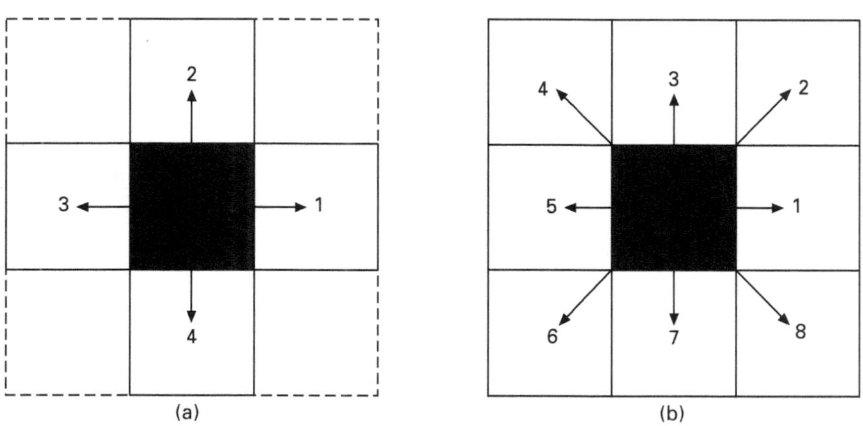

8.3 Sketches of (a) a four-way search and (b) an eight-way search. The numbers in the neighbour pixels show the searching sequence.

is performed. The black pixels with only one neighbour black pixel are recognised as noise and removed by changing their grey scale value to 255.

5. From the first black pixel on the left-most of the image on the interface between the deposit and the substrate, an eight-way search (Fig. 8.3(b)) is performed for its neighbouring black pixels to find the path from the left-most pixel to the right-most one. The searching sequence is anti-clockwise around the current pixel starting from the pixel that is on the direction from the previous pixel pointing to the current pixel. The direction of the path at the current pixel is determined by the direction from the current pixel to the last continuous black pixel in the search. For the leftmost pixels, the search sequence follows the number shown in Fig. 8.4(b) and the direction to the last black pixel is used. After each successful search, the position of the next pixel is recorded and the current black pixel is removed by changing its grey scale to 255. The search ends when it reaches to the right-most pixel and no surrounding black pixel can be found. Figure 8.4 shows an example of how this path-finding procedure works.
6. During the search, the numbers of the horizontal, vertical and diagonal move in the path (n_x, n_y and n_{xy}, respectively) are counted. For each horizontal and vertical move, the distance is recorded as 1, and for each diagonal move, the distance is counted as $\sqrt{2}$. The length of the path L_D, as the length of the interface between the deposit and the substrate, is

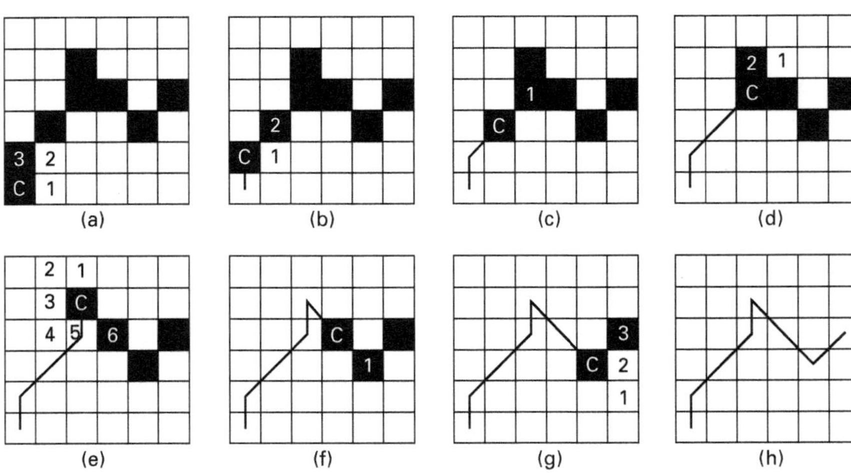

8.4 A sequence of sketches demonstrating the path searching process. The letter C in the figures indicates the current pixel and the numbers indicate the searching sequence for each current pixel. The thick line shows the result path used to calculate the length of the digital curve.

calculated by the sum of the distance of the moves. This value is then divided by the number of the horizontal pixels of the image n_h minus one to calculate the ratio of the interface length to the measured length r, and r^2 is used as the correction factor for surface area of the substrate:

$$L_D = n_x + n_y + \sqrt{2}n_{xy} \tag{8.10}$$

$$r = \frac{L_D}{n_h - 1} \tag{8.11}$$

8.2.3 The error of the length correction factor

In the above method, the length of a continuous curve is calculated by counting the number of the horizontal, vertical and diagonal connections among its digital pixel image. Due to the nature of the digitalisation, the calculated digital length is always no less than the length of the actual curve. In order to address the error that is brought in by the method, a calibration process is performed with the following procedures.

1. A set of blank images with 100 × 100 pixels are created. For each of the blank images, the grey scale of all the pixels is set to 255 (white).
2. A linear line is generated for each of the images. The slope of the line varies from 0 to 1 with a step of 0.1. All these lines cross the point (50,50) within the image. For each line, the vertical coordinate y values are calculated based on the horizontal coordinate x values of 1–100, which are the column numbers of the pixels. The y values are then rounded to their nearest integer numbers, which are used as the row numbers of the pixels that are on the line. The pixels that are determined to be on the line are set to black with grey values of 0.
3. The process in Section 8.2.2 is performed on each of the images and the digital lengths of the lines are calculated.
4. The real lengths of the lines are calculated by basic trigonometry. The error introduced by the digitalisation in each image is estimated by the error of the length of the digital line in each image against its real length from trigonometry.
5. Assume that the surface curve of the substrate acquired in Section 8.2.2 is made up by equal amount of the linear lines generated in step 2. Therefore, the average error of all the linear lines created is used as the error of the length of the surface curve of the substrate calculated digitally. Note that the calculated digital length of a curve using the method in Section 8.2.2 is always greater than its real length, so the error of the calculation is always positive.

8.2.4 Calculating the thickness and the sheet resistivity

With the calculated length correction factor and the error of the method, the actual length correction factor can be calculated. In this chapter, it is assumed that the length correction factor is the same in all directions on the surface of the substrate FR4. The area correction factor of the substrate can be estimated by squaring the actual length correction factor.

The thickness of the electroless copper deposit is calculated with the weight gain method. It is reasonable to assume that the plating rate is constant during the 20 minutes plating period so that the increase of the thickness of the deposit is linear with time. The weight gain per unit area is calculated for each substrate and plotted against time for each plating solution. The slope of the curve is calculated and used as the plating rate. The actual thickness of the deposit on each substrate is then calculated by multiplying the plating rate of each plating solution and the plating time for that substrate. By using this method, the weight change of the substrate during the preparation steps is eliminated because the four substrates for each plating solution go through exactly the same process but that of the substrates for different solutions can be slightly different.

The detailed calculation method for sheet resistivity is described in Section 8.1. The formula for sheet resistivity is used.

8.3 The error of calculating the digital length of a curve

Figure 8.5 shows the linear lines used to evaluate the error of calculating the length of digital curves. From (a) to (k), the slope of the line changes from 0 to 1. Note that in each image, the row number and column number of each pixel are used as its vertical and horizontal coordinates, and therefore the vertical axis in each image points downwards. However, the error will not change for a line with reverse slope. Due to the digitalisation, steps occur on the digital line and introduce the error in the length calculation with the method described in Section 8.2.2. Due to the replacement of a straight line with the sum of horizontal lines and diagonal lines, the calculated digital length is always no less than the real length of the analytical line. The only two situations that the calculated digital length is the same as the real length are when the slope of the line is 0 (Fig. 8.5(a)) or 1 (Fig. 8.5(k)), where the connection between pixels exactly matches the analytical line itself.

Table 8.1 lists the digital length and the real length of the curves in Fig. 8.5, and the length error in each of the images. The length ratio column shows the ratio between the calculated digital length to the horizontal length of the image, which is the horizontal pixel number minus one. This value indicates how the length of the surface cross-section changes with the fluctuation of the

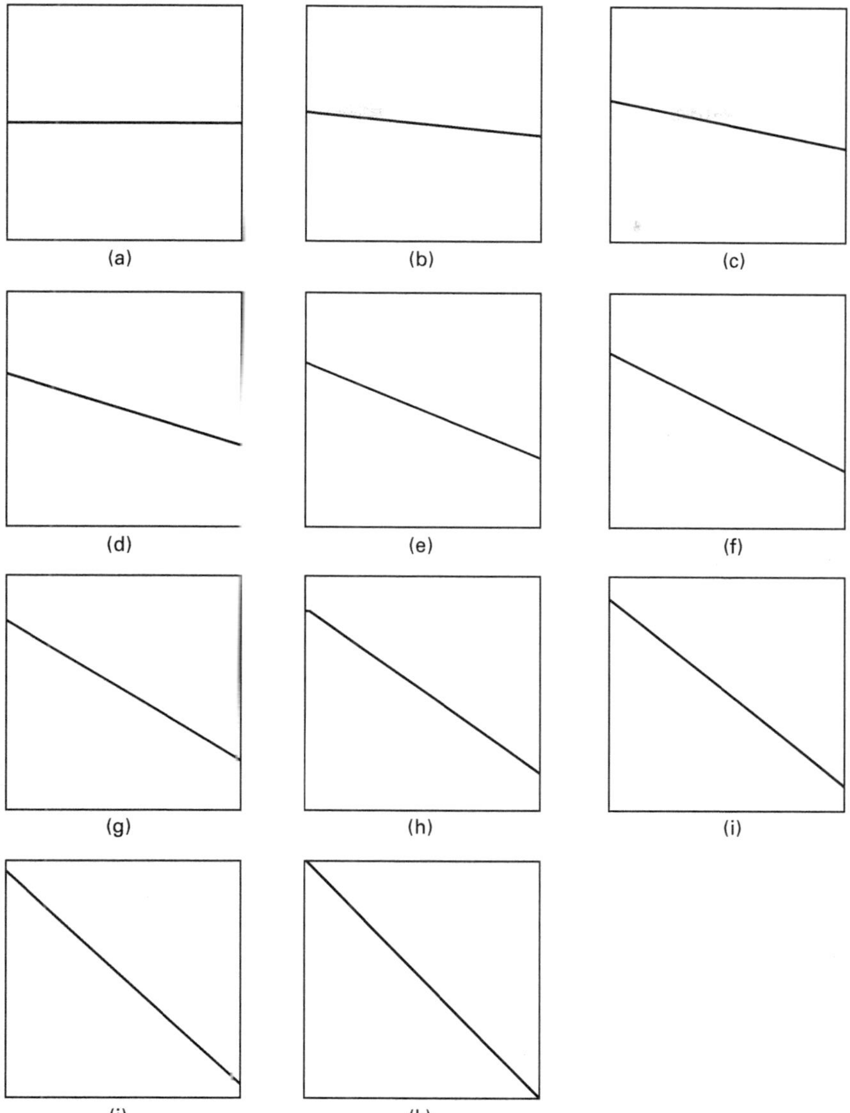

8.5 Images (100 pixels × 100 pixels) generated for the error measurement for the digital length calculation method. The slope of the line varies from (a)–(k): 0–1 with the step of 0.1. All lines cross the pixel at (50, 50) in the image.

surface. The average error is calculated by taking the average of the errors in all 11 images. From the table, it can be seen that the digital length of the lines with slope of 0 and 1 matches their real length respectively, as predicted above. The maximum error occurs when the slope of the line is around 0.4,

Table 8.1 Calculation error of the digital length of the lines shown in Fig. 8.5

Slope	Real length	Digital length	Length ratio	Error (%)
0	99.00	99.00	1.00	0.00
0.1	99.49	103.14	1.04	3.67
0.2	100.96	107.28	1.08	6.26
0.3	103.36	111.43	1.13	7.81
0.4	106.63	115.57	1.17	8.39
0.5	110.69	119.30	1.21	7.78
0.6	115.45	123.44	1.25	6.92
0.7	120.85	127.58	1.29	5.57
0.8	126.78	131.72	1.33	3.90
0.9	133.19	135.86	1.37	2.01
1	140.01	140.01	1.41	0.00
Average				4.75

and the error reaches 8.39%. The length error of all the lines is no less than 0, as predicted above. The length ratio increases with the increase of the slope of the line, which means that the greater the surface fluctuation, the greater the difference will be between the length of the surface cross-section and the direct measurement horizontally. The average error among these 11 images is 4.75%, which will be used to correct the calculated digital length of the surface cross-section curve.

8.4 The surface cross-section curve and the correction factors of FR4

The darker area at the bottom of Fig. 8.6(a) is the FR4 substrate where the brighter parts within show the glass fibres in the epoxy board. The darker area at the top is the cured epoxy resin holding the sample. The long narrow bright area in the centre of the image is the electroless copper deposit. This image shows that the fluctuation of the surface of the substrate can be nearly 10 microns. The thickness of the deposit is not even along the substrate and this can introduce large errors to the calculation of the thickness of the deposit using the weight gain method. For a deposit of no more than 20 minutes plating, a length correction factor is necessary.

The contrast of the image in Fig. 8.6(b) is increased so that the scratches on the copper deposit cross-section are removed for the convenience of the following procedure. The black and white image of Fig. 8.6(b) is shown in Fig. 8.6(c), where the edge of the surface of the substrate can be clearly observed. The edge of the surface cross-section is extracted from Fig. 8.6(c) and shown in Fig. 8.6(d). The length of the curve can then be calculated and used to evaluate the length correction ratio of the epoxy board FR4.

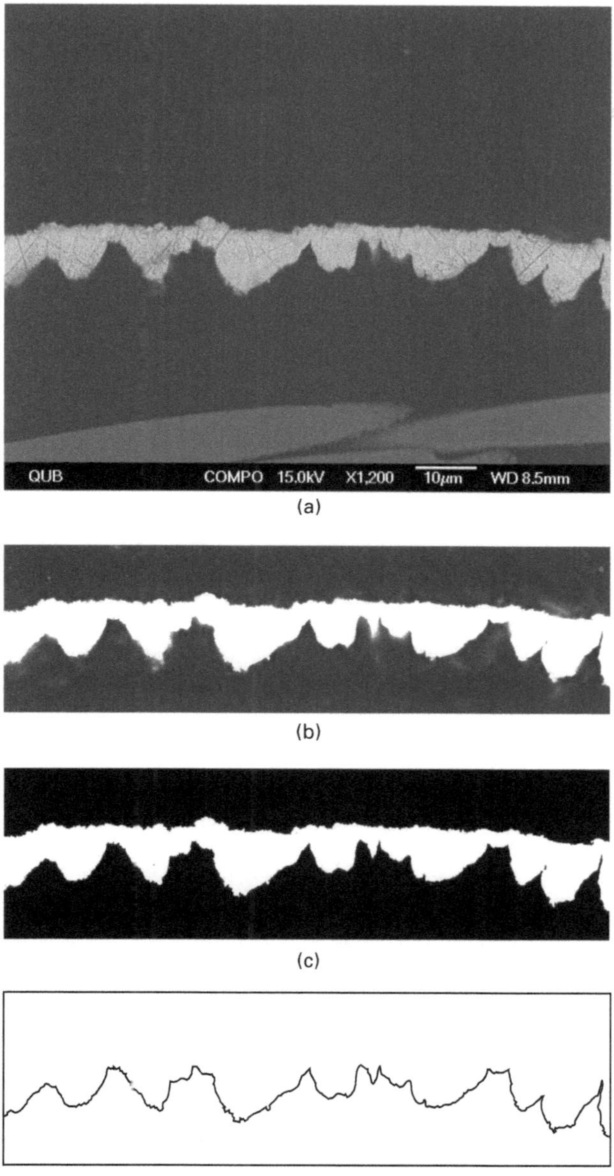

8.6 Process of extracting the surface cross-section curve from a scanning electron microscopy cross-section image. (a) A scanning electron microscopy cross-section image of a two-hour deposit under the material composition mapping mode. The bright area in the centre shows the uneven copper deposit and the edge of the substrate. (b) The cropped image of (a) with higher contrast. (c) The black-white image. The morphology of the substrate is clearly shown. (d) The surface cross-section curve that is extracted.

The calculated length correction factor of the FR4 surface from Fig. 8.6(d) is 1.98. By considering the error of the calculation method, the actual length factor of the substrate is 1.88, which means that the length of the curve of the surface cross-section of the substrate FR4 is 1.88 times the measured length. Assume that the length correction factor of the substrate is identical on the cross-section in any direction of the substrate. By taking the square of the length factor, we can get the estimated area factor of the substrate, which is 3.53, indicating that the actual surface area of the substrate FR4 is 3.53 times the measured area. Noting that it is very hard to rebuild the surface of the substrate and calculate the area directly, squaring the length factor is only an estimation of the area factor of the substrate.

8.5 The plating rates of the electroless copper deposits in different solutions

Figure 8.7 shows the increase of the weight per unit area of the substrate. All the substrates have a reasonable linear increase during the plating process. By using the least squares method, the curves can be fitted with linear functions and their slopes are the plating rates (Table 8.2).

Formaldehyde solutions in higher operating temperatures have the highest plating rates among all the plating solutions. This is one of the reasons for the popularity of formaldehyde in industry, i.e., its high plating rate. The

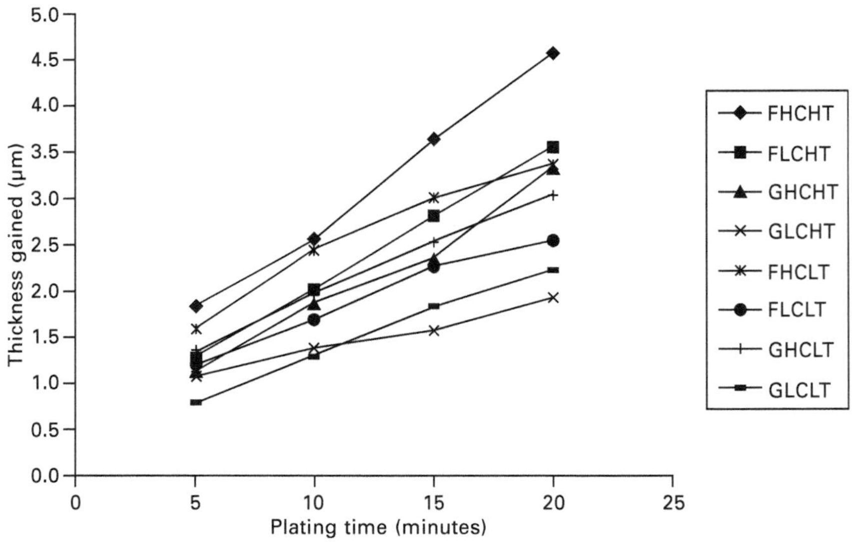

8.7 The curves of thickness of the deposit on substrate against plating time in different plating solutions. Values are not geometrically corrected.

Table 8.2 Plating rates of the deposit from different plating solutions in the form of weight gained per unit area per minute and thickness gained per minute

	$g\ m^{-2}\ min^{-1}$		$\mu m\ min^{-1}$	
Solution	No correction	With correction	No correction	With correction
Formaldehyde high concentration high temperature (FHCHT)	1.66	0.47	0.19	0.05
Formaldehyde low concentration high temperature (FLCHT)	1.35	0.38	0.15	0.04
Glyoxylic acid high concentration high temperature (GHCHT)	1.27	0.36	0.14	0.04
Glyoxylic acid low concentration high temperature (GLCHT)	0.49	0.14	0.05	0.02
Formaldehyde high concentration low temperature (FHCLT)	1.06	0.30	0.12	0.03
Formaldehyde low concentration low temperature (FLCLT)	0.83	0.23	0.09	0.03
Glyoxylic acid high concentration low temperature (GHCLT)	1.01	0.28	0.11	0.03
Glyoxylic acid low concentration low temperature (GLCLT)	0.87	0.24	0.10	0.03

rest of the solutions have similar plating rates of around 0.03 $\mu m\ min^{-1}$ among which the low concentration glyoxylic acid solution operated in high temperature has the lowest plating rate.

8.6 The sheet resistance and resistivity of the electroless copper deposit

By using the formulae in Section 8.1, the sheet resistance R_s and the resistivity ρ can be calculated.

The sheet resistance of the deposit decreases with the increase of the plating time (Fig. 8.8). This is because of the increase of the thickness of the deposits. With the increase of the plating time, the difference, between the sheet resistance of the deposits from different plating solutions, drops. Overall, the deposits from the glyoxylic acid low concentration high temperature solution have the highest sheet resistance. The reason can be the lowest plating rate of this solution, as can be seen from Fig. 8.7 and Table 8.2. The lower plating rate results in thinner deposits for the same plating time and thus higher sheet resistance.

The only difference between the two charts in Fig. 8.9 is the magnitude of the resistivity data, where the correction of the thickness changes the vertical scale. Significant decrease of the resistivity against plating time, and hence the thickness of the deposits, can be observed for some deposits. For other

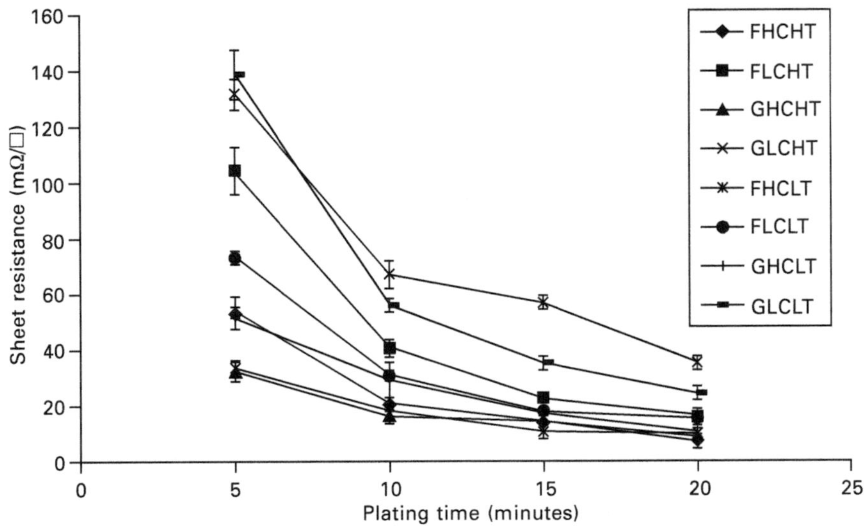

8.8 The sheet resistance of the electroless copper deposits from different plating solutions against plating time. The unit mΩ □$^{-1}$ means mΩ per square shaped sample, as for the same material, the sheet resistance is always the same for a square sized sample (with the same thickness) regardless of its dimension.

deposits, no significant decrease of resistivity against plating time is observed. The highest resistivity is discovered in the deposit from the solution where the low copper concentration solution with formaldehyde is operated at higher temperature, followed by that from the solution where the low copper concentration solution with glyoxylic acid is operated at lower temperature. The lowest resistivity is discovered in the deposit from the solution where the high copper concentration solution with formaldehyde is operated in lower temperature (formaldehyde high concentration low temperature – FHCLT). The reason for the FHCLT deposit having the lowest resistivity could be related to its crystal structure. In Chapter 2, a diamond pyramid structure in sub-micron scale is discovered on the surface of the FHCLT deposit. The analysis shows that this structure is possibly the stacking of copper {111} layers with twinning. It is known that the resistivity of the grain boundary is higher than that of the grain. Larger grain size results in better conductivity and lower resistivity because for a deposit with same size, the grain boundary area is reduced with larger grains. The resistivity of a grain with or without twinning differs little. For the FHCLT deposit, the grains with the diamond pyramid structure are bigger than on other deposits, and the twinning structure does not increase the resistivity significantly. Therefore, it is reasonable for the deposit from the FHCLT solution to have the lowest resistivity. On the formaldehyde low concentration low temperature (FLCLT) deposit, a

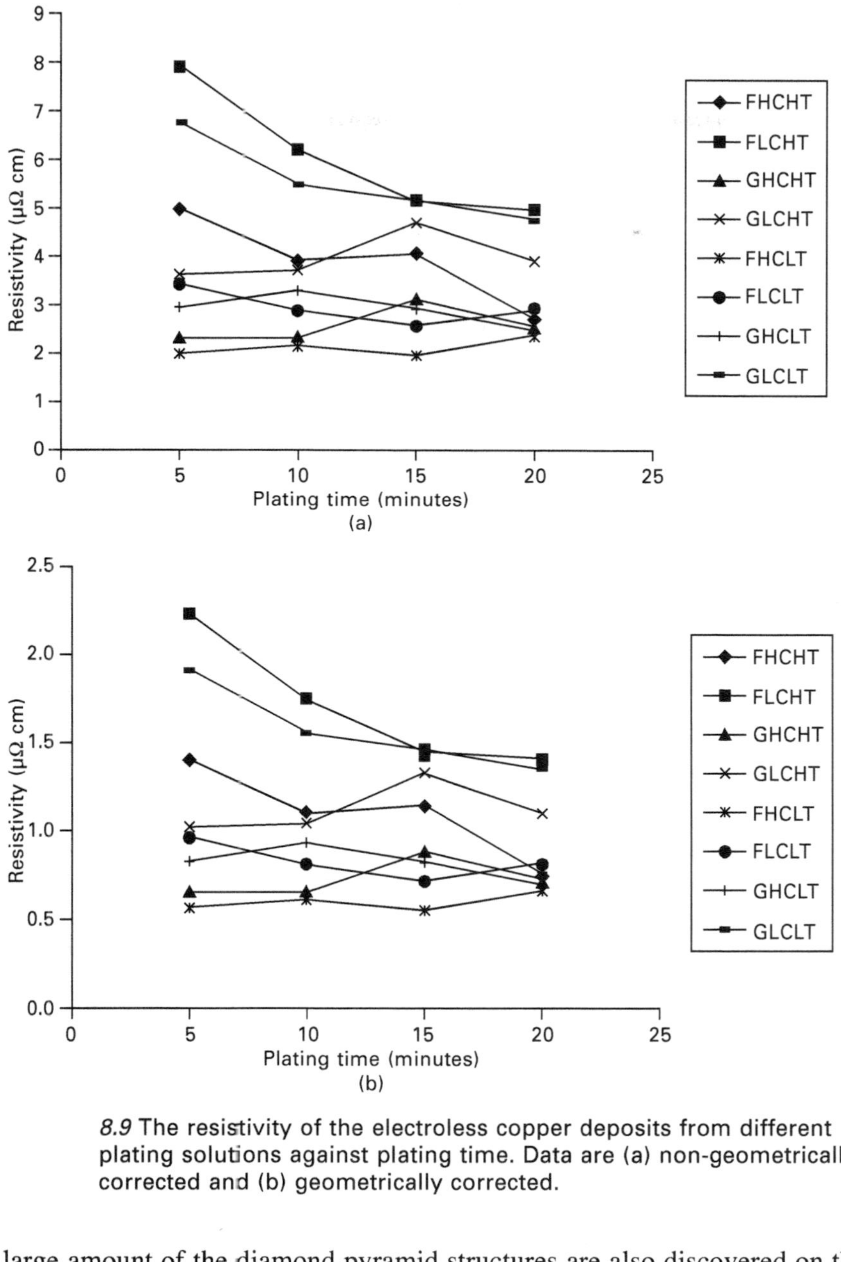

8.9 The resistivity of the electroless copper deposits from different plating solutions against plating time. Data are (a) non-geometrically corrected and (b) geometrically corrected.

large amount of the diamond pyramid structures are also discovered on the surface. These structures are even clearly visible in the early stage of the plating process, but the FLCLT sample has higher resistivity than the FHCLT one. The reason for this can be that a large amount of the diamond pyramid structures may introduce voids among grains and these voids can increase the resistivity of the deposit. In Chapter 3, the void fractions calculated from the 60-minute samples show that the FLCLT sample has the highest rate of void in the deposit.

Formaldehyde based solutions, operating in lower temperature (with higher concentration of formaldehyde) can result in a deposit with lower resistivity. For high concentration solutions (formaldehyde high concentration high temperature – FHCHT and FHCLT), the reduction of the resistivity of the deposits is up to 1 μΩ cm, and for low concentration solutions (formaldehyde low concentration high temperature – FLCHT and FLCLT), the difference is even greater, being more than 2 μΩ cm. However, for glyoxylic acid based solutions, the difference of the resistivity of the deposit is not as significant as the formaldehyde solutions. The two high copper concentration solutions (glyoxylic acid high concentration high temperature – GHCHT and glyoxylic acid high concentration low temperature – GHCLT) give the deposits with similar low resistivity. The resistivity of the two low copper concentration solution samples (glyoxylic acid low concentration high temperature – GLCHT and glyoxylic acid low concentration low temperature – GLCLT) is slightly higher than the high copper concentration ones.

The trend of the curves in Fig. 8.10 is the same as in Fig. 8.9. The only differences between Fig. 8.10(a) and (b) are the scale of the thickness of the deposits and the magnitude of the resistivity. The figures show the excellent resistivity of the deposit from some solutions, whose resistivity is at a low level even when the deposits are thin. Again, no significant reduction in resistivity against deposit thickness is discovered in most deposits.

The length correction factor used in this chapter is acquired from the scanning electron microscopy image of the cross-section of the substrate. The method is to calculate the digital length of the surface curve and compare it with the horizontal length of the image, which is the measured length. However, with the increase of the thickness of the deposit during the plating process, the surface fluctuation of the substrate is gradually covered by deposit, making the surface of the deposit smoother with less fluctuation. This means that the length and area correction factor is smaller and smaller with the increasing plating time and the increase of the thickness of the deposit. When the thickness of the deposit is much greater than the fluctuation of the substrate, no correction is needed. Therefore, the correction factor used in this chapter is actually greater than the correction factor needed for the real situation. This means that the actual thickness of the deposit and hence the resistivity are greater than the calculated values. This can also be seen from the resistivity of the deposits after correction. The known resistivity of bulk copper is 1.68 μΩ cm. In Fig. 8.9(b), where the corrected resistivity data of the deposits against plating time are illustrated, the lowest resistivity of the deposits is about 0.5 μΩ cm, which is lower than the bulk resistivity of copper. In Fig. 8.9(a), where the non-corrected data are shown, the lowest resistivity of the deposits is about 2.0 μΩ cm, which is higher than the bulk

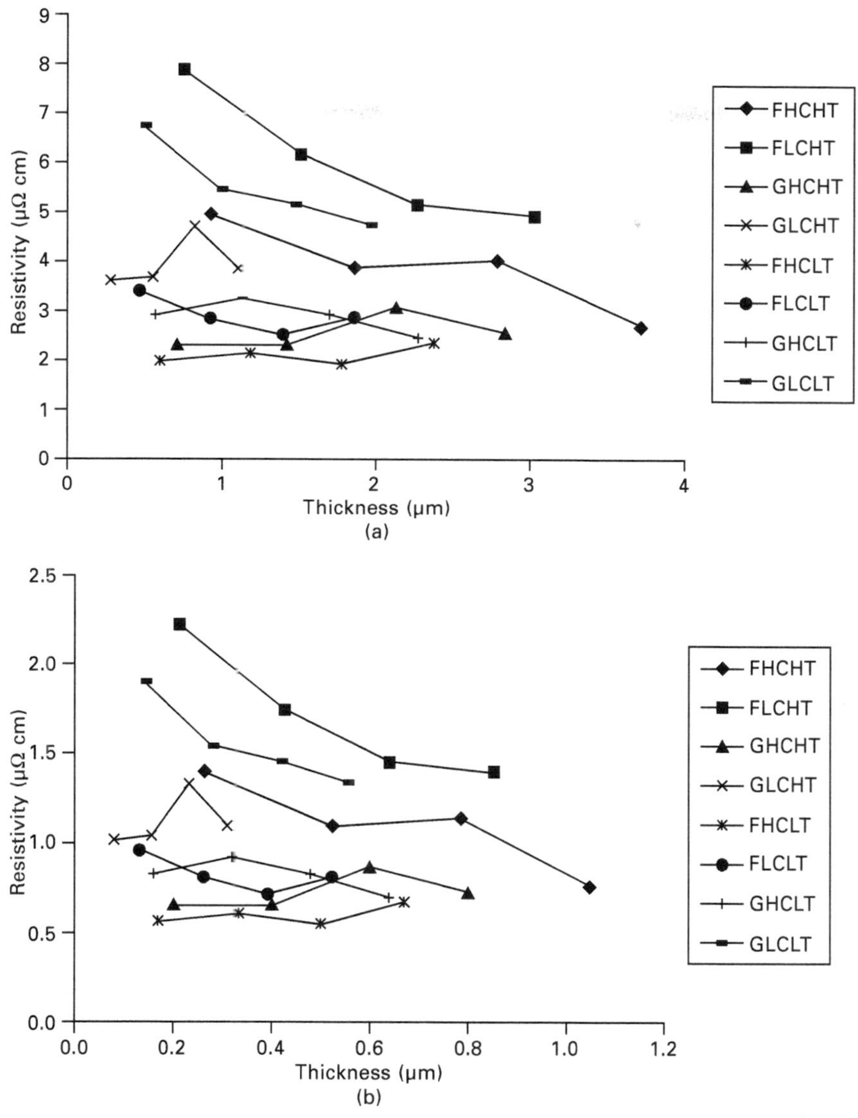

8.10 The resistivity of the deposit against film thickness (a) without geometric correction and (b) with geometric correction.

resistivity. This shows that due to the estimated correction factor being higher than the actual value, the corrected resistivity data is lower than the actual value, which should be no lower than the bulk resistivity of copper (Machlin, 2006, p. 2) but less than the non-corrected data. This makes the actual resistivity of the copper deposit close to the copper bulk resistivity.

A significant reduction of the resistivity of electroless copper deposit against the thickness of the deposit is expected. In this chapter, only the deposits from some solutions have this trend. The deposits from the rest of the solutions do not show significant reduction of resistivity against their thickness, and the level of the resistivity of these deposits remains in a very low level, close to the bulk resistivity. This means that the concentration, type of reducing agent and plating temperature can affect the resistivity of the deposit greatly.

8.7 Conclusions

In this chapter, the sheet resistance and resistivity of the electroless copper deposits from different plating solutions are compared. In order to get data that are more precise, length and area corrections are introduced. The method of calculating the length of a digital curve is also described and discussed in this chapter. The following conclusions can be given.

1. The deposit from the formaldehyde low concentration high temperature solution has the highest resistivity. It also falls sharply with the increase of plating time and thickness of the deposit.
2. Significant reduction of the resistivity against plating time is found for the deposits from some solutions. No such significant reduction is found for the deposits from the rest of the solutions.
3. The deposit from the formaldehyde high concentration low temperature solution has the lowest resistivity. The reason can be the diamond pyramid structure on the surface of the deposit. The bigger grains compared to those of the rest of the deposits and the possible twinning in the structure may lead to the low resistivity.
4. A large amount of the pyramid structures will introduce a large amount of voids in the deposits and hence increase the resistivity of the deposits. This is indicated by the higher resistivity of the deposit from the low concentration solution compared to that of the deposit from the high concentration solution of formaldehyde both operated at low temperature.
5. Deposits with fine grains and a large amount of grain boundaries have relatively high resistivity.
6. The correction factor estimated is higher than that for the actual situation, resulting in the corrected resistivity lower than the resistivity of bulk copper. The actual resistivity of the deposits should be lower than the non-corrected data and closer to that of the bulk copper.

8.8 Reference

Machlin E S (2006), *Materials Science in Microelectronics II: The effects of structure on properties in thin films* (2nd edn), Oxford, Elsevier.

9 Applications of electroless copper deposits

Abstract: A review of the background to electroless copper deposition, basic electroless copper plating procedures, methods for evaluating the mechanical and electrical properties of electroless copper deposit, and the numerical simulation methods for the electroless copper plating process is presented. The history of printed circuit boards (PCBs) is introduced followed by the application of electroless copper deposition in the PCB industry and ultra/very large-scale integration. Following that, the testing methods of mechanical properties, mainly thermal shock resistance, are summarised. In the last part, computer simulation of the electroless copper plating process is reviewed, including simulation based on a diffusion model.

Key words: application, printed circuit board, ultra/very large-scale integration, diffusion model, thermal shock resistance.

9.1 Printed circuit board (PCB) industry

The history of the printed circuit board (PCB) industry dated back to the 1950s, when Dr Paul Eisler was working to replace radio tube wiring with something less bulky. The final concept he developed was very similar to a single-sided printed wiring board.

Single-sided printed wiring boards were soon commercialised during the 1950s and 1960s. About ten years later, the plating through hole (PTH) technology was developed, which allowed copper to be plated on the walls of drilled holes. With the PTH technology, the double-sided PCB could be produced with inter-connection through the holes.

During the 1980s, with the growth in industry and technology, PCBs became more and more complex and dense with the development and commercialisation of multilayer PCBs. Today, most electric products use multilayer PCBs as the connection among parts and instruments.

In the PCB manufacturing process, electroless copper plating is an important step. It is widely used in the following two processes. One is plating onto bare laminate and the other is plating through hole, because under these two circumstances, electroplating cannot or can hardly be carried out. In the process of plating onto bare laminate, electroless copper plating plates a thin layer of copper on the bare substrate to make the substrate conductive for further electroplating. In the process of plating through hole, electroless copper plating is used to make the inner walls of the hole conductive to connect the printed circuits in different layers or the pins of the integrated chips.

For PCB industrial application, especially in the plating through hole process, more processes may be involved in the electroless plating. After holes are drilled on the epoxy board, a de-smear step is employed to remove the smear of the inner walls of the holes, and a neutralising step should follow to remove the residual chemical used in the previous de-smear step.

In the early years, compared with electroplating, electroless copper plating was lowly rated and unstable. However, the unique feature of electroless copper is that it can be plated uniformly over all surfaces, regardless of size, shape and electrical conductivity. This contributes to the application of electroless copper plating in various cases. Furthermore, the advances in electroless plating rate and stabilisation of plating solution, and an acceptance by the PCB industry of a lowering in the conductor thickness to 15 micrometers have led to a renewed interest in this manufacturing technique.

Another application of electroless copper is very- or ultra-large scale integrated (VLSI/ULSI) technology (Shacham-Diamond, 2000). Because of its lower resistivity, potentially higher resistance to electro-migration and stress induced voiding, copper has been applied to replace aluminium as the interconnect material in integrated chips. Other applications of electroless copper deposition include electromagnetic shielding and decorative plating.

9.2 Properties of electroless copper deposition and their evaluation

Before the materials can be applied in industry, their properties need to be examined. Basic properties include topological, mechanical, and electrical properties.

Although the main application of electroless copper plating is printed circuit board manufacturing, the mechanical properties of electroless copper deposits are still very important, especially when the film is expected to be thinner and appliances are getting smaller and more compact. The common mechanical properties of electroless copper deposits include the strain–stress relationship, elongation for ductility, fatigue and adhesion strength.

For electroless copper applied in the PCB industry, a unique property that needs to be characterised is the thermal shock resistance. Thermal shock testing can be used to check the reliability of electrical interconnects like plated-through holes and solder joint connections. In this test, a PCB with certain patterns of circuit (Class Y or Micro via) is prepared and sent through a sequence of repeating cycles between a low temperature (e.g. room temperature) environment and a high temperature (e.g. 260 °C) environment with a certain period of time at each temperature. The resistivity of the circuit is tested every few cycles. The number of the cycles right before the sharp increase of the resistivity is recorded as the measure of the thermal shock resistance of the copper deposit.

9.3 Diffusion based simulation

For electroless plating at low deposition rates and stationary solutions, the process can be described by a solution species transport model with the effect of only diffusion, in particular a quasi-steady-state model for electroless copper plating from the diffusion of the dominant ionic species (Cu^{2+}). The model assumes that the deposition rate at any point on the film surface is directly related to the copper ion concentration. In a 1D quasi-steady-state case, the change of the copper ion concentration with time can be written as:

$$D\frac{\partial^2 C}{\partial x^2}V - FC^r A = 0 \tag{9.1}$$

where the first term represents the diffusion contribution of copper ions in a volume V and the second term represents the reaction consumption of copper ion on a surface with area A. D in the above equation is the diffusivity of the copper ion in the solution and FC^r is the atomic flux into the film surface. The constant r can be set to 0.8. For the 2D case, the equation can be rewritten as:

$$D\frac{\mathrm{d}C}{\mathrm{d}n} - FC^r = 0 \tag{9.2}$$

The equation is solved with the finite element method by firstly creating a triangular mesh from the film topography and determining the boundary conditions (Fig. 9.1), and secondly solving the finite element equations to determine the concentration of the copper ion in the simulation region. The cross-section of the electroless deposit on a trench for VLSI application is then simulated with the model, and the morphology produced by the simulation matches well with the experimental deposits.

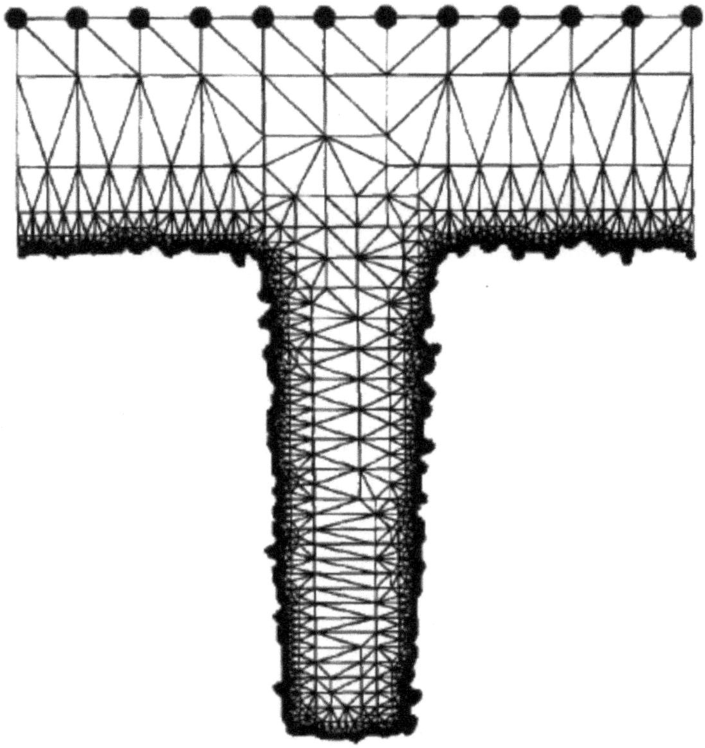

9.1 Mesh generated from film deposited into a trench (Smy *et al.*, 1997). (From Smy T, Tan L, Dew S K, Brett M J, Shacham-Diamand Y and Desilva M, 'Simulation of electroless deposition of Cu thin films for very large scale integration metallization', *J Electrochem Soc.* Reproduced by permission of ECS – The Electrochemical Society).

9.4 References

Shacham-Diamond Y Y (2000), 'Electroless copper deposition using glyoxylic acid as reducing agent for ultralarge scale integration metallization', *Electrochem Solid State Lett*, **3**, 279–82.

Smy T, Tan L, Dew S K, Brett M J, Shacham-Diamand Y and Desilva M (1997), 'Simulation of electroless deposition of Cu thin films for very large scale integration metallization', *J Electrochem Soc*, **144**, 2115–22.

Part II

Electroless nickel–phosphorus (Ni–P) depositions

10
Crystallisation of nickel–phosphorus (Ni–P) deposits with high phosphorus content

Abstract: In this chapter, experimental studies using differential scanning calorimetry (DSC) and X-ray diffraction (XRD) techniques conducted on the electroless nickel–phosphorus (Ni–P) deposits, for their crystallisation and phase transformation behaviour under linear heating conditions are described. The effects of deposit phosphorus content on the as-deposited structure, and the sequence and activation energy of crystallisation processes for the deposits are analysed. The phase transformation temperatures of the electroless Ni–P deposits in response to different phosphorus contents and heating rates are also examined. Changes in microstructure after heat treatment are examined using a high-resolution field emission scanning electron microscope.

Key words: alloys, amorphous materials, X-ray diffraction (XRD), phase transitions, calorimetry.

10.1 Introduction

Electroless nickel plating is used to deposit nickel without the use of an electric current. Electroless nickel–phosphorus (Ni–P) coating consists of an alloy of nickel and phosphorus. The autocatalytic deposition process occurs in an aqueous solution with chemical reactions caused by the catalytic reduction of the nickel ion with sodium hypophosphite in acid baths. The electroless nickel process was invented by Brenner and Riddell in the 1940s. Since the first discovery, electroless Ni–P deposits have been widely used in various industries as engineering protection coatings to protect the inner surface of pipes, valves and other parts, or as functional films.

The amount of phosphorus can be classified as high, medium and low. The microstructure of the electroless Ni–P deposits changes from having a mixture of amorphous and nanocrystalline phases to a fully amorphous phase when the phosphorus content increases. Both high (10–16 wt%) and medium (5–8 wt%) phosphorus deposits have the same microstructure or it may consist of one or more compound phases. High phosphorus deposits have very good corrosion protection and abrasion resistance. Medium phosphorus deposits can be fully amorphous or consist of a mixture of amorphous and nanocrystalline nickel phase. Low phosphorus deposits consist of nanocrystalline nickel with a small amount of amorphous phase or simply crystalline nickel phase. Microcrystalline was used in earlier publications, but we now refer to the same structure as, rightly, nanocrystalline, after the significance of nanostructure was found and nano has become trendy. At low phosphorus levels, the electroless nickel coatings have high wear resistance and excellent corrosion resistance.

As phosphorus levels increase, the hardness at as-deposited condition increases. Hardness can be improved by heat treatment that can cause the formation or precipitation of nickel phosphide. In order to achieve full hardness, the electroless nickel deposits are heat-treated at 400 °C for one hour in an inert atmosphere. High hardness and natural lubricity enable the electroless nickel coatings to have good wear resistance. The uniformity of the electroless Ni–P deposits makes them ideal wear surfaces in many sliding-wear applications. The presence of the amorphous phase increases the electrical resistivity of the deposits. Electroless nickel with high phosphorus deposits are non-magnetic as-plated. Magnetism can be increased by heat treatment. Electroless Ni–P deposits are easily soldered with a highly active acid flux. Those with low phosphorus content are more easily soldered after plating. The melting temperature decreases with an increase in phosphorus level. Low phosphorus deposits melt at approximately 1300 °C whereas medium and high phosphorus deposits melt at approximately 890 °C.

Crystallisation and phase transformation behaviour of electroless-plated Ni–P deposits during thermal processing have been the subject of various investigations (Martyak and Drake, 2000). These processes have vital roles in determining the material properties. Investigations on electroless Ni–P deposits have shown that different deposit compositions and heat treatment conditions could affect both the microstructural characteristics and the crystallisation behaviour of the deposit. Different results were reported regarding the microstructures in the as-deposited condition and the stable phases after heat treatments (Martyak and Drake, 2000).

This chapter examines the effect of the continuous heating process and phosphorus content on the crystallisation kinetics and phase transformation behaviour of electroless Ni–P deposits with high-phosphorus content. For this purpose, deposits are heated in the differential scanning calorimeter under different heating rates, and the crystallisation processes are studied. For further analysis of phase transformation, an X-ray diffractometer and a scanning electron microscope are used to identify the phases formed in the deposits heated to different end temperatures under a constant heating rate.

Electroless-plated nickel film has great potential to be used as a solder diffusion barrier in low cost flip-chip packaging. Since the barrier effect is due to the lack of grain boundaries in amorphous phase, i.e. the lack of diffusion path, the control of the formation of crystalline phase and grain size becomes crucial in such applications. The strain/stress level in the nickel film should also be controlled because of its detrimental effect on the fatigue life of the plating. Based on X-ray diffraction (XRD) profiles, line-broadening technique can be applied to estimate the grain size and microstrains in the Ni–P deposits (Martyak and Drake, 2000). Investigation of the evolution of grain size and microstrain in Ni–P deposits isothermally annealed at temperatures between 60 °C and 600 °C shows that the lattice disorder in the crystalline phase increases with increasing phosphorus content. Grain size varies from 5 to 60 nm and increases sharply above 300 °C. A study was carried out by Martyak and Drake (2000) on a Ni–P film with medium phosphorus content (5–6 wt%), isothermally annealed at temperatures between 100 °C and 600 °C. Significant grain size increase was observed at temperatures above 400 °C. Study of electrodeposited Ni–P deposits containing 1.2–15.4 wt% phosphorus using the line broadening technique shows that the grain size in the as-deposited condition decreases with the increase of phosphorus content whereas microstrain increases.

In this chapter, XRD line broadening technique is used to study the evolution of grain size and microstrain in Ni–P platings after they are subjected to different heating processes. Continuous heating is the heat treatment procedure. The influence of heating rate on grain size and microstrain is shown.

This chapter will also show the microstructures and phase transformations caused by heating, by analysing scanning electron microscopy (SEM) images. This furthers research on the heat treatment of Ni–P coatings and their characterisation using SEM, for examples by Rahimi *et al.* (2009) and Vojtěch *et al.* (2009).

10.2 Effects of phosphorus content

High phosphorus deposits are typically amorphous in the as-deposited condition, as demonstrated by the deposits of both 16 wt% P (Fig. 10.1) and 12 wt% P (Fig. 10.2). The amorphous profiles with the wide angular range of 35–55° (2θ) are around a 2θ position corresponding to the Ni {111} plane. The Fe {110}, {200} and {211} diffraction peaks in the profiles of the 16 wt% P deposit are from the mild steel substrate underneath. They are present because the coated-deposit is too thin (2.9 μm) to totally absorb the penetration of the X-ray beam. The two deposit compositions however show different microstructural characteristics after being heated (Table 10.1).

In the 16 wt% P deposit, Ni_2P and $Ni_{12}P_5$ metastable phases form after the heating processes to 350 °C and 400 °C (Fig. 10.1). The Ni_3P stable phase forms after the heating processes to between 350 °C and 800 °C. In the 12 wt% P deposit, Ni_2P and $Ni_{12}P_5$ metastable phases also form after the heating processes to 350 °C and 400 °C (Fig. 10.2). In addition, nickel phase forms in the deposit after the heating processes to between 350 and 800 °C. Hence, the effect of phosphorus content on the microstructural characteristics of the electroless Ni–P deposits is significant.

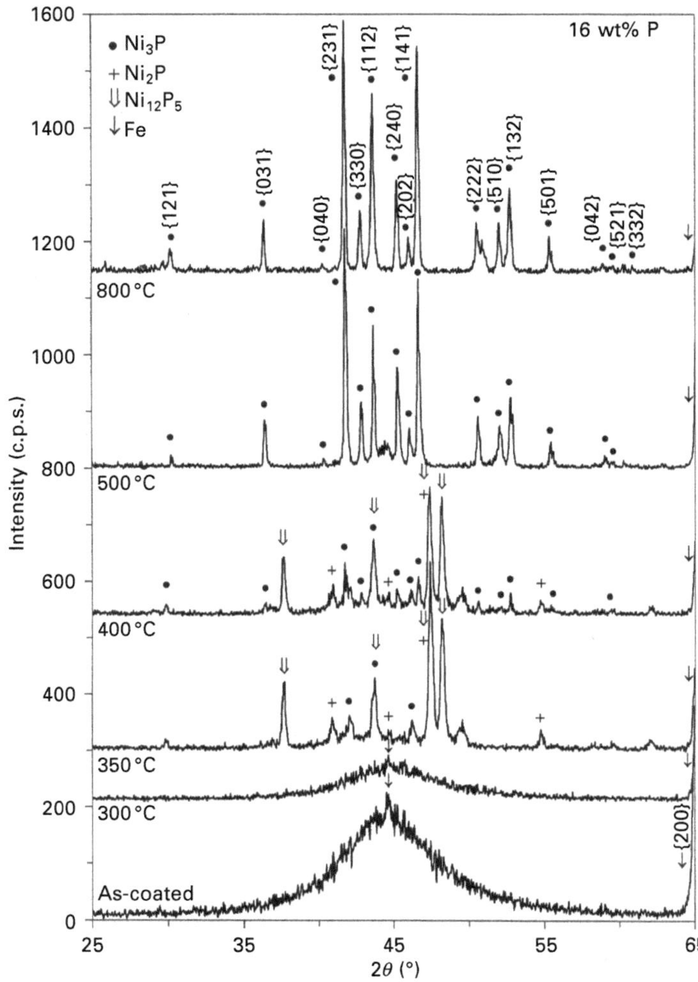

10.1 XRD profiles of the 16 wt% P deposit at as-deposited condition and after heating to the end temperatures of 300–800 °C at 20 °C min^{-1}. For the sake of clarity, the profiles are shifted arbitrarily on the vertical scale.

The phosphorus content of the electroless Ni–P deposits can also affect the phase transformation temperature of the deposits during heating process. The calorimetry curves in Fig. 10.3 show that the exothermic temperatures of the 12 wt% P deposit are higher than the 16 wt% P deposit. In addition, a shoulder is apparent right before the major exothermic peak of the DSC curves in the 12 wt% P deposit (Fig. 10.3(b)). This shoulder is, however, not contributed by the thermally activated reactions from the substrates underneath the coatings, indicated by the flat DSC curves from the substrates

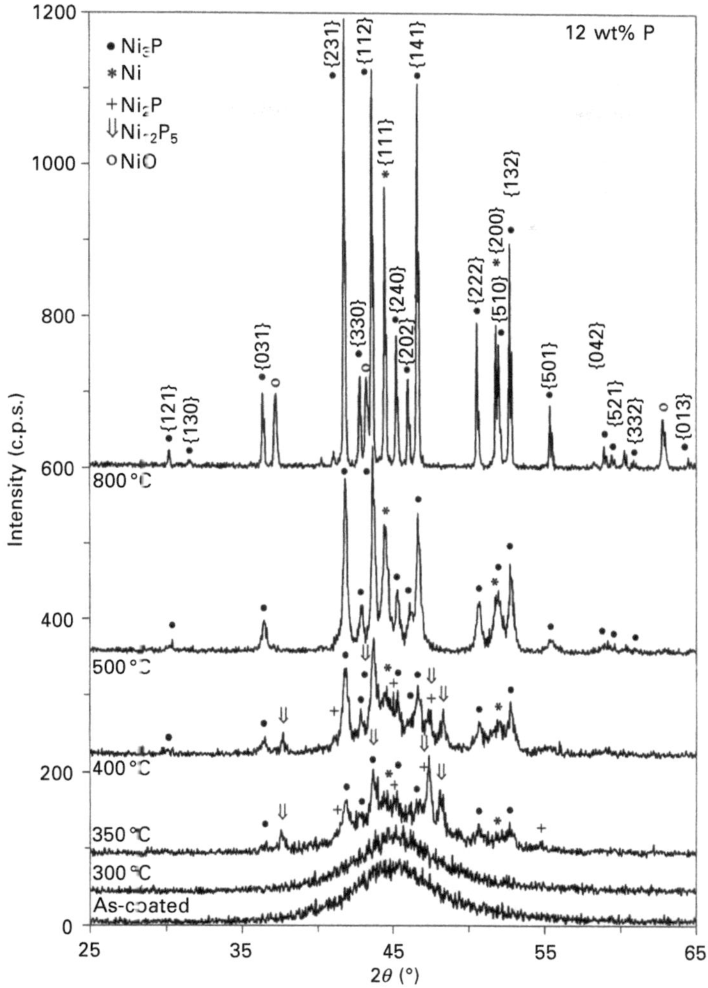

10.2 XRD profiles of the 12 wt% P deposit at as-deposited condition and after heating to the end temperatures of 300–800 °C at 20 °C min^{-1}. For the sake of clarity, the profiles are shifted arbitrarily on the vertical scale.

alone in the corresponding temperature region (Fig. 10.3(c)). The formation of shoulder in the DSC curves may be due to the first stage transformation – the short-range atomic movements and incipient crystallisation of metastable crystalline structures. The second stage transformation is the long-range atomic movements and decomposition to the stable phases, associated probably with the major exothermic peak in those DSC curves. In the 16 wt% P deposit, there may also be those processes corresponding to first stage transformation, but not showing due to the peak overlapping in the

Table 10.1 Phase compositions of the deposits of 12–16 wt% P after heating to different end temperatures at a constant heating rate 20 °C min^{-1}

Temperature (°C)	16 wt% P	12 wt% P
As-deposited	Amorphous	Amorphous
300	Amorphous	Amorphous
350	Amorphous + Ni_2P + $Ni_{12}P_5$ + Ni_3P	Amorphous + Ni_2P + $Ni_{12}P_5$ + Ni + Ni_3P
400	Amorphous + Ni_2P + $Ni_{12}P_5$ + Ni_3P	Amorphous + Ni_2P + $Ni_{12}P_5$ + Ni + Ni_3P
500	Amorphous + Ni_3P	Amorphous + Ni + Ni_3P
800	Ni_3P	NiO + Ni + Ni_3P

DSC curves (Fig. 10.3(a)). The DSC curves of the mild steel substrates alone show a broad exothermic region on each of the curves that may be in connection with recrystallisation of steel. This broad exothermic region is coincident to the second exothermic peak of the DSC curve in the 16 wt% P deposit. Nevertheless, the phase transformation temperatures increase with decreasing phosphorus content.

Crystallisation activation energies of the electroless Ni–P deposits are derived from the Kissinger plots of $\ln(T_p^2/\phi)$ versus $1000/(RT_p)$ using the peak temperatures (T_p) of crystallisation process, represented by the major exothermic peak of the DSC curves at different heating rates (ϕ) (Sha and Guo, 2009, p. 91). Based on the relationship

$$\ln\frac{T_p^2}{\phi} = \frac{Q}{RT_p} + \ln\frac{Q}{RK} + \ln C \qquad 10.1$$

a linear regression line could be obtained from the plots (Fig. 10.4). The slope of the plot yields the activation energy (Q).

In Eq. 10.1, R is the gas constant (8.314 J mol^{-1} K^{-1}), and K and C are constants. The crystallisation peak temperatures of the 16 wt% P and the 12 wt% P deposits range from 325–354 °C and 346–375 °C, respectively (Fig. 10.3). The calculated crystallisation activation energy of the two deposits (210 ± 17 and 225 ± 20 kJ mol^{-1}, respectively, for the 16 wt% P and the 12 wt% P deposits) is about the same. The calculated activation energies can be compared with the self-diffusion activation energy of nickel, 289 kJ mol^{-1} (Wilkinson, 2000, p. 242). The calculations of standard deviation (±) for the activation energies (the slope of Kissinger plots) are explained below.

The standard deviation (S_β) for the slope of a linear regression plot is given by:

$$S_\beta = \sqrt{\frac{1}{(n-2)}\frac{S_{yy}}{S_{xx}}(1-r^2)} \qquad 10.2$$

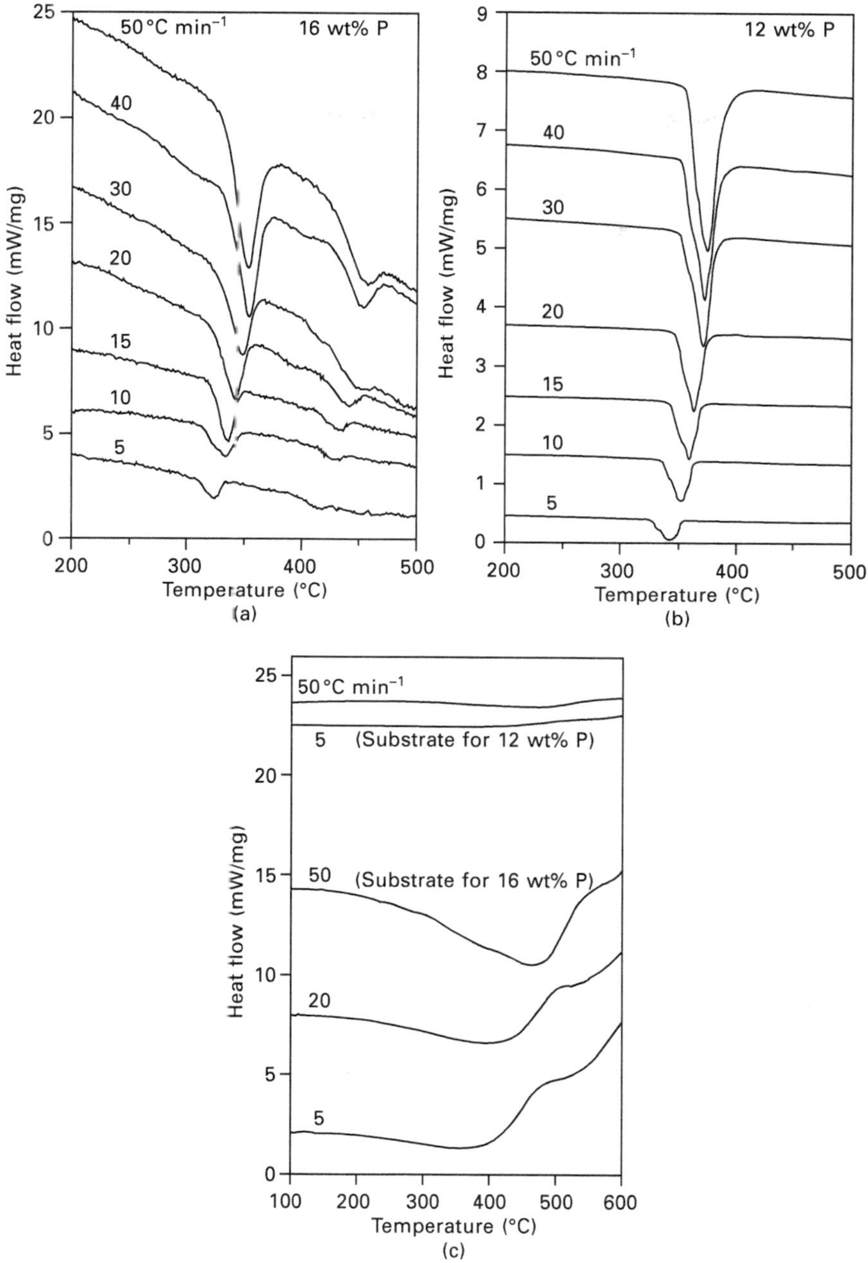

10.3 Differential scanning calorimetry (DSC) curves of (a) 16 wt% P coating, (b) 12 wt% P coating and (c) mild steel substrates at the heating rates of 5–50 °C min^{-1}. For the sake of clarity, the calorimetric curves are shifted arbitrarily on the vertical scale.

where n is number of (x, y) input data sets,

$$S_{yy} = \sum_{i=1}^{n} (y_i - \bar{y})^2 \text{ and } S_{xx} = \sum_{i=1}^{n} (x_i - \bar{x})^2 \qquad 10.3$$

where $\bar{x}$ and $\bar{y}$ are the mean of x and y values, respectively, and r^2 is the coefficient of determination given as $(S_{yy} - \text{SSE})/S_{yy}$.

The sum of squares for error (SEE) is defined as:

$$\text{SEE} = \sum_{i=1}^{n} (y_i - \hat{y})^2 \qquad 10.4$$

where $\hat{y}$ denotes the predicted value of y obtained from the fitted plot.

For the Kissinger plots, the slope of the plots equals the crystallisation activation energy (Q) of the electroless Ni–P deposits. Hence, the standard deviation of Q is the same as that calculated for S_β.

10.3 Effects of the heating process and degree of phase transformation

Both the phase composition and phase transformation behaviour of the electroless Ni–P deposits depend on the heating temperatures. The effect of heating rate on the phase composition and transformation behaviour is also considerable. However, the cooling rate has no significant effect on the phase composition from the heating temperature that is far higher than the crystallisation temperature of the deposit.

The XRD profiles of the 12 wt% P deposit heated to 300–800 °C show different phase compositions (Fig. 10.2). After heating to 300 °C, there is no change, with only a single broad amorphous profile. This is confirmed by the DSC curve of the deposit (Fig. 10.3(b)), which shows no major crystallisation reaction up to that temperature. At 350 °C and 400 °C, the diffraction peaks corresponding to the metastable Ni_2P and $Ni_{12}P_5$, and stable fcc nickel and Ni_3P phases in the XRD profiles of the deposit suggest the ongoing of phase transformation processes. This is in agreement with the major exothermic peak of crystallisation process in the DSC curves of the deposit at the same temperature range. The XRD profiles of the deposit also show the amorphous phase remaining. Hence, the microstructure of the 12 wt% P deposit is the mixture of amorphous and crystalline phases. At 400 °C and 500 °C, the diffraction peaks corresponding to the fcc nickel and Ni_3P phases are intensified with a decrease of the amount of amorphous phase. The Ni_2P and $Ni_{12}P_5$ metastable phases are decomposed completely after the heating process to 500 °C. Consequently, the crystallisation process in the 12 wt% P deposit is near to completion. At 500 °C and 800 °C, numerous minor diffraction peaks develop in the 12 wt% P deposit with increased

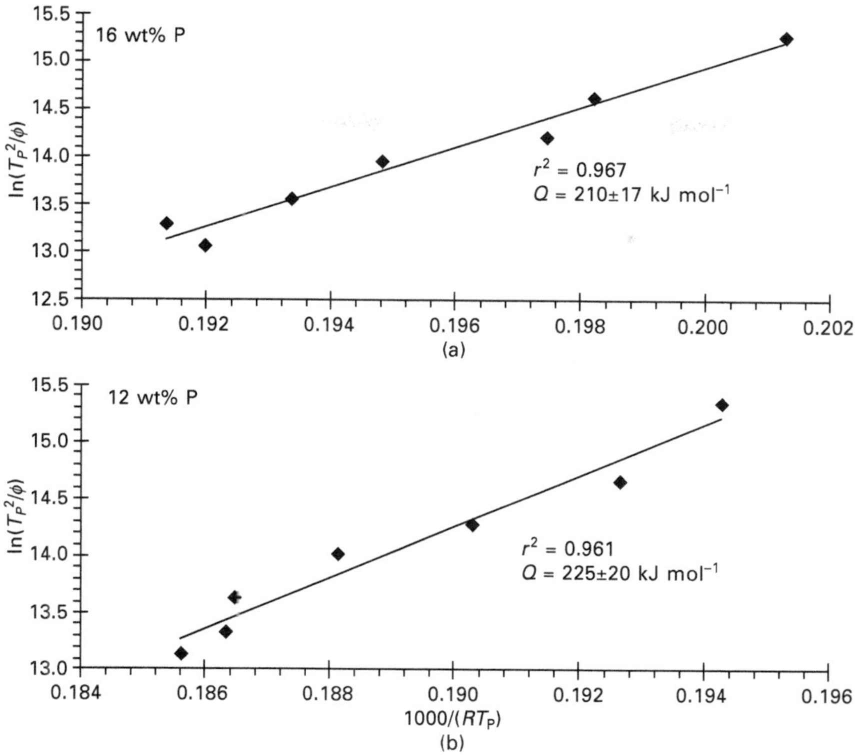

10.4 Kissinger plots of $\ln(T_P^2/\phi)$ versus $1000/(RT_P)$ using crystallisation peak temperatures of (a) the 16 wt% P deposit and (b) the 12 wt% P deposit.

intensities and further refinement of the existing diffraction peaks. The fcc nickel and Ni_3P phases are the only two dominant resultant phases at this stage. Besides, a small amount of nickel oxide (NiO) is in the XRD profile at 800 °C, possibly as the purity of helium (99.996 vol. %) and the base vacuum level in the heat treatment furnace are not high enough to prevent oxidation at high temperature Nonetheless, the 12 wt% P deposit has completed the crystallisation process and developed a crystalline microstructure. In the 16 wt% P deposit, the heating temperatures affect the microstructural properties and phase transformation behaviour of the deposit in a similar way (Figs 10.1 and 10.3(a)), except that there is no nickel phase in the deposit.

The effect of the heating rate on the electroless Ni–P deposits during the heating process can be studied through the XRD and DSC analyses. The XRD profiles of the 12 wt% P deposit after heating to 400 °C, at different heating rates of 5 °C and 20 °C min^{-1} are compared in Fig. 10.5(a). Ni_2P and $Ni_{12}P_5$ metastable phases are in the deposit heated at the higher heating rate. The 12 wt% P deposit used in these XRD analyses is heated to a temperature

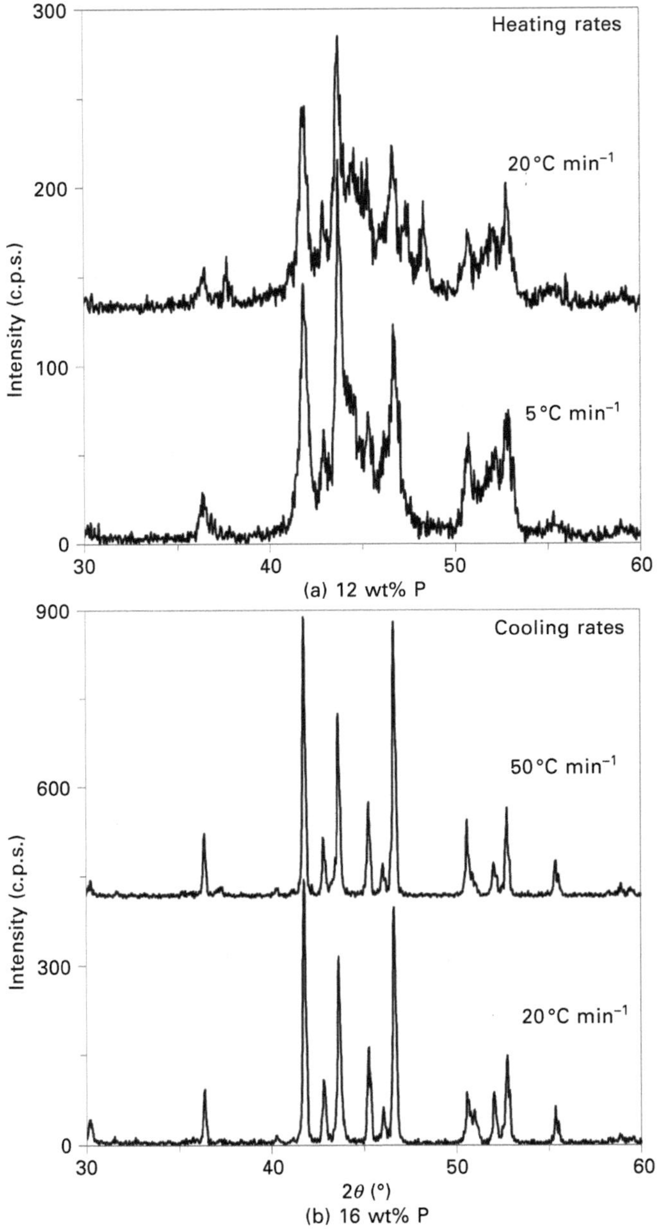

10.5 XRD profiles showing the effects of (a) heating and (b) cooling rates on the crystallisation of the 12 wt% P and the 16 wt% P deposits, respectively. The heating end temperatures are 400 °C for the 12 wt% P deposit, and 800 °C for the 16 wt% P deposit at 20 °C min^{-1}. For the sake of clarity, the XRD profiles are shifted arbitrarily on the vertical scale.

(400 °C) not much higher than the peak crystallisation temperature (346 °C and 366 °C, respectively, at 5 °C and 20 °C min^{-1}). The effects of the heating rate on the microstructural property of the deposits to temperatures far before and after the major crystallisation reaction may not be identical to the effect shown in Fig. 10.5(a). In addition to this, the DSC curves of the 12 wt% P deposit in Fig. 10.3(b) show that the major crystallisation reaction, indicated by the major exothermic peak, is delayed by the increase of heating rates. A similar effect is in the DSC curves of the 16 wt% P deposit (Fig. 10.3(a)). The increase in heating rates also correspondingly increases the area of exothermic peaks in the DSC curves. This phenomenon is caused by the higher rate of exothermic heat flow and shorter crystallisation time at high heating rate.

The effects of the cooling rate on the resultant microstructure are not significant. The deposits are heated to 800 °C at a constant rate of 20 °C min^{-1}. They are then cooled down at different cooling rates of 20 °C and 50 °C min^{-1}. The XRD profiles of the deposits cooled at different cooling rates are similar (Fig. 10.5(b)). However, this indicates only the effect of cooling rate on the microstructural property of the deposit that has been heated to a temperature (800 °C) much higher than the crystallisation temperatures of the deposit. The effects of cooling rate on the microstructural property of the electroless Ni–P deposits during the major crystallisation reaction may be different.

The integrated intensities of amorphous, metastable, nickel and Ni_3P phases can be calculated from the corresponding reflections in the XRD profiles using standard computer programs. The ratio of the integrated intensity of amorphous phase (I_{amorp}) to the total integrated intensity (I_{total}) shows the relative proportion of the amorphous phase remaining in the deposit. Consequently, the degree of transformation in the electroless Ni–P systems can be represented by the variation of this integrated intensity ratio with the heating temperature. In Fig. 10.6, the integrated intensity ratios of the amorphous phase (I_{amorp}/I_{total}) remain constant to 300 °C, indicating that no significant crystallisation reaction had occurred in both deposits. From 300 °C to 500 °C, the I_{amorp}/I_{total} decreases sharply to a very low level in both deposits. The major crystallisation reactions in the deposits take place at this temperature range. From 500 °C to 800 °C, the I_{amorp}/I_{total} of both deposits further decreases to zero, indicating that the complete crystallisation is achieved at 800 °C.

10.4 X-ray diffraction (XRD) data analysis procedures

Calculation of grain size and microstrain values requires well-separated reflections. However, the reflections from nickel are severely overlapped

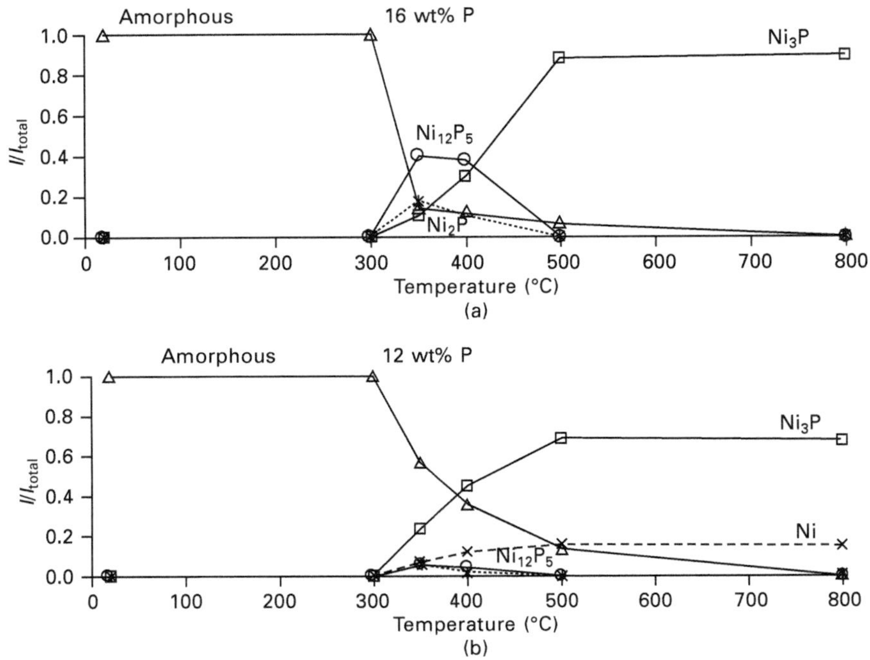

10.6 Variation of the integrated intensity ratio (I/I_{total}) over heating temperature in (a) the 16 wt% P deposit and (b) the 12 wt% P deposit.

either by an amorphous profile in the as-deposited state or by reflections from the phosphide precipitates in the heat-treated condition. Profile refinement is required to separate the individual profiles from a cluster of overlapping reflections. A linear combination of Cauchy and Gaussian functions, namely, the pseudo-Voigt function, describes the X-ray diffraction profiles best. A split-type pseudo-Voigt function is adopted as the profile function to accommodate the asymmetry in the profiles. Based on the above considerations, a computer program is used for profile refinement. In this program, a number of parameters, such as integrated intensity, peak maximum position, full width at half maximum (FWHM), and integral breadth (β), are obtained using the least squares technique by minimising the squared difference between the experimental and theoretical intensities. These parameters are then used for further analysis.

In the as-deposited and some of the heat-treated conditions, the {111} reflection of nickel can only be recorded along with a heavily overlapped amorphous profile. To separate the amorphous profile using the computer program, initial values of FWHM of nickel {111} and amorphous profile should be provided. A proper determination of the peak parameters depends upon a judicious choice of these values. A procedure is proposed to separate the overlapped peaks in an accurate manner. In the XRD profile, a broad {222} reflection peak can also be recorded. The overlapping of the amorphous profile and the crystalline reflection is confined to the {111} reflection of nickel alone. From the integral breadth of the {222} reflection, that of the {111} can be directly calculated using the Scherrer formula and this value can be given as the initial parameter corresponding to the FWHM of the {111} reflection. This initial parameter is then subject to further least squared refinement by the computer program. Such procedures are followed in the peak separation whenever the XRD profile allows an accurate determination of the {222} peak.

Assuming the shape of the line profiles to be of Gaussian form and both the crystalline grain size and microstrain contribute to the line broadening, the integral breadth, β, of the structurally broadened profile is given by:

$$\beta^2 = (K\lambda/(D\cos\theta))^2 + (4e\tan\theta)^2 \qquad 10.5$$

where K is the Scherrer factor, ≈ 1, λ is the wavelength and θ is the angular position of a certain reflection. Plotting $\beta^2\cos^2\theta/\lambda^2$ against $\sin^2\theta/\lambda^2$ for the two orders of reflection ({111} and {222}) and extrapolating the line to $\sin^2\theta/\lambda^2 = 0$ allow the estimation of $1/D^2$ and the slope of the line yields $16e^2$, from which the grain size D and microstrain e are calculated.

10.5 Grain size and microstrain in the platings

In this section, the reflections of two parallel planes of nickel, {111} and {222}, are used to calculate the grain size and microstrain values following the procedures described in Section 10.4.

In the as-deposited condition, the material is fully amorphous. Even when heated up to 300 °C, the reflections of crystalline nickel are unclear. Clear nickel reflections are first observed when the temperature reaches 350 °C. Similar to other Ni–P platings, the grain size of the crystalline nickel increases with increasing temperature, with a dramatic increase at temperatures above 500 °C (Table 10.2). After being heated up to 800 °C, the grain size is estimated as 48.9 nm when microstrain is assumed zero.

The microstrain in the platings becomes smaller and smaller when the temperature increases, especially when the temperature is over 400 °C. If one assumes the microstrain to be zero, the grain size can be calculated from each nickel reflection peak, using a simplified Eq. 10.5, i.e. $\beta = K\lambda/(D\cos\theta)$ (Fig. 10.7). The crystalline nickel may be regarded as largely randomly orientated. The grain size is a strong function of temperature. It should be noted that microstrain, which is larger at lower temperatures, is ignored in the calculation. The omission of microstrain causes the grain size to be underestimated which can be understood through Eq. 10.5. The results here are qualitatively in agreement with Martyak and Drake (2000).

Table 10.2 Grain size and microstrain values in the 12 wt% P plating after being heated up to different temperatures at 20 °C min^{-1}

Temperature (°C)	Size *D* (nm)	Microstrain *e* (×10^{-2})
As-deposited	Fully amorphous	
300	Fully amorphous	
350	7.0	1.4
365	11.1	1.4
400	12.7	1.9
500	17.9	0.57
600	51.9	0.56
800	48.9	–

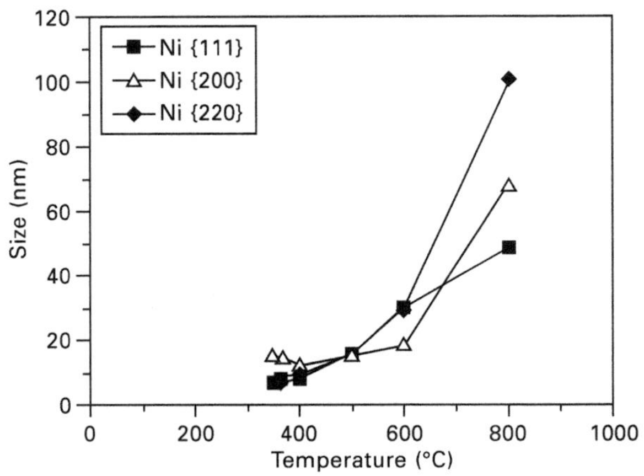

10.7 Grain size values of the 12 wt% P deposit calculated from different nickel reflections {111}, {200} and {220} at different temperatures.

The estimated grain size is smaller than that by Martyak and Drake (2000) which is primarily because of the difference in heat treatment: continuous heating here versus isothermal holding for one or two hours by Martyak and Drake (2000). Such a difference in heat treatment also explains the smaller microstrain measured by Martyak and Drake (2000).

10.6 Scanning electron microscopy (SEM) and electron microprobe analysis

The thicknesses of the coated-deposit on both sides of the substrate are remarkably uniform (2.9 μm on each side, Fig. 10.8). The appearance of the deposits is homogeneous although they were plated through the repeated plating process. There are numerous voids (arrow in Fig. 10.8(a)) of different shapes and sizes in the mild steel substrate.

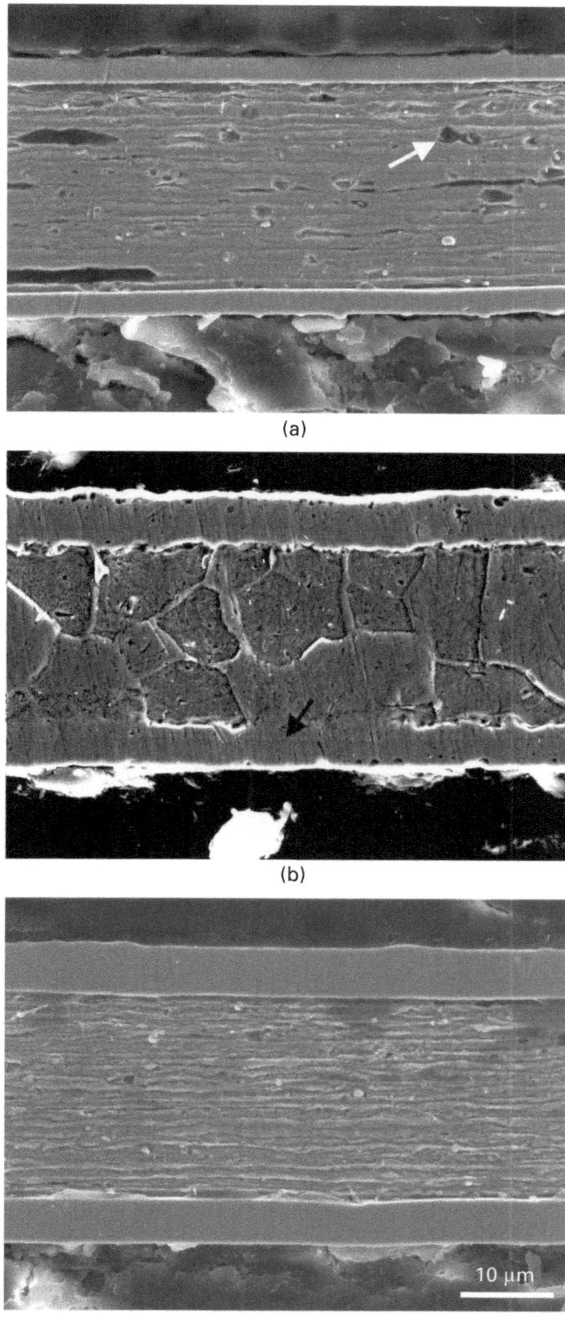

10.8 SEMs of the cross-section of the 16 wt% P deposit at (a) as-deposited condition, and after heating to (b) 400 °C and (c) 800 °C.

From Fig. 10.8(b) and (c), it is evident that a thermally activated diffusion process takes place between the coated-deposits and the mild steel substrate underneath. The deposit heated to 400 °C shows an increase in the iron content with a relative decrease of Ni content (Table 10.3). Ni has diffused into the substrate during the diffusion process, indicated by the increase of Ni level in the diffusion layer (arrow in Fig. 10.8(b)). However, there is neither Ni nor P in the substrate at the point far from the deposit-substrate interface, suggesting that the diffusion process does not involve the entire mild steel substrate. After heating to 800 °C, the Fe content in the deposit appears to have reduced to the level about half of the as-coated deposit, though the contribution from sample contamination is hard to quantify. Again, any possible diffusion process is believed to not involve the entire mild steel substrate, as there is still absence of Ni and P in the substrate at the point far from the deposit-substrate interface.

Film growth after the heating processes (Fig. 10.8(b) and (c)) has increased the film thicknesses to 4.8 and 5 μm, respectively. The increase might be related to the phase transformation in the deposits as indicated by the XRD profiles of the 16 wt% P deposit (Fig. 10.1). The same kind of diffusion process could be present in the deposit-substrate interfaces of other deposits.

10.7 Microstructural evolution of nickel–phosphorus (Ni–P) deposits with heat treatment

10.7.1 16 wt% P coating

The thickness of this coating is about 3 μm. The material above the coating cross-section in the micrograph of Fig. 10.9(a) is the mild steel substrate, and the material below it is the mounting resin for metallographic specimens. There is a very good bonding between the coating and the substrate. There is not much contrast feature within the coating; those features with small contrast being likely the inevitable sample preparation defects. Such a largely featureless structure agrees with the amorphous nature of the coating, as identified using XRD in earlier sections of this chapter.

After heating at 100 °C, the coating microstructure remains largely

Table 10.3 Electron microprobe analyses of the average compositions in the 16 wt% P deposit at as-deposited condition and after heating to 400 °C and 800 °C

Location	Ni	P	Fe
As-deposited	81.2±1.1	15.5±0.2	3.4±0.2
400 °C deposit	78.4±1.0	16.5±0.2	5.1±0.3
400 °C diffusion layer	26.6±0.7	0.4±0.1	73.0±0.8
800 °C deposit	82.7±1.1	15.5±0.2	1.9±0.2

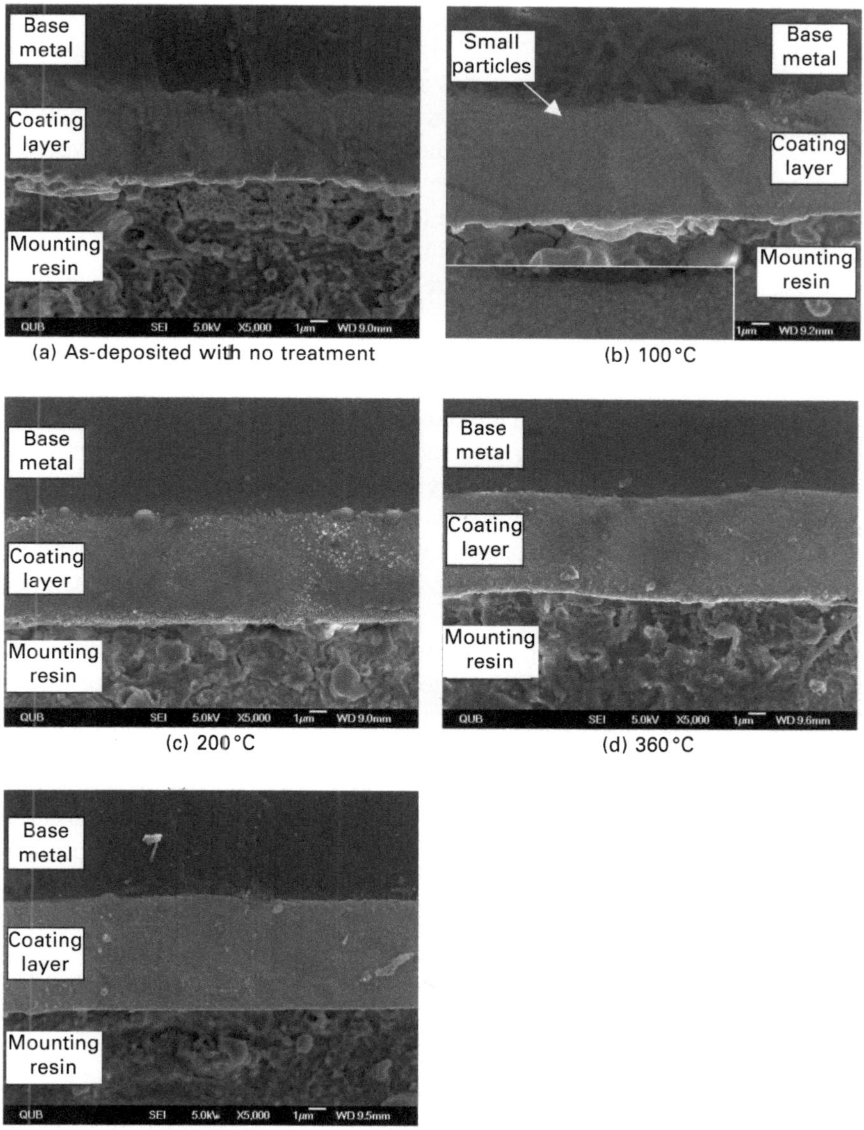

(a) As-deposited with no treatment

(b) 100 °C

(c) 200 °C

(d) 360 °C

(e) 450 °C

10.9 Cross-section SEM images showing microstructure of the Ni–16P coating after heat treatment at different temperatures for one hour. (a) As-deposited with no treatment; (b) 100 °C, with magnified inset; (c) 200 °C; (d) 360 °C; (e) 450 °C.

unchanged, but there are small particles that have formed within the coating, especially near the coating/substrate interface, some as indicated by the arrow in Fig. 10.9(b) and shown in the magnified inset. The size of these small particles is well below the resolution of chemical analysis by the electron microprobe attached to the SEM, so it is not possible to establish the variation in composition of the small particles shown in heat-treated coatings to the surrounding matrix. Increase of the heating temperature to 200 °C significantly increases the number and contrast of these small particles, though the volume fraction is actually rather low and the visual effect may make them appear more abundant than they actually are (Fig. 10.9(c)). Further increase of the heating temperature to 360 °C results in the further increase of the numbers of these particles, covering the entire surface, with some growth of their sizes as well (Fig. 10.9(d)).

XRD has shown crystallisation of this coating upon heating to above 300 °C (earlier sections in this chapter); the crystallisation products being the metastable $Ni_{12}P_5$ and Ni_2P and the stable Ni_3P. In general, the microstructure of the coatings is significantly affected by post-coating heat treatment that causes the precipitation of nickel crystallites and various forms of nickel phosphides (Taheri *et al.*, 2005). The particles observed here are certainly the crystallisation products upon heating, but their appearance at temperature as low as 100 °C (though faintly, Fig. 10.9(b)) and, clearly, at 200 °C (Fig. 10.9(c)) is not detected using XRD. However, conventional XRD is not good at detecting phases with very low volume fractions, especially for the kind of compounds in the Ni–P system that have many diffraction peaks spreading the intensity. These are opposite to the strong diffraction peaks from the high symmetry elemental metals, which are move easily detectable. Therefore, it is very conceivable that XRD has missed the crystallisation at low temperatures, due to the low volume fraction of the crystallisation products. Ordered clusters as crystalline nuclei in amorphous electroless Ni–P alloy have been reported before (Gao *et al.*, 2005).

After further increase of the heating temperature to 450 °C, when the crystallisation process is (almost) completed, the particle contrast in the coating is diminished (Fig. 10.9(e)), because the structure is now almost a single-phase structure consisting only of the stable crystallisation product Ni_3P (see earlier sections). This is because the 16 wt% P content in the alloy is near the stoichiometric Ni_3P composition.

The good bonding between the coating and the mild steel substrate is evident throughout the heating process at different temperatures.

10.7.2 12 wt% P coating

This deposit was made by Lea Manufacturing, instead of lab-made, and it is much thicker (50 μm, Fig. 10.10(a)). The straight, inclined lines in Fig. 10.10(b) are scratch marks made during sample preparation.

No structural changes are apparent in the deposits heated at 100–300 °C for one hour. Upon heating at 330 °C, however, at places, material shrinkage results in large voids, as shown in Fig. 10.10(c). Such heat-shrinkage induced voids appear to be at where the layering is the most intense in the as-deposited coating, which is reasonable, because the layering is a kind of defect with voids already existing before the heat treatment.

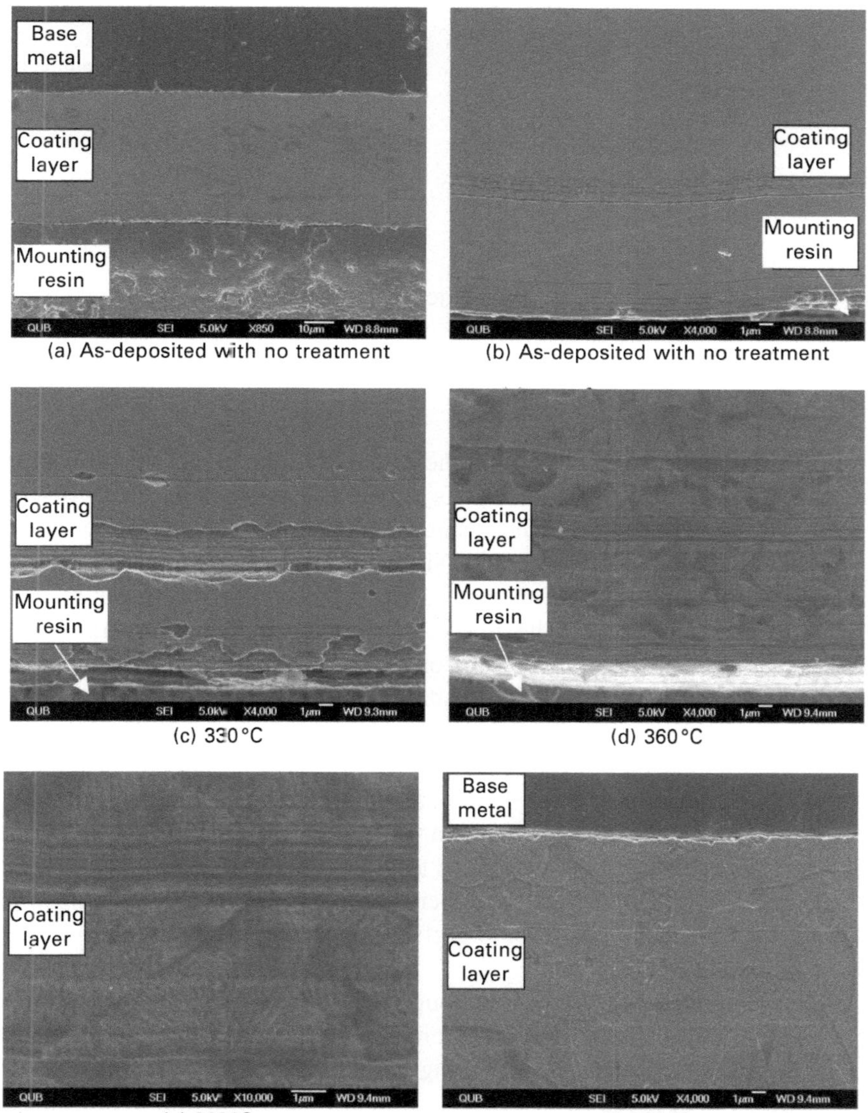

10.10 Cross-section SEM images showing microstructure of the Ni–12P coating after heat treatment at different temperatures for one hour. (a), (b) As-deposited with no treatment; (c) 330 °C; (d), (e) 360 °C; (f) 400 °C; (g) 450 °C; (h) 500 °C.

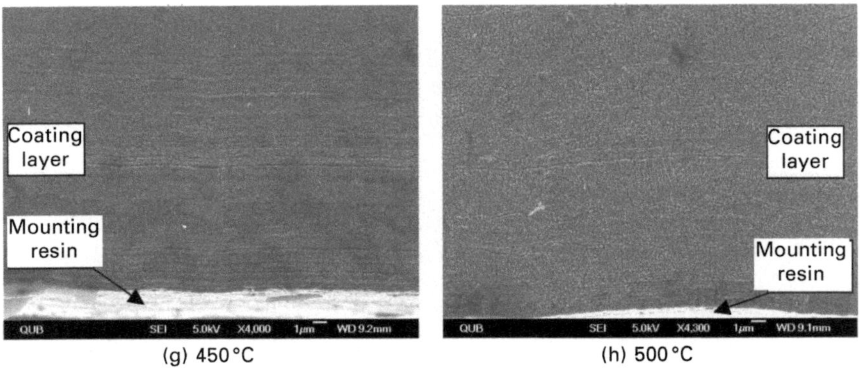

10.10 Continued

With an increase of the heating temperature to 360 °C, there are very significant microstructural changes. There are obvious phase transformation (crystallisation) products, in the form of regular, straight-sided shapes (Fig. 10.10(d)). The phase transformation is supported by DSC analysis (Fig. 10.3(b)) (Section 10.2).

There are finer structures within these shapes in the higher magnification picture of Fig. 10.10(e). 360 °C is about the mid-temperature of the crystallisation process (see earlier sections), and the phase composition should be about 50% amorphous and 50% Ni_3P. The straight-sided, darker shapes are Ni_3P, with no obvious inner structure, which is in amorphous matrix. The amorphous matrix is in the process of decomposition and crystallisation, and its inner structures are a sign of these. It should be noted that the crystallisation process is studied in the earlier sections during continuous heating, while the deposits here are under isothermal heating for the duration of one hour at each temperature. So, all transformations should happen at a lower temperature in deposits here compared to in the earlier sections, or, equivalently, it may be said that the transformation should be more advanced at a same temperature here compared to in the earlier sections. The deposit for Fig. 10.10(d) is not exactly 90° cross-section, but is slightly tilted, as can be seen from the coating top surface toward the bottom of the image not being a single line.

After heating at 400 °C for one hour, the decomposition appears to have progressed (Fig. 10.10(f)), but the microstructural feature looks similar to in the 360 °C treated deposit. Note that Fig. 10.10(f) shows the coating/substrate interface, instead of the coating free surface as in previous micrographs. It is not clear whether the apparent debonding occurred during the heat treatment process or is a result of sample preparation that included grinding, polishing and etching.

With an increase of the heating to the higher temperature of 450 °C, similar to the trend in the 16 wt% P coating, the decomposition contrast has receded, because of the growth and homogenisation (Fig. 10.10(g)). For this alloy, at 450 °C, the transformation is complete or nearly complete (see earlier sections). The product at this temperature is mainly Ni_3P, but also with some Ni and possibly some amorphous structure remaining. Such a product mixture agrees in principle with the microstructure shown in Fig. 10.10(g).

Finally, after heating at the highest temperature (500 °C) used in this experiment, we have received a homogeneous structure with clear grain boundaries (Fig. 10.10(h)). The phase composition should be mainly Ni_3P, plus a small fraction of the Ni phase. Amutha *et al.* (2006) give support to this by showing that the heat-treated Ni–P film at 500 °C gives rise to Ni and Ni_3P phases with increased crystallinity. It is not entirely clear whether the apparent fine structure, within each grain in Fig. 10.10(h), is real fine structure or a polishing or etching artefact, but it is more likely the former, because of its directional nature, i.e., having different weak alignment directions in different grains.

This section has concentrated on microstructural aspects. Taheri *et al.* (2004) have investigated the effect of post-deposition heat treatment on the properties of low and high phosphorus electroless Ni–P coatings. Post-deposition heat treatment increases brittleness of the coatings occasioned by the crystallisation of nickel crystals and nickel phosphide precipitates. In addition, according to the depth profile analysis, the annealing drives diffusion of both Ni and P toward the substrate phase. This strongly changes corrosion resistance of the plating layer (Wang *et al.*, 2009).

10.8 Conclusions

Electroless Ni-P deposits with high phosphorus content are amorphous at as-deposited condition. The sequence of phase transformation in the 16 wt% P deposit is amorphous phase → co-existence of stable Ni_3P, and metastable Ni_2P, $Ni_{12}P_5$ and unidentified phases → stable Ni_3P phase. The sequence of phase transformation in the 12 wt% P deposit is amorphous phase → co-existence of stable Ni_3P, fcc nickel, and metastable Ni_2P, $Ni_{12}P_5$ and unidentified phases → stable Ni_3P and fcc nickel phases. Phase transformation temperatures of the deposits increase with increasing heating rate, and decreasing phosphorus content. Effect of heating rate on the microstructural property is significant for the heating temperature that is close to the crystallisation temperature range. Effect of cooling rate on the phase constitution is not significant after heating to the temperature far beyond the crystallisation process. There is film growth after heating processes.

Differential scanning calorimetry and XRD analysis can be employed to study the crystallisation and phase transformation behaviour in electroless

Ni–P platings with phosphorus contents varying in a large range. X-ray line broadening technique can be used to estimate the grain size and microstrain in the Ni–P platings. The grain size of fcc nickel increases in the platings when temperature increases, significantly at temperatures above 400 °C. The microstrain in the platings decreases with increasing temperature. The crystalline nickel formed in the Ni–P platings has largely random orientation.

SEM using a field-emission source has excellent resolution and is a suitable technique for examining microstructure changes in the electroless Ni–P deposition because of heat treatment. Microstructural features characteristic of these coatings, including interface, and different stages of crystallisation and grain growth of the coatings with different phosphorus levels are revealed by the microscopy, including the microstructural effects of the formation of Ni–P compounds such as Ni_3P. The detailed microscopy data in the main agrees with crystallisation kinetics study using XRD.

10.9 References

Amutha K, Shakkthivel P, Sathiyamoorthi R and Vasudevan T (2006), 'Electroless Ni–P alloy on zincated aluminium substrates from tartrate bath', *Bull Electrochem*, **22**, 169–75.

Gao J Q, Wu Y T, Lei L, Bin S and Hu W B (2005), 'Crystallization temperature of amorphous electroless nickel–phosphorus alloys', *Mater Lett*, **59**, 1665–9.

Martyak N M and Drake K (2000), 'Peak-profile analysis of electroless nickel coatings', *J Alloy Comp*, **312**, 30–40.

Rahimi A R, Modarres H and Abdouss M (2009), 'Study on morphology and corrosion resistance of electroless Ni–P coatings', *Surf Eng*, **25**, 367–71.

Sha W and Guo Z (2009), *Maraging Steels: Modelling of microstructure, properties and applications*, Woodhead Publishing, Cambridge.

Taheri R, Oguocha I N A and Yannacopoulos S (2004), 'Effect of coating parameters and heat treatment on the adhesive properties of electroless Ni–P coatings', *Can Metall Quart*, **43**, 363–70.

Taheri R, Oguocha I N A and Yannacopoulos S (2005), 'Microstructure characterization of electroless NiP alloys', *Can Metall Quart*, **44**, 13–21.

Vojtěch D, Novák M, Zelinková M, Novák P, Michalcová A and Fabián T (2009), 'Structural evolution of electroless Ni–P coating on Al–12 wt% Si alloy during heat treatment at high temperatures', *Appl Surf Sci*, **255**, 3745–51.

Wang K, Hong L and Liu Z L (2009), 'Exploring the water-soluble phosphine ligand as the environmentally friendly stabilizer for electroless nickel plating', *Ind Eng Chem Res*, **48**, 1727–34.

Wilkinson D S (2000), *Mass Transport in Solids and Fluids*, Cambridge University Press, Cambridge.

11
Crystallisation of nickel–phosphorus (Ni–P) deposits with medium and low phosphorus content

Abstract: The crystallisation process in electroless nickel phosphorus (Ni–P) platings during continuous heating is investigated using the X-ray diffraction (XRD) analysis and the differential scanning calorimetry (DSC). The X-ray line broadening technique can estimate the grain size and the microstrain with the aid of software to separate the reflections of crystalline nickel from the amorphous phase. In the as-deposited condition, platings with 5–8 wt% phosphorus are mixtures of amorphous and nanocrystalline materials. In all the platings after continuous heating, the grain size increases sharply when the temperature increases above 400 °C. The microstrain in the platings decreases with increasing temperature.

Key words: transition metal alloys, microstructure, X-ray diffraction, amorphous materials, mechanical properties.

11.1 Calorimetric study and the major exothermal peak

This chapter is concerned with electroless nickel–phosphorus (Ni–P) deposits with medium and low phosphorus contents. The research topics are similar to those of Chapter 10 for the electroless Ni–P deposits with high phosphorus content. The same experimental techniques, namely the differential scanning calorimetry (DSC) and X-ray diffraction (XRD), are used. The medium- and low-phosphorus deposits are heated in the differential scanning calorimeter under different heating rates, and the crystallisation processes are studied. An X-ray diffractometer identifies the phases formed in the deposits heated to different end temperatures for further analysis of the phase transformation.

The effect from the mild steel substrate on the resulting DSC curves is insignificant, indicated by the flat DSC curves of the substrate alone (Fig. 10.3(c)). Crystallisation peak temperatures (T_p) of the three deposits determined from the major exothermic peaks of the DSC curves (Fig. 11.1) are 330–360 °C, 332–362 °C and 386–406 °C, respectively. These increase with decreasing phosphorus content. Though this tendency appears to be the same as for the DSC analyses of the deposits in Chapter 10, it should be noted that there is a dip of crystallisation temperature for the 8 wt% P

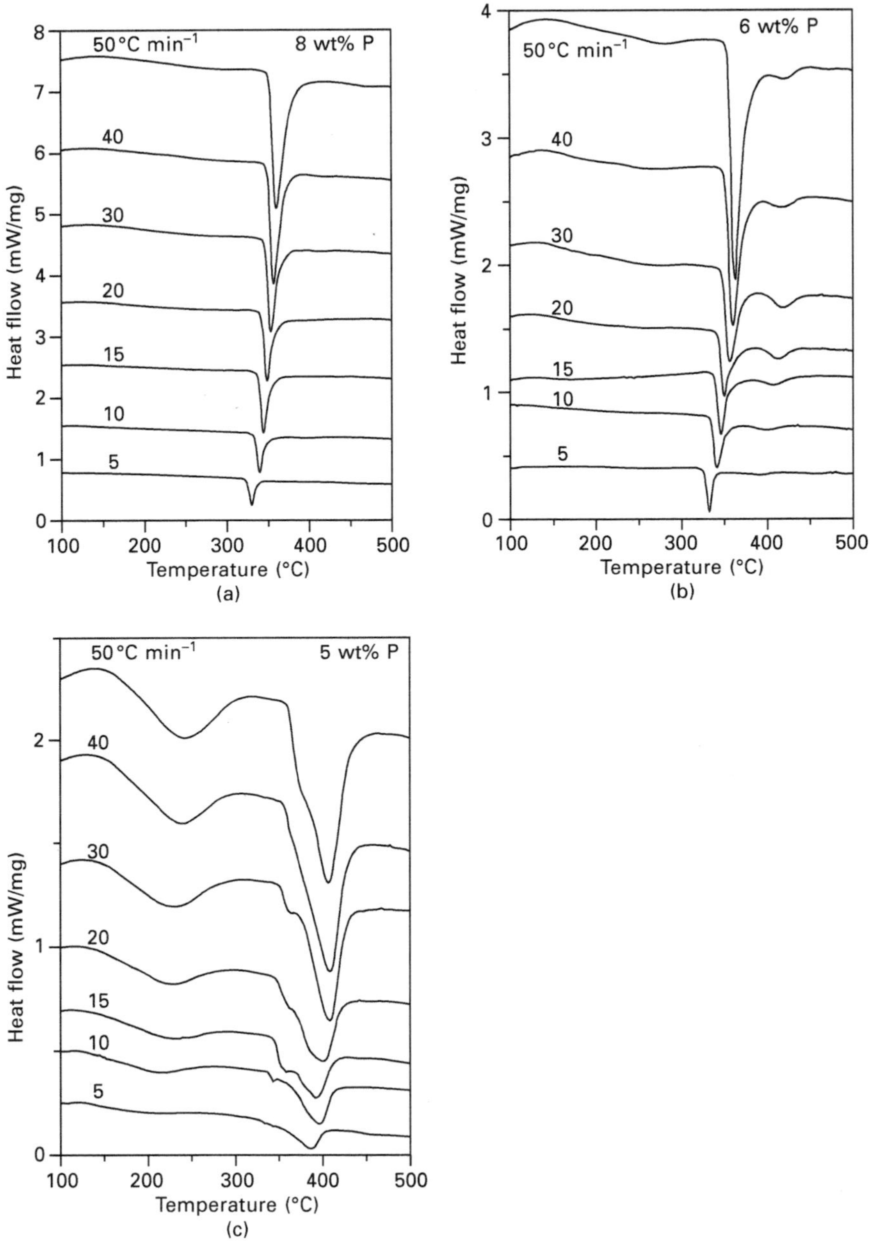

11.1 DSC curves of (a) 8 wt% P, (b) 6 wt% P and (c) 5 wt% P coatings at the heating rates of 5–50 °C min^{-1}. For the sake of clarity, the calorimetric curves are shifted arbitrarily on the vertical scale.

deposit compared to the 12 wt% P one. Other DSC studies on the deposits with 5.3–11 wt% P, and 2.3 and 9.2 wt% P (Xiang *et al.*, 2001) show the same trend.

In the 8 wt% P deposit, only one exothermic peak is in each of the DSC curves. No significant shoulder(s) (secondary peak) coexists with the major exothermic peak. In the 6 wt% P deposit, there is a shoulder (secondary peak) after the major exothermic peak. The presence of these shoulders in the DSC curves indicates the final stage of crystallisation process or the relaxation of lattice strain energy during phase separation. In the 5 wt% P deposit, the DSC curves show a small exotherm between 150 °C and 300 °C. The presence of a small exotherm before the major crystallisation exothermic peak is also reported in the deposit with 5–6 wt% P by Martyak and Drake (2000). The small exotherm is related to the release of microstrain, associated with microstructural changes before the major crystallisation reaction. As in the 6 wt% P deposit, there is a shoulder, but this time right before the major exothermic peak in the DSC curves (Fig. 11.1(c)). The presence of these shoulders can be related to the higher crystallisation temperatures in the 5 wt% P deposit. The formation of these shoulders may be related to the short-range atomic movements, according to Martyak and Drake (2000), as previously described for the DSC curves of the 12 wt% P deposit in Chapter 10 (Section 10.2). The major exothermic peak in the DSC curves corresponds to the long-range atomic movements causing precipitation to the stable phases such as fcc nickel and Ni_3P. In the 5 and 6 wt% P deposits, the onset (T_S) and the end (T_E) temperatures of the major exothermic peak in some DSC curves are difficult to determine due to the peaks overlapping. As these two deposits contain a large amount of nanocrystalline Ni (Section 11.3), the energy released by Ni grain growth may have also contributed to the exothermic peaks.

The exothermic temperatures of the major crystallisation processes of the electroless Ni–P deposits increase with increasing heating rates. The reason for this is that the effects of thermal ageing are less intense when the heating rate is high, as the time for crystallisation is shorter. In addition, the increase of heating rate correspondingly increases the areas of the DSC exothermic peaks of the deposits. The DSC exothermic peaks become sharper and deeper with increasing heating rates. The area of the DSC exothermic peak is equivalent to the total amount of heat flow during the crystallisation process. Hence, at higher heating rates, the rate of heat release will be higher, as the time for crystallisation is shorter. In the same deposit, the enthalpy of crystallisation process calculated from the area of the DSC exothermic peak between the T_S and T_E is, however, about the same regardless of the different heating rates (Table 11.1). The partial area of the DSC curves between the T_S and T_P indicates that the major crystallisation process achieves less than 50% completion at the peak temperature. For the 5–6 wt% P deposits, the

Table 11.1 DSC peak parameters of the 8 wt% P deposit

Heating rate (°C min^{-1})	T_S (°C)	T_P (°C)	T_E* (°C)	Enthalpy (J g^{-1})	Area from T_S to T_P (%)
5	325	330	335	34.4	44.6
10	334	340	346	33.9	36.3
15	338	344	354	34.2	33.9
20	342	349	358	33.9	37.6
30	347	354	367	34.3	36.4
40	351	356	372	35.0	24.8
50	353	360	380	+	26.8

*Accurate determination of the T_E is difficult as the DSC peaks end too gradually.
+The 50 °C min^{-1} heating rate is too fast to allow accurate measurement of the enthalpy.

interference of a secondary peak causes difficulty in measuring the T_S and T_E in some of the DSC curves.

Similar to the deposits in Chapter 10, the crystallisation activation energies (Q) of the deposits can be calculated using the peak temperatures (T_P) of crystallisation process at different heating rates (ϕ). Plots of $\ln(T_P^2/\phi)$ versus $1000/(RT_P)$ (Fig. 11.2) from the Kissinger method (Sha and Guo, 2009, p. 91) yield the crystallisation activation energies of 239±5, 230±6 and 322±54 kJ mol^{-1}, respectively, for the 8, 6 and 5 wt% P deposits. These values are relatively close to the self-diffusion activation energy of nickel (289 kJ mol^{-1}) (Wilkinson, 2000, p. 242). Nevertheless, the crystallisation activation energy of the 5 wt% P deposit (low-phosphorus deposit) is higher than the 6–8 wt% P deposits (medium-phosphorus deposits). Though the term 'crystallisation activation energy' is used, as mentioned before, the 'crystallisation peak' may include grain growth contribution, especially for the 5–6 wt% P deposits.

11.2 X-ray diffraction (XRD) analysis

11.2.1 Microstructures at as-deposited condition

The as-deposited structures of the electroless Ni–P deposits are the mixture of amorphous and nanocrystalline nickel phases (Figs 11.3–11.5). The wide angular range profile at the 2θ position of 35–55° indicates the presence of amorphous phase in the deposits. For all the deposits, there is one major and one weak broad-peak diffraction at the 2θ positions corresponding to Ni {111} and {200} planes, superimposed with the amorphous diffraction in the XRD profiles. The as-deposited medium (Martyak and Drake, 2000) and low phosphorus deposits consist of a semi-amorphous structure.

Crystallinity, texture and perhaps composition inhomogeneity can happen in the electroless Ni–P deposits at the as-deposited condition. Similar to the crystallinity of deposit, the presence of texture in these deposits is probably

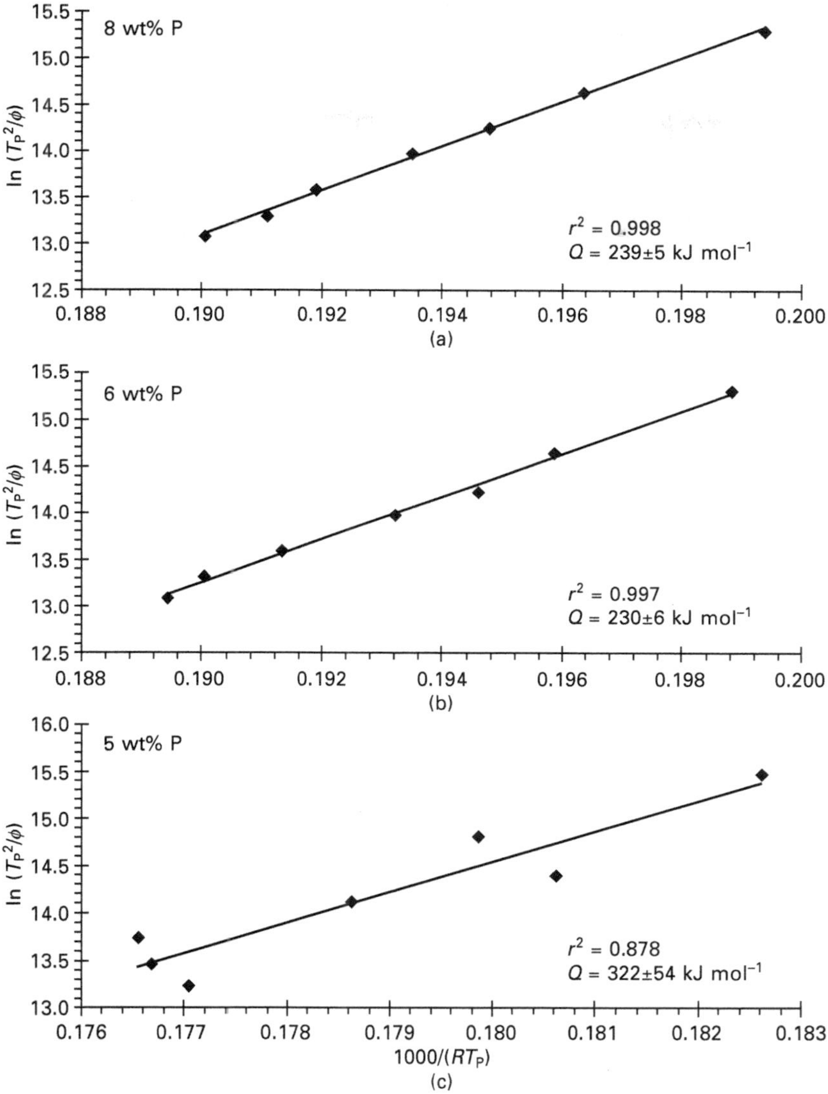

11.2 Kissinger plots of $\ln(T_P^2/\phi)$ versus $1000/(RT_P)$ using crystallisation peak temperatures of (a) the 8 wt% P deposit, (b) the 6 wt% P deposit and (c) the 5 wt% P deposit.

predetermined by the initial phosphorus content of deposits, as well as the deposition process. In Fig. 11.6, the phenomenon (inhomogeneity) is illustrated. The two as-deposited profiles shown in the figure are measured under the same condition, but they display different diffraction intensity distribution. The specimens used for the XRD measurement are cut from different locations of the deposit on a large piece of substrate. An additional

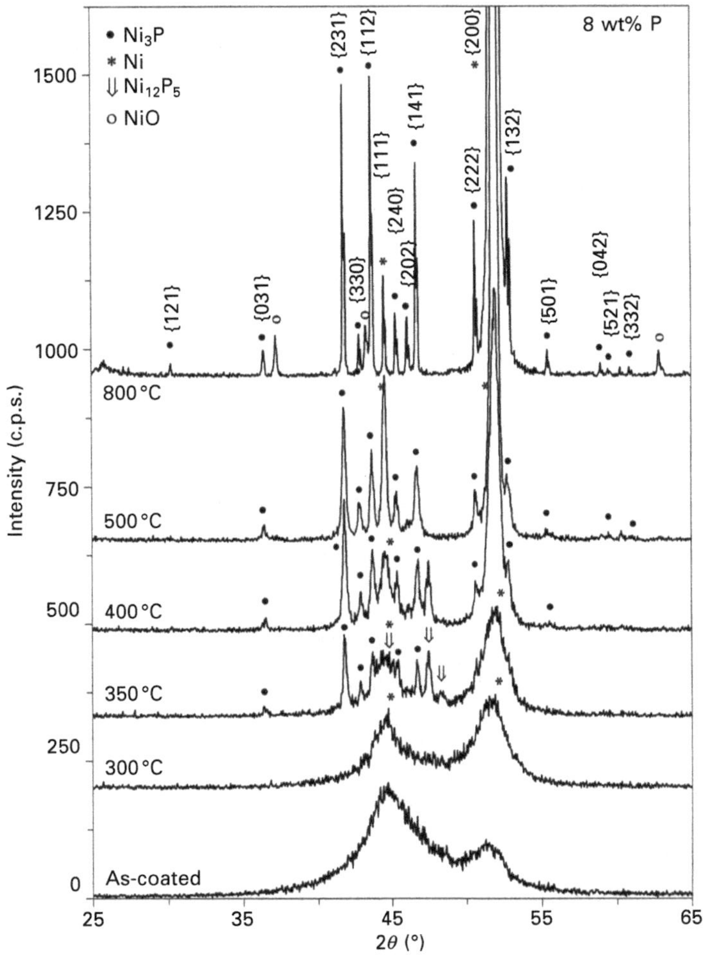

11.3 XRD profiles of 8 wt% P deposit at as-deposited condition and after heating to the end temperatures of 300–800 °C at 20 °C min^{-1}. For the sake of clarity, the profiles are shifted arbitrarily on the vertical scale.

XRD profile, from a third specimen after heating to 300 °C (Fig. 11.3), is included in the figure for comparison with the as-deposited profiles, for the location (2θ) and tendency of their diffractions. The effect of such phenomena (i.e. crystallinity, texture and composition inhomogeneity) in the as-deposited state should not affect significantly the overall phase transformation process of the deposit to the heating temperature of 300 °C or higher. The average of those two as-deposited profiles is used for the 8 wt% P deposit in Fig. 11.3. The preferred orientation of the nickel phase tends to disappear with increasing phosphorus content.

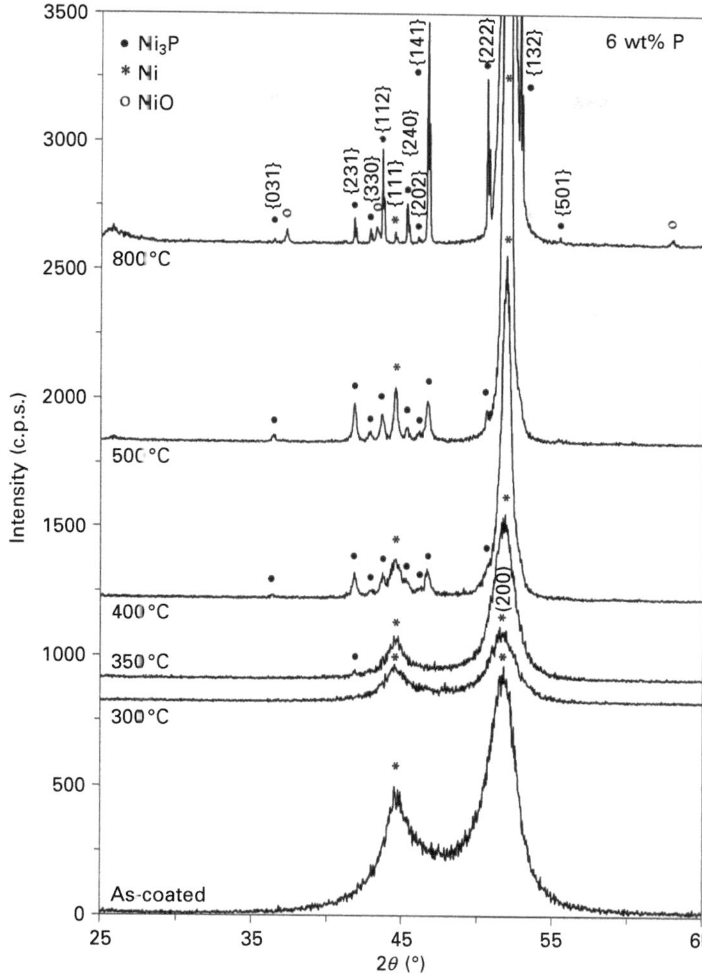

11.4 XRD profiles of 6 wt% P deposit at as-deposited condition and after heating to the end temperatures of 300–800 °C at 20 °C min^{-1}. For the sake of clarity, the profiles are shifted arbitrarily on the vertical scale.

11.2.2 Microstructural changes due to continuous heating processes

After the deposits are heated to the end temperatures of 300–800 °C, they exhibit different phase compositions in respect to different heating temperatures. In the 8 wt% P deposit, the heating end temperature of 300 °C has caused the broad-peak reflections that correspond to the 2θ positions of Ni {111} and {200} planes, to develop into the relatively broad Ni {111} and {200} diffraction peaks, respectively, with limited height (intensity) (Fig. 11.3).

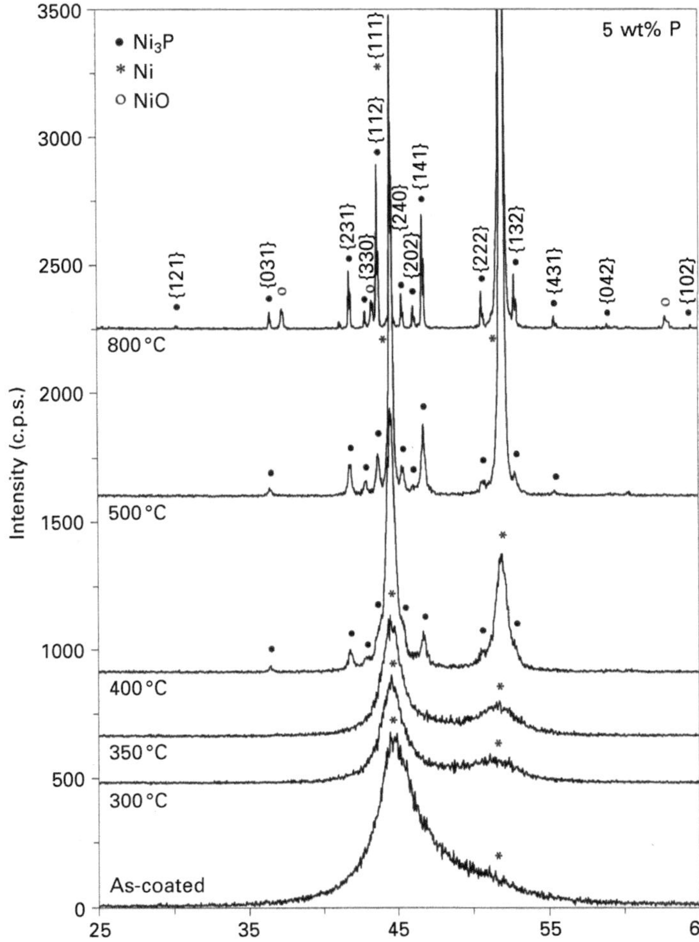

11.5 XRD profiles of 5 wt% P deposit at as-deposited condition and after heating to the end temperatures of 300–800 °C at 20 °C min^{-1}. For the sake of clarity, the profiles are shifted arbitrarily on the vertical scale.

The intensity of the amorphous profile decreases, and the Ni {311} reflection is in the XRD profile. However, the major crystallisation process has not yet happened, indicated by the absence of numerous diffraction peaks in the XRD profile. This suggestion is confirmed by the flat DSC curve recorded at the same temperature (Fig. 11.1(a)). After heating to 350 °C, the presence of a number of diffraction peaks from nickel, metastable $Ni_{12}P_5$, and stable Ni_3P phases in the XRD profile indicates the ongoing of major crystallisation reaction in the deposit. This is in agreement with the major exothermic peak in the DSC curve recorded (Fig. 11.1(a)). Besides, the Ni {200} reflection

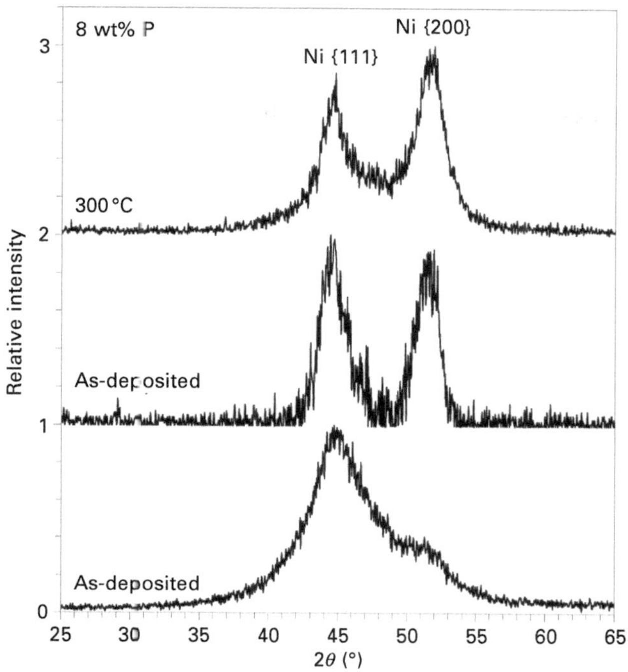

11.6 XRD profiles (with normalised intensity) of 8 wt% P deposit at as-deposited condition and after continuous heating to 300 °C under a constant heating rate of 20 °C min^{-1}. Two profiles from different specimens at the as-deposited condition are presented to show the specimen-to-specimen variation in diffraction intensity distribution at that state.

becomes the most intense and there is some amorphous phase left in the material. The 2θ positions of metastable $Ni_{12}P_5$ diffraction peaks in the XRD profile do not match exactly the data from the powder diffraction file. The possible reasons for this shift are the presence of internal stresses in the deposit and the deviation from stoichiometric composition. After heating to 400 °C, the amorphous phase remains but further decreases in amount. The $Ni_{12}P_5$ metastable phase is absent. Nickel and Ni_3P phases are present in the deposit. In addition, there is an unidentified reflection (at plane spacing d = 1.915 Å), which does not match any of the existing known Ni–P phases in the XRD profile. In addition, the number of diffraction peaks from Ni_3P phase increases dramatically, but they are low and wide, possibly caused by their small (grain) size and the residual stresses of the deposit. The Ni {200} peak is still the most intense diffraction peak in the XRD profile at this stage. The phase transformation process has not yet completed. After heating to 500 °C, there is only a small fraction of amorphous phase remaining in the deposit, indicating nearly complete crystallisation. There are only the diffraction peaks that match either the nickel or stable Ni_3P phase. In addition, the diffraction peaks have increased intensity and are refined into narrower shape, implying homogeneity, grain growth and absence of internal stresses. The first three major diffraction peaks are the Ni {200} and {111}, and the Ni_3P {231} peaks. With further increase in the heating end temperature to 800 °C, the amorphous phase completely disappears from the deposit, indicating complete crystallisation process. All the peaks from both the Ni and Ni_3P phases are further refined, which indicates better crystallinity. In addition, new minor diffraction peaks from the Ni_3P phase emerge, and a small amount of nickel oxide (NiO) is in the XRD profile. The presence of NiO can be related to the purity of helium gas and the lack of high vacuum during heat treatment as explained in Chapter 10 (Section 10.3) during the discussion on the XRD profile of the 12 wt% P deposit. Again, the Ni {200} reflection is the most intense among the XRD peaks.

In the 6 wt% P deposit, heating to 300 °C brings small effect on the Ni {111} and {200} reflections (Fig. 11.4). There is emergence of Ni {311} reflection at 92.9° (2θ). After heating to 350 °C, the Ni {200} reflection becomes sharper with considerably increasing intensity. The Ni_3P {231} reflection is also in the XRD profile. The phase transformation process is at the expense of the amorphous phase. The major exothermic peak (T_P = 349 °C) in the DSC curve measured at the same heating rate (20 °C min^{-1}) is related to these (Fig. 11.1(b)). Heating to 400 °C leads to further sharpening and intensifying of the Ni {200} reflection, and precipitation of nickel phosphide, Ni_3P, in the 2θ region around the Ni {111} reflection. Numerous minor reflections from the Ni_3P phase are in the XRD profile at this heating condition. The further growth of nickel phase and formation of Ni_3P can be related to the small exotherm (shoulder, onset at about 400 °C) in the DSC curve measured at 20 °C min^{-1}.

The emergence of the Ni_3P phase in the deposit also relates to the crystallisation process. However, a small amount of amorphous phase remains in the deposit after the heating process. No amorphous phase was found by Martyak and Drake (2000) in the electroless Ni–P medium phosphorous deposit (5–6 wt%) that was heat treated at 400 °C, but held at this temperature for one hour. After heating to 500 °C, the Ni and Ni_3P reflections further sharpen and strengthen into well-defined diffraction peaks. However, the reflections from the Ni_3P phase are considerably wide and short compared to the nickel reflections, indicating that further refinement of phase reflections is likely at higher temperatures. There is still a relatively small fraction of amorphous phase in the deposit as identified after profile fitting, though it is not obvious when viewing the diffraction pattern. The deposit is near to complete crystallisation. After heating to 800 °C, the Ni and Ni_3P reflections further refine and intensify into narrower and sharper peaks. No amorphous phase remains after the heating process, indicating complete crystallisation process in the deposit. As in the 8 wt% P deposit, a small number of NiO reflections are in the XRD profile, indicating the presence of oxygen in the DSC heating chamber. The Ni {200} reflection stays as the strongest reflection throughout the heating processes.

In the 5 wt% P deposit, heating to the temperature 300 °C slightly sharpens the Ni {111} and {200} reflections (Fig. 11.5). This can be related to the small exotherm (at about 240 °C) in the DSC curve measured at the same heating rate (Fig. 11.1(c)). In addition, the Ni {311} and {222} reflections are at the higher 2θ values (not shown in Fig. 11.5). After heating to 350 °C, the nickel reflections are slightly further strengthened, but no new phases have been observed so far. Heating to 400 °C shows considerable strengthening and sharpening of the Ni {111} and {200} reflections. The reflections from the Ni_3P phase are also observed, but most are shadowed by the overlapped strengthened Ni reflections. The formation of the Ni_3P phase in the deposit also indicates the ongoing of major crystallisation process. This agrees with the major exothermic peak in the DSC curve measured at the same heating rate (Fig. 11.1(c)). The amorphous phase remains, but only in a relatively small amount. After heating to 500 °C, the nickel and Ni_3P reflections have further sharpened. The Ni {200} reflection becomes the strongest but at the expense of the Ni {111} reflection. The number of Ni_3P reflections has also increased, but they are considerably short and wide. The transformation of amorphous phase into the crystalline phases has nearly completed. Heating to 800 °C has further sharpened and strengthened the nickel and Ni_3P reflections. The Ni_3P reflections become narrow and increase in numbers. As in the 6–8 wt% P deposits, the presence of oxygen in the heating process is manifested by the NiO reflections in the XRD profile. Complete crystallisation process in the deposit has been achieved after the heating process.

Overall, the crystallisation and phase transformation process of the deposits is affected by the phosphorus content and heating end temperatures. For example, the presence of intermediate phases is evident in the 8 wt% P deposit (at 350 °C and 400 °C) but not in the two other deposits. In addition, heating to the higher end temperatures (at 500 °C and 800 °C) will bring about the intensive development of Ni_3P stable phase. Besides, the Ni {200} reflection remains as the strongest peak in the 6–8 wt% P deposits throughout the heating processes, but not in the case of the 5 wt% P deposit. In the 5 wt% P deposit, the Ni {111} reflection is apparently dominant at the heating end temperatures of 300–400 °C (Fig. 11.5).

11.2.3 Peak separation, grain size and microstrain, and Ni–5P heated at different rates

One example of peak separation is shown here to demonstrate the performance of the X-ray pattern analysis software (Fig. 11.7). The structure of the plating in this condition is a mixture of an amorphous phase and crystalline nickel.

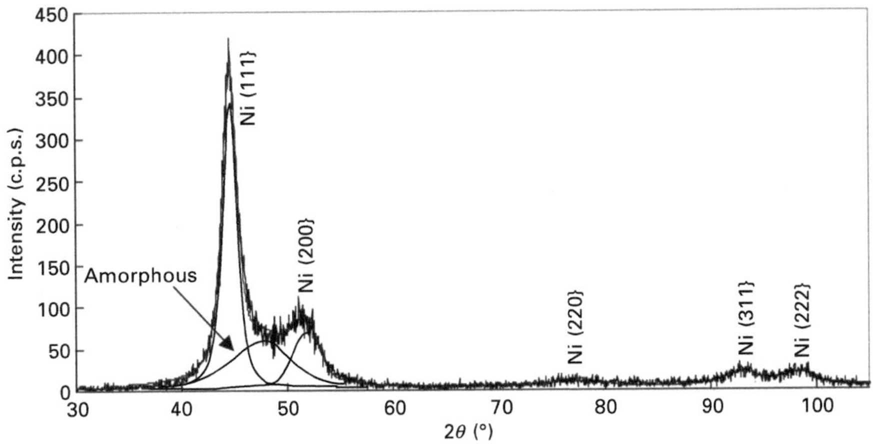

11.7 Separated amorphous and crystalline nickel profiles, using 5 wt% P deposit heated up to 300 °C at 20 °C min^{-1} as an example.

Table 11.2 Grain size and microstrain values in 5 wt% P plating after being heated up to different temperatures at 20 °C min^{-1}

Temperature (°C)	Size *D* (nm)	Microstrain *e* ($\times 10^{-2}$)
As-deposited	2.6*	–
300	6.9	3.7
350	6.4	2.5
400	13.3	0.97
500	28.5	0.04
800	74.5*	–

*The values are obtained from {111} peak of nickel when microstrain is assumed to be zero.

The reflections of fcc nickel and the broad peak between 40 and 55° of the amorphous phase are clearly separated.

The grain size increases with increasing temperature, whereas microstrain in the plating decreases (Table 11.2). The grain size increases dramatically at 400 °C and above (Martyak and Drake, 2000). Because the {222} peak of nickel in the as-deposited condition is not clearly observed, the separation of the influences of microstrain and grain size on line broadening cannot be carried out. The grain size in the as-deposited state is an underestimation since the microstrain in this condition may be larger than 3.7% (Table 11.2). For the 5 wt% P plating heated up to 800 °C, the plot obtained following the procedures described in Section 10.4 gives a negative slope ($16e^2$). Nevertheless, it is reasonable to assume the microstrain in this condition to be zero.

Since crystalline structures with grain sizes below 100 nm are considered as nanocrystalline, these platings are a mixture of an amorphous phase and nanocrystalline material under the as-deposited condition (Tables 11.2–11.4).

The phosphorus content plays an important role in determining the structure of the Ni–P plating in the as-deposited condition, as well as the intermediate stages of the phase transformation during the heating process before the stable Ni_3P phase forms (Martyak and Drake, 2000). The tendency that higher phosphorus results in finer grain size is not observed in the 5–6 wt% P deposits may be because their compositions are very close (Figs 11.8–11.10).

Two Ni–P deposits with 5 wt% phosphorus are stopped during heating at the termination point of the major DSC peak at heating rates of 5 °C and 40 °C min^{-1} (Fig. 11.11). The termination temperatures of the DSC peak are 410 °C for 5 °C min^{-1} and 450 °C for 40 °C min^{-1}, as shown in Fig. 11.1. In the DSC curve at 40 °C min^{-1} heating rate, in addition to the major DSC peak at 350–450 °C, there is a smaller exothermal peak between 150 °C and 300 °C. This is related to the release of microstrain associated with microstructural changes before the major crystallisation reaction (Martyak and Drake, 2000). The calculation of grain size and microstrain follows the procedures described in Section 10.4. The grain size in the deposit heated at

Table 11.3 Grain size and microstrain values in 6 wt% P plating after being heated up to different temperatures at 20 °C min^{-1}

Temperature (°C)	Size *D* (nm)	Microstrain *e* (×10^{-2})
As-deposited	6.7	4.7
300	4.4	1.8
350	5.4	0.64
400	9.5	0.71
500	28.8	0.51
800	110.0	0.12

Table 11.4 Grain size and microstrain values in 8 wt% P plating after being heated up to different temperatures at 20 °C min^{-1}

Temperature (°C)	Size *D* (nm)	Microstrain *e* (×10^{-2})
As-deposited	1.1*	–
300	6.5	4.4
350	13.1	2.4
400	9.3	0.26
500	68.1	0.86
600	66.6	0.47
800	103.7*	–

*The values are obtained from {111} peak of nickel when microstrain is assumed to be zero.

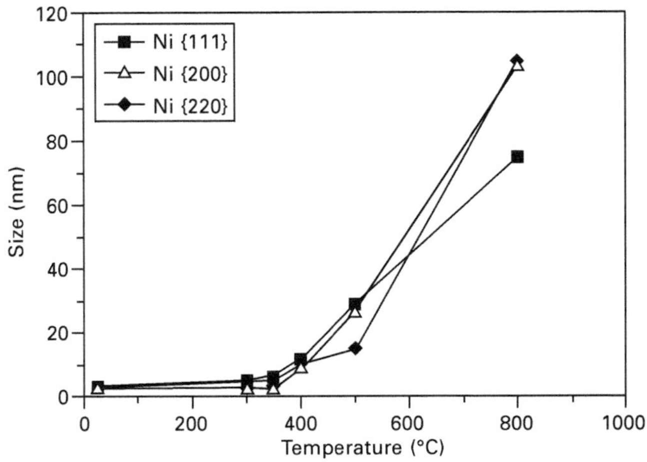

11.8 Grain size values of 5 wt% P deposit calculated from different nickel reflections {111}, {200} and {220} at different temperatures.

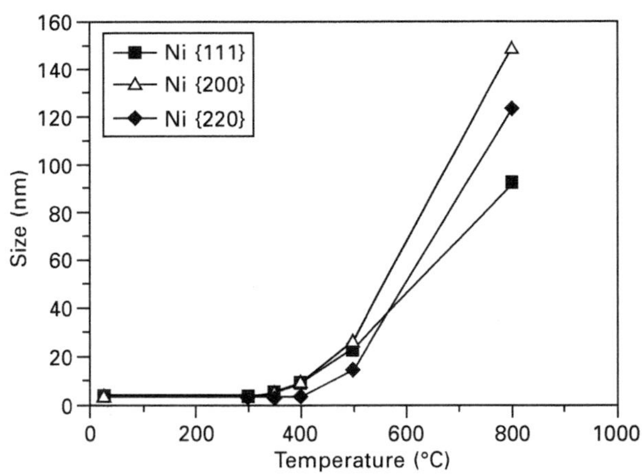

11.9 Grain size values of 6 wt% P deposit calculated from different nickel reflections {111}, {200} and {220} at different temperatures.

40 °C min^{-1} is slightly larger than that of the deposit heated at 5 °C min^{-1}, although the difference may be within the error range of the calculation (Table 11.5). On the other hand, the microstrain in the deposit heated at 40 °C min^{-1} is much larger than that in the deposit heated at 5 °C min^{-1}.

11.3 Degree of phase transformation

The integrated intensities of amorphous, nickel and Ni_3P phases are calculated from the corresponding XRD reflections using a computer program (Fig.

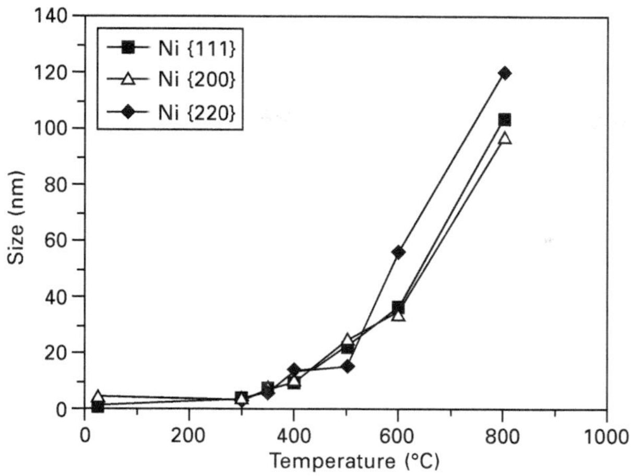

11.10 Grain size values of 8 wt% P deposit calculated from different nickel reflections {111}, {200} and {220} at different temperatures.

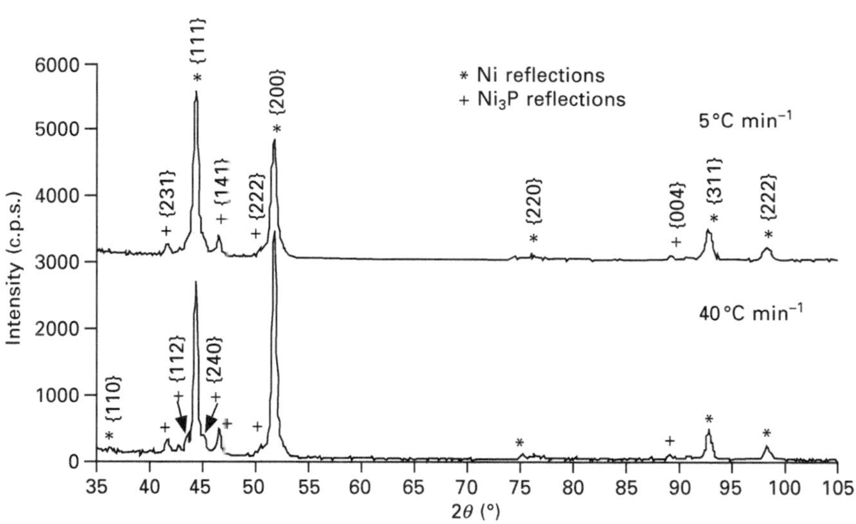

11.11 XRD profiles of 5 wt% P deposit heated up to the termination of the DSC peaks at 5 °C and 40 °C min^{-1}.

11.12). Similar to the deposits in Chapter 10 (Section 10.3), the degree of phase transformation in the electroless Ni–P deposits can be represented by the variations of relative proportion of the amorphous phase remaining in the deposit over the heating temperature, and of the fcc nickel and Ni_3P phases formed. The relative proportion of the amorphous phase can be calculated from the ratio of integrated intensity of amorphous phase (I_{amorp}) to the total integrated intensity (I_{total}) of the deposit.

Table 11.5 Grain size and microstrain values in 5 wt% P plating after being heated up to the termination temperature of the DSC peaks at 5 °C min^{-1} (410 °C) or 40 °C min^{-1} (450 °C), together with the fraction of Ni_3P formed

Heating rate (°C min^{-1})	Size D (nm)	Microstrain e (×10^{-2})	I_{Ni_3P}/I_{total} (%)
5	14.2	0.08	14
40	18.2	0.23	13

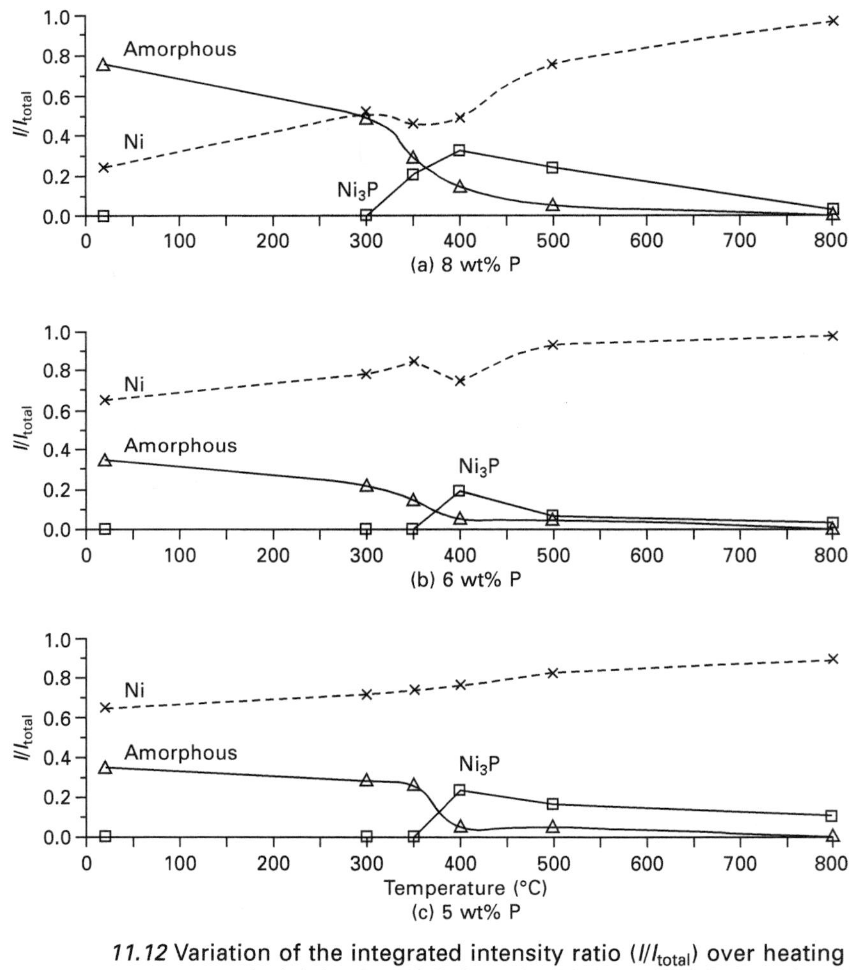

11.12 Variation of the integrated intensity ratio (I/I_{total}) over heating temperature in (a) the 8 wt% P deposit, (b) the 6 wt% P deposit and (c) the 5 wt% P deposit.

In the 8 wt% P deposit, the ratio of I_{amorp}/I_{total} declines after heating to 300 °C, accompanied by an increase of integrated ratio of nickel phase (Fig. 11.12(a)). However, no indication of major crystallisation reaction is apparent in the DSC curve to the same temperature (Fig. 11.1(a)). The reason for this can be that the precipitation of nanocrystalline nickel phase from the amorphous phase, without the formation of any other new phases, is not accompanied by measurable heat evolution. From the end temperatures of 300–400 °C, the ratio of I_{amorp}/I_{total} declines quite sharply to a low level. Figure 11.12(a) shows that the decline is accompanied by the rapid increase of integrated intensity ratio of Ni_3P stable phase. This can be the reason for the non-increase of integrated intensity ratio of nickel phase in the deposit at this temperature range. At the same time, a small amount of $Ni_{12}P_5$ metastable phase (less than 2%) forms in the deposit at 350 °C. Consequently, the deposit is undergoing major crystallisation reaction at this temperature range. This is confirmed by the presence of a major exothermic peak in the DSC curve at the same temperature range (Fig. 11.1(a)). From the end temperatures of 400–500 °C, the decline of the ratio of I_{amorp}/I_{total} becomes gradual, and it reaches zero at the end temperature of 800 °C. Accompanied with this are the massive increase in the ratio of I_{Ni}/I_{total} that is largely contributed by the dominant Ni {200} reflection, and the corresponding drop in the ratio of I_{Ni_3P}/I_{total} corresponding to Ni_3P stable phase. The Ni_3P phase has highly developed at the temperature of 400 °C, with only peak-refinement process at the higher temperatures (Fig. 11.3). The deposit has become crystalline and rich in nickel phase at the end temperature of 800 °C.

In the 5–6 wt% P deposits, the ratio of I_{amorp}/I_{total} decreases gradually to the end temperature at 300 °C indicating the absence of major crystallisation reactions. From 300 °C to 400 °C, the ratio of I_{amorp}/I_{total} drops to a very low level, due to the ongoing major crystallisation reactions at this temperature range. The ratio of I_{Ni_3P}/I_{total} has also increased sharply from 350 °C to 400 °C due to the formation of Ni_3P phase. From 400 °C to 800 °C, the ratio of I_{amorp}/I_{total} further reduces to zero, a sign of the complete crystallisation process. The fcc nickel is the dominant phase from the beginning to the complete crystallisation process (Figs 11.4 and 11.5). The amount of nickel phase increases at the expense of the amorphous and Ni_3P phases in the deposits.

In the 5 wt% P deposit, the ratio of the integrated intensity of the Ni_3P phase to the total integrated intensity of the deposit is 14 wt% in the deposit heated at 5 °C min^{-1} and 13 wt% in the deposit heated at 40 °C min^{-1}, respectively. About the same amount of Ni_3P phase has been formed at the termination point of the DSC peak, regardless of the heating rate. Therefore, the DSC peaks at the two heating rates correspond to similar process(es). This is the fundamental basis of the non-isothermal analysis based on experiments performed at different heating rates. Though it appears that heating at 5 °C min^{-1} favours the {111} orientation, whereas 40 °C min^{-1} heating favours the {200} reflection (Table 11.6), the sample-to-sample variation cut from the coated panel has a strong contribution.

Table 11.6 Integrated intensity of each individual reflection of nickel as a proportion of the total integrated intensity of the nickel phase (I_{Ni})

Heating rate (°C min^{-1})	$I_{Ni\{111\}}/I_{Ni}$	$I_{Ni\{200\}}/I_{Ni}$	$I_{Ni\{220\}}/I_{Ni}$	$I_{Ni\{311\}}/I_{Ni}$
5	0.44	0.37	0.017	0.12
40	0.35	0.51	0.011	0.081

As a whole, the integrated intensity ratios of individual phases indicate that the amorphous phase of the 8 wt% P deposit is decreasingly dominant with the increase of fcc nickel phase up to the heating end temperature of 300 °C. At the higher end temperatures, fcc nickel phase is dominating with the further expense of amorphous phase into crystalline phases. In the 5–6 wt% P deposits, the fcc nickel phase is dominant throughout the entire range of heating end temperatures (up to 800 °C). Compared to the high phosphorus deposits described in Chapter 10 (Fig. 10.6 in Section 10.3), the medium- and low-phosphorus deposits are mainly dominant by the fcc nickel phase, while those high-phosphorus deposits are dominated by the Ni_3P stable phase at the higher heating end temperature ranges.

11.4 The major exothermal peak

Many attempts have been made to calculate the activation energy of the crystallisation process using the Kissinger method based on the major exothermal DSC peaks obtained at different heating rates. Only when the process(es) involved in the exothermal peak is fully understood should the activation energy be used to describe the crystallisation and phase transformation process. The crystallisation process is essentially the formation of crystalline nickel, Ni_3P and other intermediate phases. However, the precipitation of crystalline nickel may not be accompanied by significant heat evolution, since no obvious heat release is observed before the onset of the major DSC peak when a considerable amount of nickel has crystallised already. Martyak and Drake (2000) proposed that the formation of metastable phases has taken place prior to the major DSC peak where the formation of Ni_3P takes place. However, the possible participation of metastable phases such as Ni_2P and $Ni_{12}P_5$ in the major DSC peak cannot be completely excluded.

Research usually employs XRD and DSC to study the crystallisation and phase transformation behaviour in Ni–P platings (Martyak and Drake, 2000). XRD is normally carried out on isothermally annealed specimens, whereas DSC is in a continuous-heating mode. The difference in the heating process of XRD and DSC experiments can affect the accuracy when using the phase structure characterised by XRD to interpret the DSC profile due to kinetic factors. For instance, it is widely accepted that the amorphous phase in Ni–P platings of high phosphorus content will disappear after isothermal holding at 300 °C. However, the amount of amorphous phase is still detectable when the temperature reaches 400 °C, or even 500 °C in continuous-heating experiments. When isothermal annealing treatment is carried out, the difference in annealing time may partly contribute to the discrepancies between the resulting phase structures (Martyak and Drake, 2000).

Within the temperature range of the major DSC peak, significant increase in crystalline grain size is observed. Such increase in grain size may also result in heat release, which makes the interpretation of the exothermal DSC peak even more complicated. There is always grain size increase accompanying temperature increase. This process will release energy due to the reduction of the area of interface. When the average grain size is D, the ratio of surface area over volume can be estimated as $2/D$. Assuming the interface energy to be 0.5 J m^{-2} (a typical value used for grain boundary interface) and the deposit density to be 8×10^3 kg m^{-3} (the density of pure nickel), the energy release by the 5 wt% P deposit from 350 °C to 500 °C (this temperature range covers the exothermal peak in the DSC curve at 20 °C min^{-1}), i.e. from 6.4 to 28.5 nm in grain size, is 15 J g^{-1}. One concern here is the magnitude of interfacial energy, which can vary within a range of 10^{-4}–1 J m^{-2}. If 0.5 J m^{-2} is a reasonable value, compared with the integrated area of the DSC peak (34 J g^{-1}), the heat release due to size increase is a significant contribution. In the above estimation, the material is assumed to have fully crystallised. The size increase in the four Ni–P platings approximates parabolic shape, which implies that the grain growth is a diffusion-controlled process. Clearly, the major DSC exothermal peak involves all the above mentioned processes. Determining which of these is the dominant process is needed.

11.5 Conclusions

Electroless Ni–P deposits with medium- and low-phosphorus contents have the mixture of amorphous and nanocrystalline nickel phases at the as-deposited condition. The sequence of phase transformation in the 8 wt% P deposit is amorphous phase + nanocrystalline nickel → co-existence of fcc nickel and Ni_3P stable phases, and metastable $Ni_{12}P_5$ and unidentified phases → fcc nickel and Ni_3P stable phases. The sequence of phase transformation in the 5–6 wt% P deposits is amorphous phase + nanocrystalline nickel → stable Ni_3P and fcc nickel phases. Phase transformation temperatures of the deposits increase with decreasing phosphorus content and increasing heating rate. Activation energies of the 6–8 wt% P deposits (medium phosphorus) are lower than that of the 5 wt% P deposit (low phosphorus). Preferred orientation has developed at the Ni {200} plane of the deposits after the continuous heating processes.

There is no significant obvious dependency of grain size on the amount of phosphorus in the platings. The 5 wt% P deposit heated up to the termination point of the DSC peak at 40 °C min^{-1} has a larger microstrain than that at 5 °C min^{-1}. The amount of Ni_3P phase formed at this point is the same regardless of the heating rate. The increase of grain size may significantly contribute to the exothermal peak in the DSC curve.

11.6 References

Martyak N M and Drake K (2000), 'Peak-profile analysis of electroless nickel coatings', *J Alloy Comp*, **312**, 30–40.

Sha W and Guo Z (2009), *Maraging Steels: Modelling of microstructure, properties and applications*, Woodhead Publishing, Cambridge.

Wilkinson D S (2000), *Mass Transport in Solids and Fluids*, Cambridge University Press, Cambridge.

Xiang Y, Hu W, Liu X and Luo S (2001), 'Evaluation of the phosphorus content of electroless nickel coatings by XRD & DSC', *Plat Surf Finish*, **88**, 96–9.

12
Modelling the thermodynamics and kinetics of crystallisation of nickel–phosphorus (Ni–P) deposits

Abstract: Modelling based on the Johnson–Mehl–Avrami (JMA) theory is carried out on the crystallisation kinetics of the electroless Ni–P deposits. The degree of transformation and the differential scanning calorimetry curves corresponding to the major crystallisation processes of those deposits are simulated using the JMA models and compared to the experimental results. The models could make reasonable estimations on the major crystallisation processes of the electroless Ni–P deposits with amorphous as-deposited structure, but in the case of non-isothermal crystallisation, not of those with partial crystalline structure. In addition, thermodynamic modelling of the crystallisation products of amorphous solids is described.

Key words: modelling, phase transformation, amorphous materials, thermodynamic modelling, kinetics.

12.1 Introduction

A significant advance in thermodynamic databases was the completion of SGTE (Scientific Group Thermodata Europe) SOLution database, SSOL. In this chapter, thermodynamic calculations for several amorphous systems are carried out, and the results are compared with experimental data. The type of materials investigated is mainly amorphous coating produced using an electroless process or electrodeposition. The aim is to assess the viability of using a thermodynamic calculation to evaluate the crystallisation products upon heating of the coating. Various authors have carried out a large amount of experimental work over the years to identify such products; it is envisaged that much of this can be replaced, or at least guided, with much less laborious calculations. The calculations are conducted with Thermo-Calc version L.

In the main parts of this chapter, we will model the non-isothermal crystallisation kinetics of electroless Ni–P deposits, based on the classical Johnson–Mehl–Avrami (JMA) approach, using experimental data obtained by differential scanning calorimetry (DSC). Although good modelling results in agreement with experimental data are obtained, the linear heating to high temperatures necessary for completing the crystallisation heat-release peak dictates that the crystallisation and grain growth processes cannot be separated. Therefore, the modelling is for the entire crystallisation process, which in itself usually has several stages involving various metastable phases, plus the grain growth immediately after formation of the crystals.

Some excellent and elegant work by Révész *et al.* (2001a,b,c) included DSC studies of the isothermal crystallisation kinetics of Ni–P and $Fe_{40}Ni_{40}P_{14}B_6$ amorphous alloys. The isothermal temperatures were chosen to be just below the onset temperature for crystallisation during linear heating. In this way, the incubation period of crystallisation was clearly demonstrated, and the crystallised grains did not grow significantly. Nanosized grains resulted and they were maintained.

The middle part of this chapter uses the data obtained and published by Révész *et al.* (2001a,b,c) to develop computer models for simulation of the crystallisation kinetics of amorphous materials. It differs from the later sections in that the later sections concern non-isothermal heating. As stated above, isothermal annealing below the crystallisation onset temperature during linear heating can separate crystallisation from the subsequent grain growth process. In all but one case, published curves have been digitised and readings obtained from scanned diagrams. Small errors are possible during this process, and the data points are vastly reduced compared with the original DSC output file. The background relating to applications of these amorphous alloys can be found in earlier chapters and in the papers by Révész *et al.* (2001a,b,c).

In the remaining and major parts of this chapter, the objective is to model the non-isothermal crystallisation kinetics of electroless Ni–P deposits based on a theoretical approach using a computer-based model. For this purpose, the model has been developed based on the JMA theory, using the non-isothermal DSC data in Chapters 10 and 11. Both the degree of transformations and the DSC experimental curves related to the major crystallisation process of the deposits are simulated in the modelling. The applicability of the JMA model to these deposits is discussed, and the corresponding JMA kinetic parameter is used for describing the crystallisation kinetics of the deposits. In previous books, the use of the JMA theory and the incorporation of DSC experimental data into the theory for such modelling have been described by Sha and Malinov (2009) and Sha and Guo (2009).

12.2 Thermodynamic analysis of crystallisation in amorphous solids

We will start the modelling with the most common electroless amorphous plating systems, Ni–B and Ni–P. For Ni–B, a typical B content of 4 wt% is used. The equilibrium phases for temperatures ranging from 250 °C to 375 °C are consistently 74 mol% Ni_3B and almost pure Ni. In the case of Ni–P, for 6.8 wt% P, the equilibrium at 300 °C consists of almost pure Ni and 48.6 mol% Ni_3P.

In the ternary Fe–Ni–P system, for the $Fe_{80}Ni_8P_{12}$ alloy, at 350 °C, the equilibrium constitution is 52 mol% bcc with a composition of Fe–1.77at%Ni–0.02at%P and 48 mol% $Fe_{60}Ni_{15}P_{25}$ ($(Fe,Ni)_3P$), in agreement with the result from the very time-consuming experimental programme involving isothermal heat treatment and phase identification using X-ray diffraction. We now extend the calculations to more complex systems (Table 12.1). In a further alloy, $Fe_{32}Ni_{36}Cr_{14}P_{12}B_6$ (Metglass 2826A), the equilibrium between 375 °C and 600 °C consists of four phases. These are Cr_2P (12 mol% at 375–400 °C reducing to 11.3 mol% at 600 °C), CrB (12 mol%), Ni_3P (32 mol% at 375–400 °C increasing to 32.6 mol% at 600 °C), and fcc. In the fcc phase, the concentrations of Ni, Fe and Cr change with temperature (Fig. 12.1). The solubility of phosphorous in fcc increases from 0.004 at% (375 °C), to 0.006 at% (400 °C), and to 0.2 at% (600 °C). The solubility of boron remains very small for the entire temperature range, as expected. For the Ni_3P phase, although its mole fraction remains almost constant, the iron level changes from 40.3 at% at 375 °C, to 39.7 at% at 400 °C, and to 33.1 at% at 600 °C.

Electrodeposition is used to produce Ni–W–P alloy coating and the phase

Table 12.1 Equilibrium phase compositions (at%) in $Fe_{40}Ni_{40}P_{14}B_6$ (Metglass 2826) alloy at 350/500 °C: Thermo-Calc thermodynamic calculation results

Phase	Ni	Fe	P	B	Mol%
fcc	59.8/56.7	40.2/43.3	0.002/0.04	0	26
Ni_3P	43.2/43.6	31.8/31.4	25*	0	56
Fe_2B	1.6/4.6	65.1/62.1	0	33.3*	18

*The concentrations of these are always stoichiometric in the database.

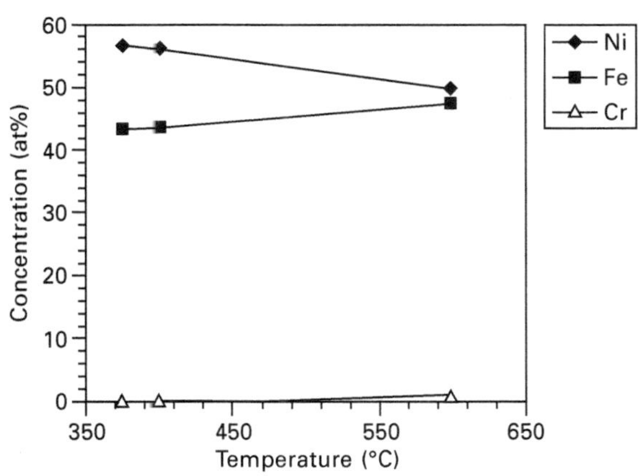

12.1 The variation of equilibrium Ni, Fe and Cr concentrations in fcc phase in the $Fe_{32}Ni_{36}Cr_{14}P_{12}B_6$ alloy as a function of temperature.

evolution upon heating can be investigated using transmission electron microscopy, X-ray diffraction and microhardness measurements. For the composition of the coating of Ni–31wt%W–1wt%B, the phase equilibria at 400 °C and 600 °C are both fcc Ni, Ni_3B and Ni_4W. At 400 °C, there are, respectively, 26 and 40 mol% of Ni_3B and Ni_4W in the equilibrium. The W content in fcc Ni is 11.5 at% and the B content is virtually zero. At 600 °C, the amount of Ni_4W has decreased to 34 mol%, accompanied by an increase in the W content in fcc Ni to 12.7 at%. The solubility of B in Ni is only 0.001 at%, so there is no change of the amount of Ni_3B in equilibrium. Ni_4W was also identified experimentally by the techniques mentioned in the beginning of this paragraph.

During solder reaction-assisted crystallisation of electroless Ni–P, a reaction equation is proposed:

$$\mathrm{Ni_{85}P_{15}} + \frac{160}{3}\mathrm{Sn} \rightarrow 15\mathrm{Ni_3P} + \frac{40}{3}\mathrm{Ni_3Sn_4}$$

Using Thermo-Calc, when an alloy composition as in the left-hand side of the equation is used, the equilibrium reached matches exactly the right-hand side.

It is reasonable that good results are obtained by applying equilibrium calculations to what starts as a very non-equilibrium system. One of the tenets of thermodynamics is that the equilibrium state can be reached via a number of different routes and it matters not at all that one is travelling through metastable intermediaries. Indeed using a highly distorted or highly stressed intermediary state is a common way of increasing the kinetics of sluggish reactions.

12.3 Application of Johnson–Mehl–Avrami (JMA) theory in isothermal crystallisation

The kinetics of transformation can be described in the theoretical framework of JMA theory, with the following formula:

$$f(t)/f^{\max} = 1 - \exp\left(-kt^{n}\right) \qquad 12.1$$

where $f(t)$ is the amount of product after time t, $f^{\max}$ is the maximum amount of product, k is the reaction rate constant and n is the Avrami index. In the present case of modelling crystallisation, the DSC technique measures only the heat signal of the transformation, and does not distinguish the source of the heat, i.e. transformation products. When several transformation products are formed simultaneously, the heats from them are measured collectively. Therefore, the formation of individual crystalline products cannot be modelled separately based on DSC data alone if these products are produced at the same time. Here, the crystalline products are considered collectively, effectively as one phase, the crystalline state of the solid. Therefore the value of f^{max}, the maximum amount of product, is 1 (in fraction of the total material), because at any temperature, the amorphous solids are in a thermodynamically unstable state and should eventually transform to a crystalline state entirely. The JMA equation is now:

$$f(t) = 1 - \exp(-kt^n) \tag{12.2}$$

The isothermal modelling in the chapter is based on this equation. The parameter f is also termed the degree of transformation. By plotting $\ln\{\ln[1/(1-f)]\}$ versus $\ln(t)$, the Avrami index n can be obtained from the gradient of the linear regression line, and from the intercept, the k value can be calculated.

The first step of modelling is to obtain the f versus t curve for the given transformations. The original DSC data are presented as heat flow versus annealing time t. A basic assumption in using the DSC data is that the heat flow produced is proportional to the amount of product phase, in this case, the crystalline phases collectively. Strictly, this assumption is only correct if the different phases give the same amount of heat per unit mass and there is no interaction between different phases. These two conditions, in fact, are not met, as the enthalpy for the transformations is always different for different phases, and in the case of crystallisation, there is transformation between different metastable phases formed in the initial stages of crystallisation. Therefore, modelling based on heat signals should be used only to describe the heat signals, and not the amounts of phases. Nevertheless, such modelling of the heat signals is useful in that:

1. Although not precisely as explained above, it does give a good estimation of the amount of crystalline phases formed.
2. Using the model developed, heat signals can be predicted for heating conditions not covered by experiments, and as such, they can be related back to products when the relationship between heat signals and phase products is established.

If the heat signal can be described by the JMA equation, it is a good indication that the heat is directly proportion to the phases formed. Later sections include further discussion of these aspects. It should be noted that the initial materials are purely amorphous for the alloys studied here, but the crystallisation product after treatment is nanophase. The grain size of the crystallised structure after heat treatment will have an effect on the JMA kinetics, but it is difficult to tell whether the grain size at the end of the crystallisation treatment is a result of significant grain growth or not. This would require numerous transmission electron microscopy (TEM) investigations at intermediate stages of crystallisation to examine the change in grain size.

12.4 Electroless and melt quenched nickel–phosphorus (Ni–P)

Several Ni–P alloys have been studied experimentally by Révész *et al.* (2001a,b,c) including an electroless alloy (Révész *et al.*, 2001a) and several melt quenched alloys (Révész *et al.*, 2001b). The electroless alloy has varying composition across the deposit thickness, from near eutectic composition (19 at% P; at% is used throughout the isothermal modelling sections) on the bath side approximately linearly down to 10 at% on the substrate side. The films produced by melt spinning have better controlled composition: at eutectic 19 at%P and just below eutectic 18 at%P.

12.4.1 Electroless alloy

The first step of modelling is to plot the curve of transformed fraction versus time based on the assumption or approximation that the heat recorded is proportional to the amount of the crystalline phases formed. This conversion is done by partial integration of the peak area of the curve for heat flow versus time recorded by DSC (Fig. 12.2). The original DSC curves are published by Révész *et al.* (2001a), and are not reproduced here. The reader should consult the above for the original data. The curve for 590 K is plotted using the DSC computer data file provided by an author of Révész *et al.* (2001a) (AR), while the other three curves, and all the remaining curves of fraction versus time used in the isothermal modelling sections, are plotted using data points in the published heat flow curves. For this reason, the curve for 590 K

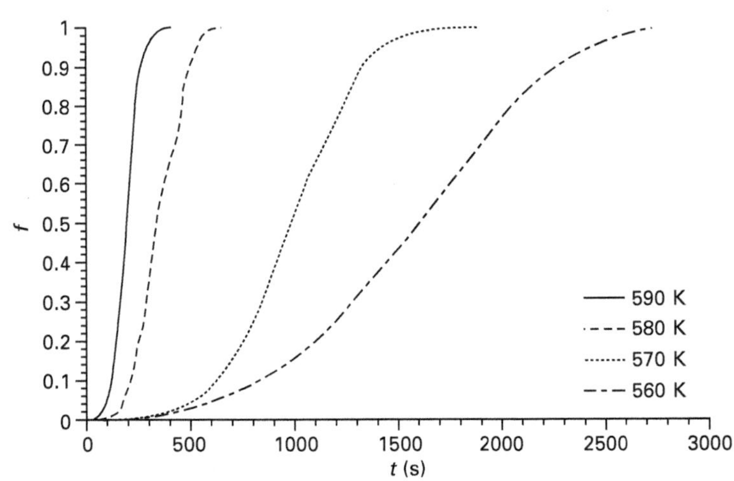

12.2 Degree of transformation *f* for electroless Ni–P coatings under isothermal annealing at 590, 580, 570 and 560 K, obtained from published DSC heat flow data (Révész *et al.*, 2001a): *t* time.

includes many more data points than do the other curves in Fig. 12.2. This is better revealed by other figures in these isothermal modelling sections.

It should be particularly noted that the JMA equation does not consider the incubation period of the transformation. In other words, time t in the JMA equations begins (when $t = 0$) at the start of the transformation. Experimentally, the DSC data show clearly that there is an incubation period for crystallisation, the length of which depends on how much the isothermal annealing temperature is below the crystallisation start temperature during linear heating. In Fig. 12.2, this incubation period is subtracted from the time in the original DSC heat flow curves published by Révész *et al.* (2001a) so that t in Fig. 12.2 always starts at 0. Such an incubation period is best illustrated using a classical time–temperature–transformation (TTT) diagram. This is shown in Fig. 12.3, where the 50% transformation (t_{50}) and transformation completion (t_f) points are also presented. In Fig. 12.3, t on the horizontal axis is the actual time during isothermal annealing, and this time equals the start time t_s plus the time t shown in Fig. 12.2. Exclusion of the induction period in JMA is a feature of the theory, and does not mean that the induction period is ignored. To model this period, it is expected that some considerations have to be given to atomic rearrangements, as there is no new phase formed.

The three stages of crystallisation are clearly revealed (Fig. 12.4). However, the first stage, although appearing to be long in Fig. 12.4 owing to the logarithmic scale, occurs only for a short period in the initial stage of the crystallisation process, within the first 15 s. The third stage is the completion stage of crystallisation, where the increase of transformed fraction becomes

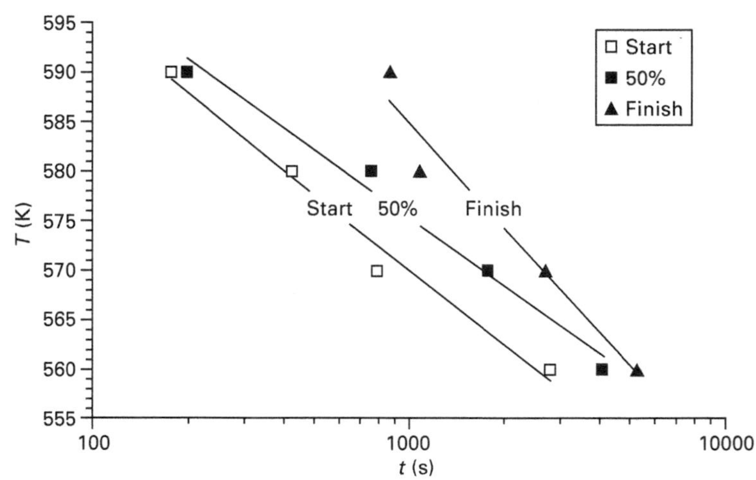

12.3 TTT diagram for crystallisation process of amorphous electroless Ni–P alloy.

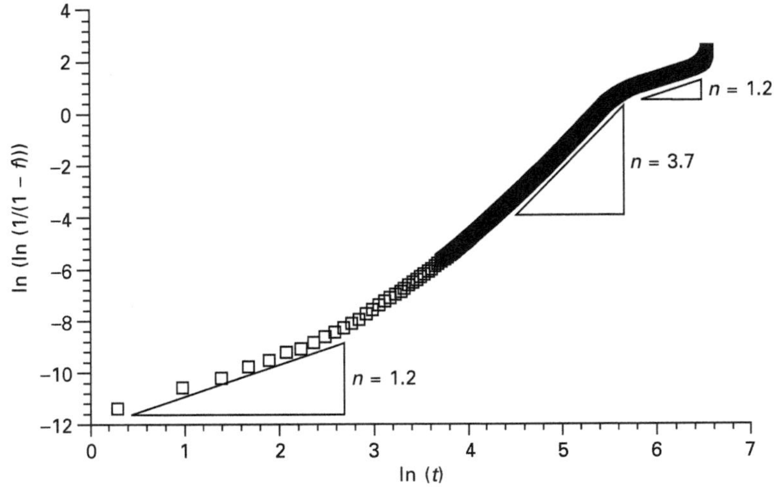

12.4 A detailed plot of ln{ln[1/(1–*f*)]} versus ln (*t*) for the case of isothermal annealing at 590 K of amorphous electroless Ni–P alloy: *n* Avrami index.

very slow (see curve for 590 K in Fig. 12.2). These starting and finishing stages are usually associated with large errors in DSC signals, because of the limitations in instrumental sensitivities caused by the inhomogeneity of the sample temperature inside the crucible (pan). Such temperature inhomogeneity will cause a widening or broadening of the heat peak, and the initial and final long 'tails' can be a result of this peak widening or broadening effect, rather than the actual transformation heat. However, without special techniques, the two effects cannot be separated. This is discussed in later sections. On many occasions, it is even difficult to determine exactly the starting and finishing points of a peak when they are very gradual. Therefore, the first and last stages of the DSC curves and the derived JMA results should be treated with caution. Although their relationships to the transformation mechanisms are in principle correct, the exact values of JMA parameters may well be inaccurate, and may not be suitable for quantitative interpretation and further predictive modelling. For this reason, only the major crystallisation stage, the second stage illustrated in Fig. 12.4, is modelled.

Although there are some variations, essentially the linear regression lines have similar fitting equations (Fig. 12.5). It should be remembered that the fitted JMA parameters have errors associated with them. The Avrami index n varies within a small range from 3.1 to 3.8, indicating that the mechanism of transformation remains the same for the range of isothermal annealing temperatures. It corresponds to a three-dimensional nucleation and growth process in the bulk of the deposits.

12.4.2 Melt quenched eutectic alloy and $Ni_{82}P_{18}$

The approach presented above is extended to two melt quenched Ni–P alloys during their primary crystallisation process for the first isothermal heat treatment of the amorphous state (Fig. 12.6). In Fig. 12.6(b), there are two stages of crystallisation. In this alloy, the first stage is more pronounced than in the case of the electroless alloy (Fig. 12.4) as it extends to 132 s, and therefore probably cannot be ignored. The first stage corresponds to a surface nanocrystallisation. It may be noted that the melt quenched ribbon has a thickness of 10 μm, compared with the 15 μm thickness of the electroless

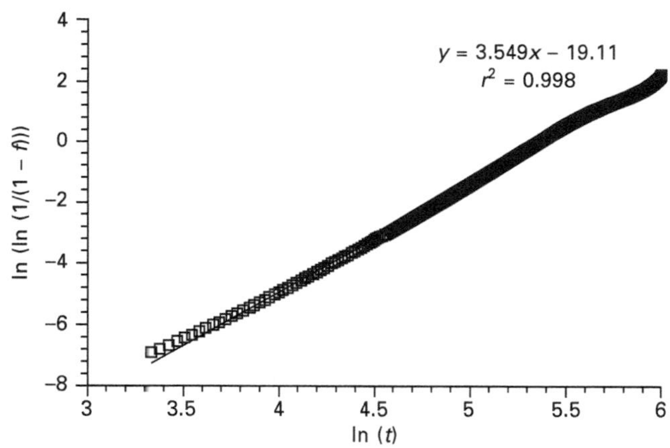

(a)

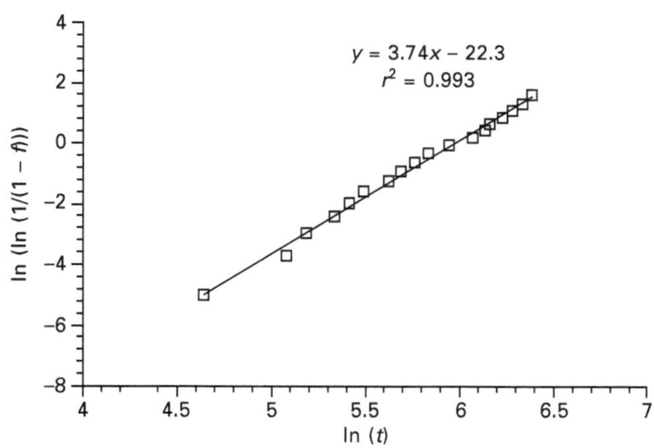

(b)

12.5 ln{ln[1/(1–f)]} versus ln(t) plots for the major crystallisation stage in the case of electroless Ni-P deposits under isothermal annealing at given temperatures derived from f versus t curves in Fig. 12.2: standard deviations on Avrami index are 0.009, 0.08, 0.08, 0.07 for (a)–(d), respectively. (a) 590 K; (b) 580 K; (c) 570 K; (d) 560 K.

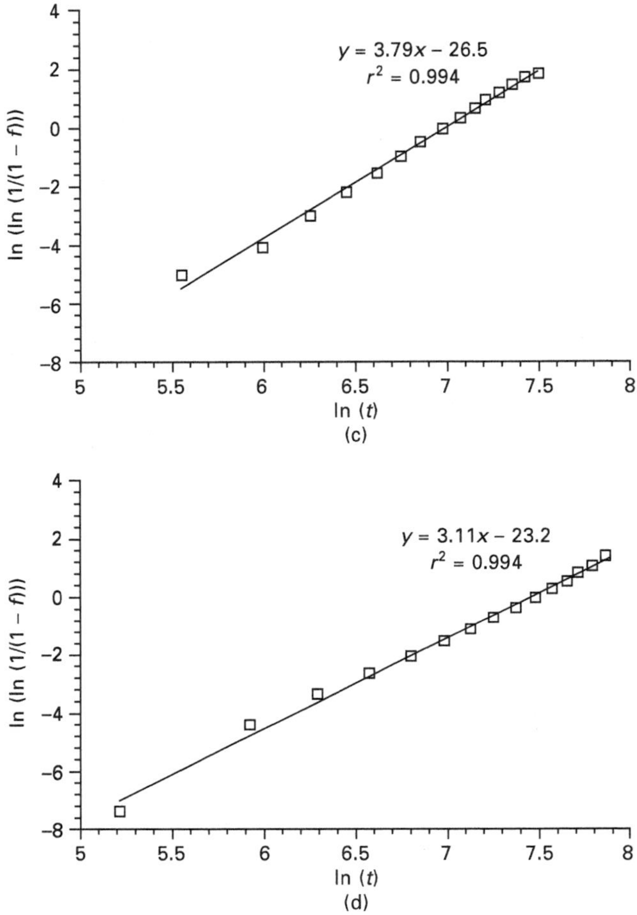

12.5 Continued

foil. The Avrami index for the second stage is abnormally high, indicating a fast increasing nucleation rate of the crystalline grains.

When the melt quenched eutectic alloy annealed at 600 K for 1200 s is subjected to a second isothermal anneal at 660 K, there is an abnormal grain growth process overlapping with the normal grain growth. The abnormal grain growth is similar to recrystallisation. It can be described using the JMA theory (Fig. 12.7). Compared with the crystallisation process, this abnormal grain growth has a much lower enthalpy, indicated by the very small peak in DSC curve.

The DSC result for the slightly lower phosphorus $Ni_{82}P_{18}$ alloy is reported by Révész *et al.* (2001b). The crystallisation process at 565 K of this alloy is characterised by two peaks in the DSC curve. This is because the alloy has a lower phosphorus concentration than the eutectic level, and therefore nickel is

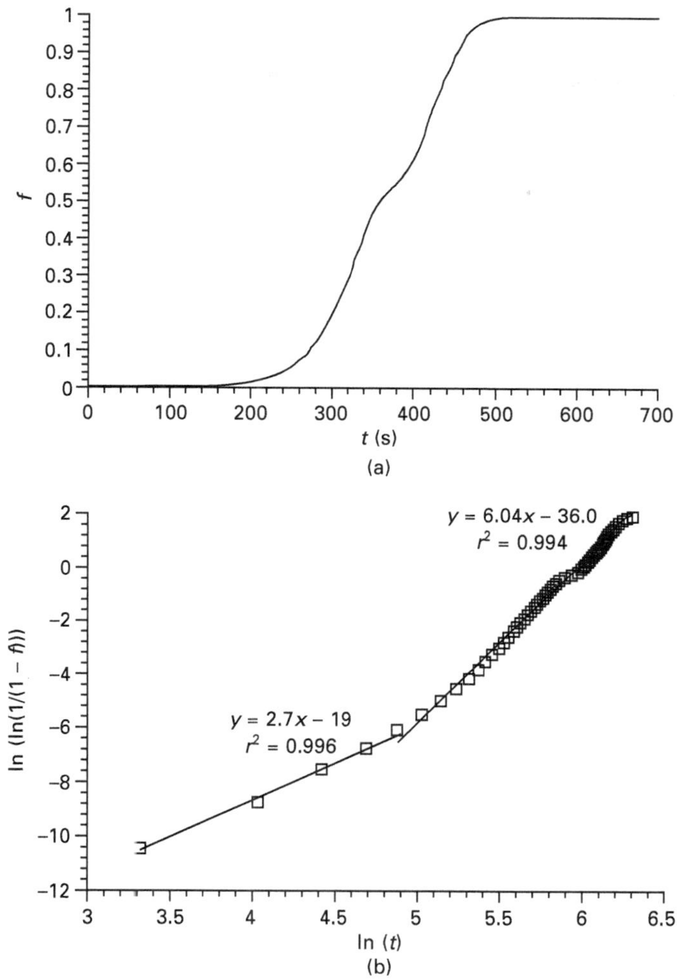

12.6 Given plots for melt quenched eutectic $Ni_{81}P_{19}$ during isothermal crystallisation at 600 K: standard deviations on Avrami index are 0.1 and 0.06 for first and second regression equations, respectively. (a) Degree of transformation *f* versus *t*; (b) ln{ln[1/(1–*f*)]} versus ln(*t*).

precipitated first from the amorphous alloy, which is a clearly separate process from the subsequent simultaneous eutectic formation of Ni and Ni_3P phases from the amorphous structure. As the phosphorus-level is only slightly below the eutectic level, the first peak is rather small compared with the major peak. After the first, small crystallisation process, the remaining amorphous matrix should have a composition close to the eutectic composition. Both DSC peaks are modelled according to the JMA theory (Figs 12.8 and 12.9). The major crystallisation process, i.e. the second crystallisation peak, modelled in Fig.

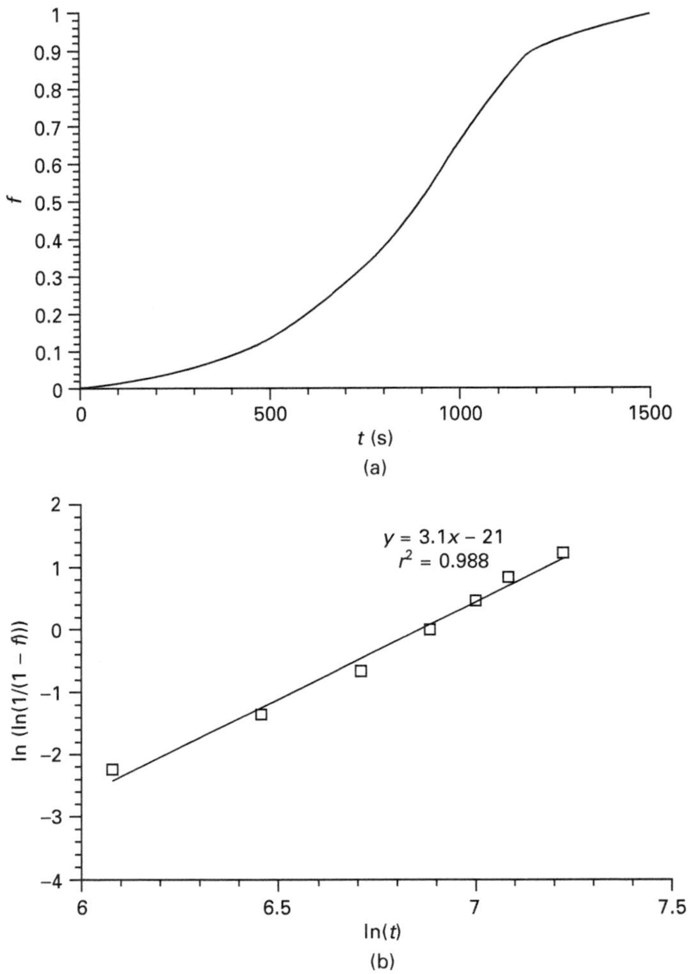

12.7 Given plots for abnormal grain growth of melt quenched eutectic $Ni_{81}P_{19}$ at 660 K of nanocrystals formed in first anneal at 600 K for 1200 s: standard deviation on Avrami index is 0.1. (a) Degree of transformation *f* versus *t*; (b) ln{ln[1/(1–*f*)]} versus ln(*t*).

12.9(b), shows a similar two stage process to that of the eutectic alloy (Fig. 12.6(b)). Therefore, there are significant similarities of the crystallisation process from the Ni–P eutectic composition in the two alloys, although in the $Ni_{82}P_{18}$ alloy this eutectic composition itself is a product of the initial precipitation of nickel from a matrix richer in nickel. The different JMA parameters demonstrate quantitatively the different crystallisation behaviours between this $Ni_{82}P_{18}$ and the eutectic $Ni_{81}P_{19}$ alloy. Incidentally, the first and shorter stage of the major crystallisation process in $Ni_{82}P_{18}$ (Fig. 12.9(b))

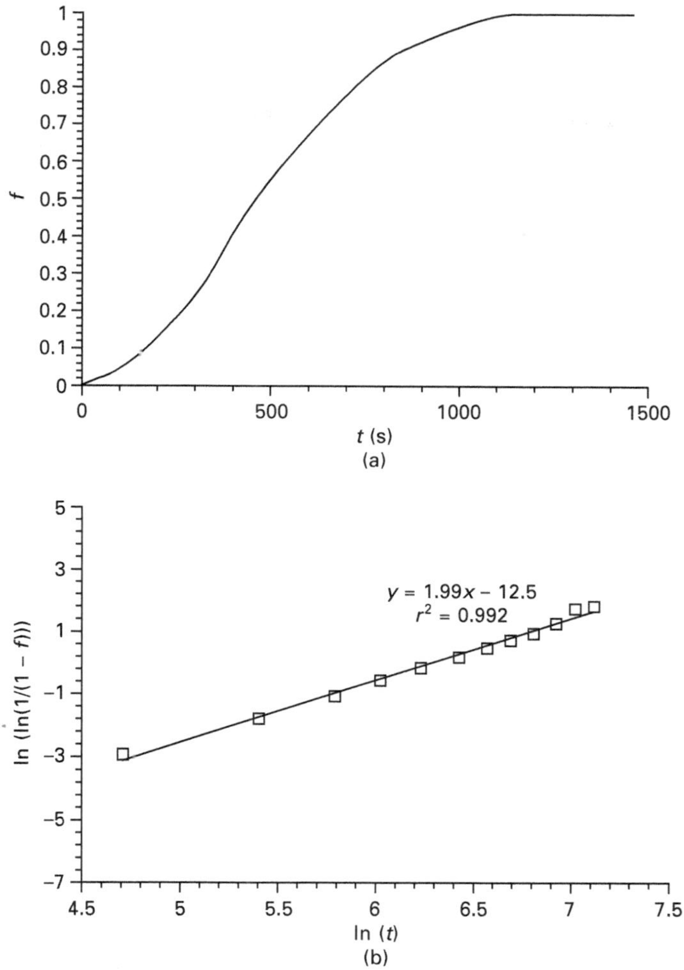

12.8 Given plots of first and small crystallisation peak for melt quenched $Ni_{82}P_{18}$ during isothermal crystallisation at 565 K: standard deviation on Avrami index is 0.06. (a) Degree of transformation *f* versus *t*; (b) ln{ln[1/(1−*f*)]} versus ln(*t*).

has an Avrami index close to that of the nickel precipitation process (Fig. 12.8(b)).

12.5 Melt quenched $Fe_{40}Ni_{40}P_{14}B_6$

The applicability of the JMA theory to this alloy can be used to testify the range of suitability of the theory for the nanocrystallisation process in different amorphous alloys. Similar DSC data to those for the electroless Ni–P are presented by Révész *et al.* (2001c), in which study isothermal crystallisation

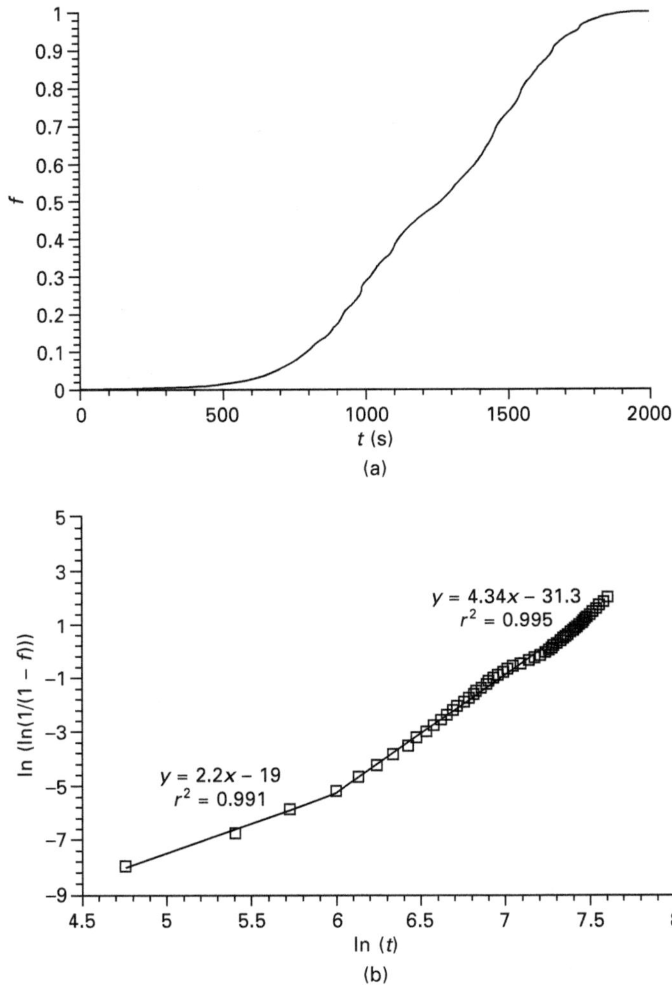

12.9 Given plots of second and major crystallisation peak for melt quenched $Ni_{82}P_{18}$ during isothermal crystallisation at 565 K: standard deviations on Avrami index are 0.1 and 0.05 for first and second regression equations, respectively. (a) Degree of transformation f versus t; (b) $\ln\{\ln[1/(1-f)]\}$ versus $\ln(t)$.

is conducted at five different temperatures below the crystallisation onset temperature during linear continuous heating. Again, the incubation period is evident from the isothermal thermograms, where the crystallisation peak starts after a finite time, depending on the temperature. The lower the temperature is, the longer the incubation period is. These results are summarised in Fig. 12.10, where the crystallisation start, 50% and finishing times are plotted against the isothermal treatment temperature, using a traditional semi

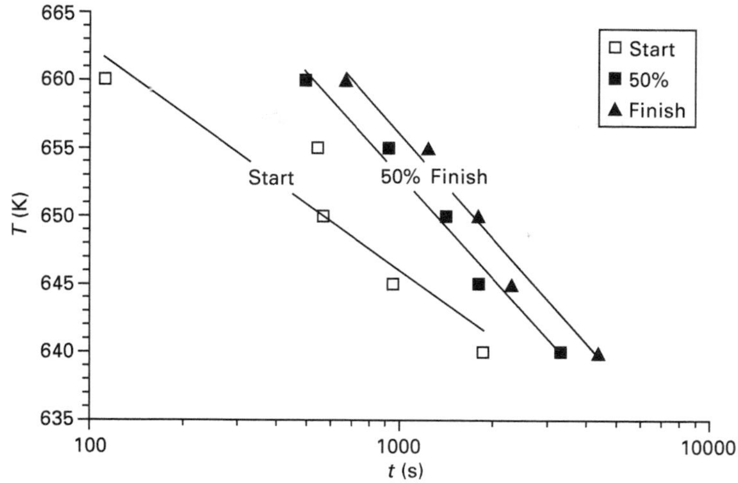

12.10 TTT diagram for crystallisation process of melt quenched $Fe_{40}Ni_{40}P_{14}B_3$ alloy.

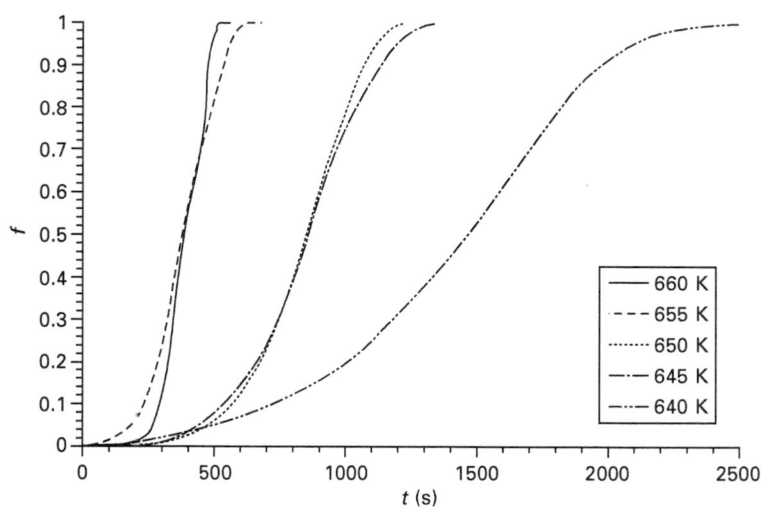

12.11 Degree of transformation *f* for melt quenched $Fe_{40}Ni_{40}P_{14}B_6$ alloy under isothermal annealing at 660, 655, 650, 645 and 640 K, obtained from published DSC heat flow data (Révész *et al.*, 2001c).

logarithmic scale. Similar behaviour compared with the electroless Ni–P is demonstrated.

Following the same procedure for the analysis of electroless Ni–P, Fig. 12.11 shows the crystallised fraction versus time plots at various isothermal temperatures. As for all other such plots in these sections, *t* is the time after the start of crystallisation, i.e. the isothermal holding time after subtraction of t_s in Fig. 12.10. The kinetic parameters including the Avrami index n and rate parameter k can be obtained from the regression equations shown in each graph in Fig. 12.12. At intermediate temperatures, the Avrami index is close to 4. At the highest temperature tested, 660 K, the Avrami index is very high, indicating a different crystallisation mechanism with fast kinetics and increasing nucleation rate. Note also the large $\ln(k)$ value, the constant term in the regression equation. At the lowest temperature tested, 640 K, there are two fitting lines for two different stages of crystallisation. The second stage is similar to the crystallisation kinetics at intermediate temperatures, indicated by the similar regression equations. The first stage represents slower crystallisation, which is reasonable considering the lower temperature. In the five graphs in Fig. 12.12, the scales of the horizontal x and the vertical y axes are such that the slopes of the regression lines are visually comparable. Similar consideration is given in deciding the scales for the graphs for Ni–P alloys, except for Fig. 12.4, so the slopes in these plots of $\ln\{\ln[1/(1-f)]\}$ versus $\ln(t)$ are visually comparable as the scales have the same y range and x range ratios.

12.6 Johnson–Mehl–Avrami kinetic modelling of non-isothermal crystallisation

An alternative form of the JMA equation to Eq. 12.2 is

$$f = 1 - \exp[-(kt)^n] \qquad 12.3$$

Equation 12.3 can be used to describe many phase transformations for their characteristics, such as the nucleation and growth mechanisms as well as the growth morphology of crystal. It must be mentioned that, however, the equation was derived based on the following assumptions:

- isothermal transformation conditions,
- random nucleation (for both homogeneous and heterogeneous),
- a nucleation rate that is either constant, maximum at the beginning of transformation and decreasing during the course of transformation, or being negligible (site saturation) during the transformation,
- that growth rate of new phase depends just on temperature, and is independent of time.

Accordingly, the use of the JMA equation in describing non-isothermal processes is not well based from a theoretical point of view (Málek, 2000). Besides, it should be realised that there is no generally valid theoretical justification for use of the JMA theory to describe solid-state transformations (like the crystalline transition of amorphous phase) at either isothermal or non-isothermal conditions. The reason for this is that the JMA theory refers only to simple processes, whereas solid-state transformations are usually

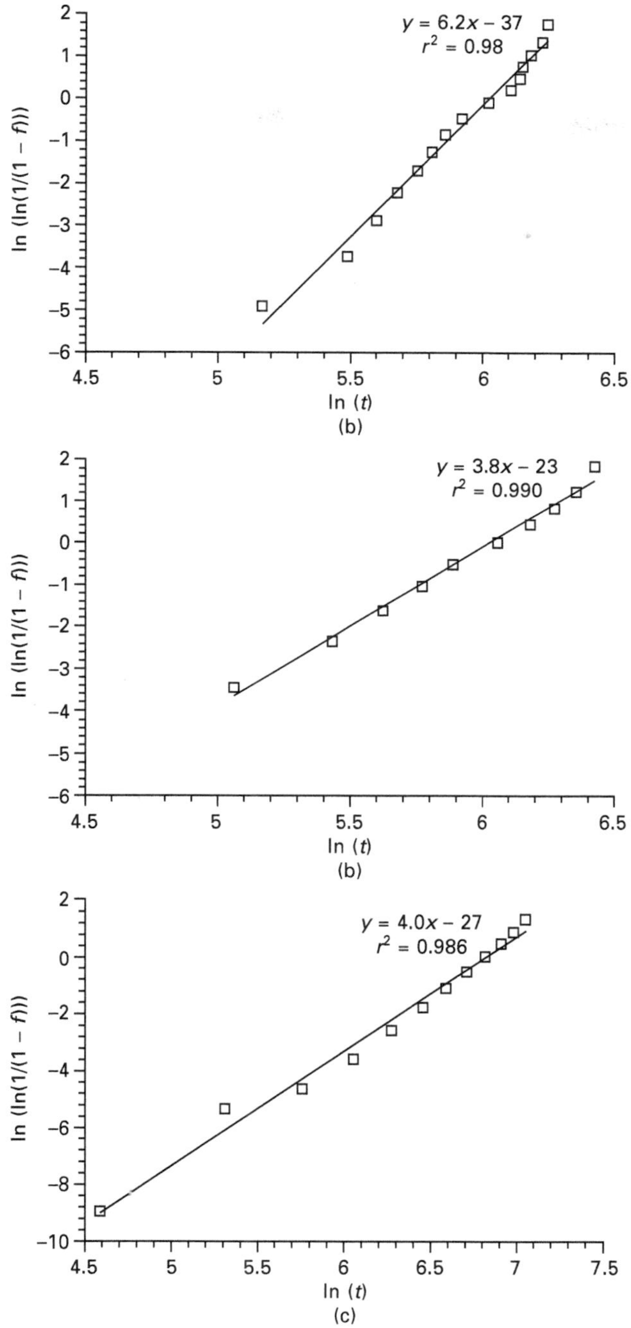

12.12 ln{ln[1/(1–*f*)]} versus ln(*t*) plots for melt quenched $Fe_{40}Ni_{40}P_{14}B_6$ alloy under isothermal annealing at given temperatures derived from *f* versus *t* curves in Fig. 12.11: standard deviations on Avrami index are 0.2, 0.1, 0.2, 0.1, 0.06 (first)/0.09 (second) for (a)–(e), respectively. (a) 660 K; (b) 655 K; (c) 650 K; (d) 645 K; (e) 640 K.

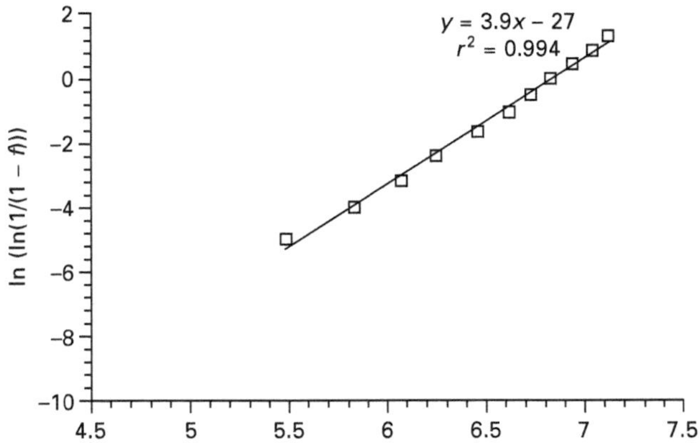

(d)

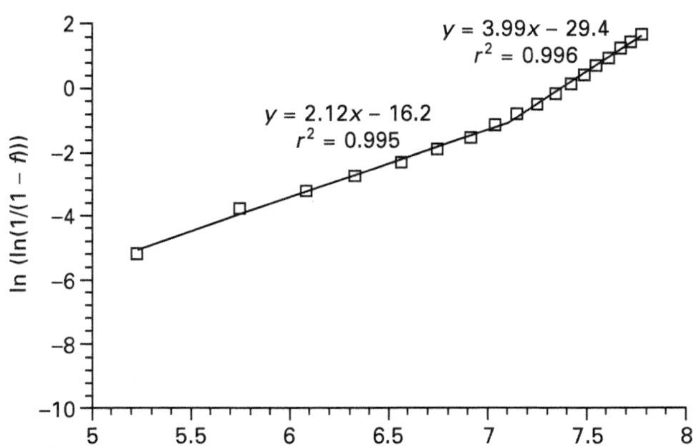

(e)

12.12 Continued

complex and could involve different simultaneous mechanisms as the process progresses. Nonetheless, the JMA theory can be applied in non-isothermal conditions if the site saturation condition is fulfilled. Besides, the JMA theory can present a relatively good approximation in a phenomenological way for the interpretation of transformation kinetics of a system governed by the nucleation-growth mechanism. Accordingly, the adoption of JMA theory for the modelling of non-isothermal crystallisation kinetics of electroless Ni–P deposits can give a reasonable estimation of the kinetics interpretation related to the major crystallisation process of the deposits.

Over the years, various approaches have been proposed based on the JMA theory to derive expressions for describing the crystallisation and phase transformation processes at non-isothermal conditions. All these approaches have been generally justified in describing the kinetics of transformation at such conditions (Sha and Malinov, 2009). The approaches used for the kinetics modelling based on the JMA theory include the direct integration of the generalised form of the JMA equation, and the introduction of a new state variable. In this chapter, the latter approach is adopted in view of its simple expression and acceptance in general. The approach is based on the concept that the degree of transformation during non-isothermal processes is not a state function of time and temperature (Sha and Malinov, 2009). It is justified, therefore, to introduce a state variable that is fully determined by the path of process described in a time–temperature (t–T) diagram.

Based on the temperature–time diagram given in Fig. 12.13, a sample specimen experiencing a thermally activated phase transformation under a non-isothermal condition proceeds from state-1 (t_1, T_1) to state-2 (t_2, T_2) via either path-a or b. Although the time to proceed from state-1 to state-2 is the same for both paths, the higher temperatures operating along path-b cause a stage of transformation in state-2 for path-b which is more progressive than that reached along path-a. Hence, the stage of transformation in state-2 depends, in general, on the path followed. Accordingly, for non-isothermal analyses, t and T are not state variables for the stage (degree) of transformation.

Accordingly, the degree of transformation is fully accommodated by the state variable. The JMA equation with the new state variable can be written as:

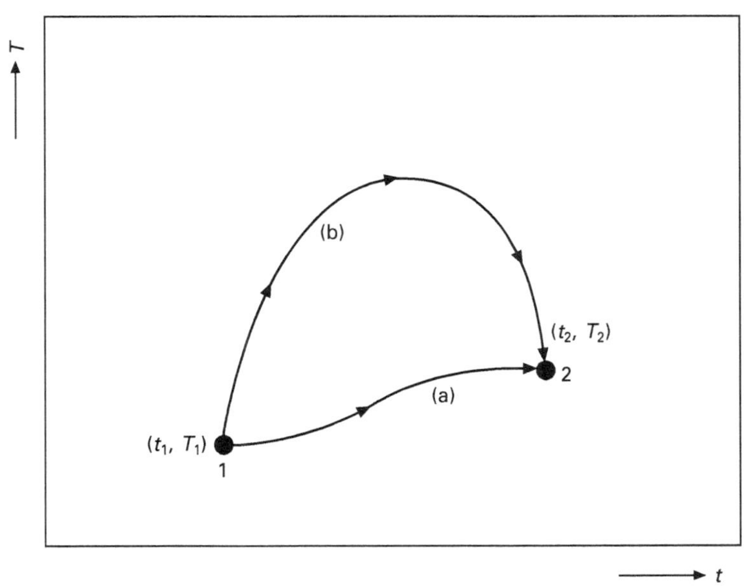

12.13 Temperature (T) – time (t) diagram (Mittemeijer, 1992) (With kind permission from Springer Science+Business Media: *Journal of Materials Science*, Analysis of the kinetics of phase transformations, 27, 1992, 3977, E. J. Mittemeijer, Fig. 2, © 1992 Chapman & Hall).

$$f = 1 - \exp(-\beta^n) \tag{12.4(a)}$$

where f is the degree of transformation, n is the Avrami index, and the state variable β is:

$$\beta = \int_0^t k(T)\mathrm{d}t \qquad k(T) = k_0 \exp\left(\frac{-E}{RT}\right) \tag{12.4(b)}$$

where t is the time (s) and $k(T)$ is the reaction rate constant at any given temperature T. The reaction rate constant follows an Arrhenius type equation and it is temperature dependent. The denotation for k_0 is the pre-exponential factor, Q is the crystallisation activation energy, and R is the gas (Boltzmann) constant. The values of Q for the deposits (Table 12.2) are derived from the Kissinger plots of $\ln(T_P{}^2/\phi)$ versus $1000/(RT_P)$ using the peak temperatures (T_P) of crystallisation process, represented by the major exothermic peak of the DSC curves at different heating rates (ϕ) (Sections 10.2 and 11.1).

Based on JMA kinetic modelling, the degrees of transformation f for the deposits are simulated using Eq. 12.4. Based upon these f values, the DSC curves of the major crystallisation process of the deposits are calculated. Discussion on the corresponding kinetic parameter n values, and comparison between the simulated and experimental results are presented in the following sections.

12.7 Determination of degree of transformation

As used by Sha and Malinov (2009) from their non-isothermal DSC analyses, the fraction of transformation heat released is proportional to the degree of transformation, f, at any given time t (or temperature T):

$$f(t) = \frac{\int_{t_S}^{t} \frac{\partial h}{\partial t}\mathrm{d}t}{\int_{t_S}^{t_E} \frac{\partial h}{\partial t}\mathrm{d}t} = \frac{\int_{t_S}^{t} \Delta H\,\mathrm{d}t}{\int_{t_S}^{t_E} \Delta H\,\mathrm{d}t} \quad \text{or} \quad f(T) = \frac{\int_{T_S}^{T} \frac{\partial h}{\partial T}\mathrm{d}T}{\int_{T_S}^{T_E} \frac{\partial h}{\partial T}\mathrm{d}T} = \frac{\int_{T_S}^{T} \Delta H\,\mathrm{d}T}{\int_{T_S}^{T_E} \Delta H\,\mathrm{d}T} \tag{12.5}$$

where t_S and t_E are the transformation onset and end times respectively, T_S and T_E are the transformation temperatures corresponding to t_S and t_E respectively, and ΔH is the measured transformation heat released at any given time t (or temperature T).

Table 12.2 Crystallisation activation energy (Q)

Deposit	16 wt% P	12 wt% P	8 wt% P	6 wt% P	5 wt% P
Q (kJ mol^{-1})	210±17	225±20	239±5	230±6	322±54

Taking into account the expression in Eq. 12.5, the degree of transformation for the major crystallisation process of the electroless Ni–P deposits can be calculated from the respective DSC experimental curves, and plotted as a function of time, $f(t)$ (or temperature, $f(T)$). Figure 12.14 shows the plots (indicated by solid lines) of the deposits calculated by Eq. 12.5 using the experimental data of the DSC curves in Figs 10.3 and 11.1 (in Sections 10.2 and 11.1, respectively). The plots also enable the path of the processes at different heating rates to be assessed. Comparison between the plots and those simulated from the JMA modelling will be commented on later in Section 12.9.

In Figs 10.3 and 11.1, the presence of noticeable overlapping peaks in some of the DSC curves implies that the major crystallisation process of electroless Ni–P deposits could involve more than one crystallisation stage. These are probably attributed to several different reactions, such as the structural relaxation process, the formation and the decomposition of metastable phases, and the development of different stable phases. These crystallisation stages could be affected by the as-deposited condition of the electroless Ni–P deposits. Here, the deposits are fully (the 12–16 wt% P deposits) or partially (the 5–8 wt% P deposits) amorphous at as-deposited condition. In the 8–16 wt% P deposits heated to 350 °C, there are metastable phase(s), like the Ni_2P and $Ni_{12}P_5$ phases, in the major crystalline transition process. The 5–12 wt% P deposits have two stable phases (Ni and Ni_3P), while the 16 wt% P deposit has only one of those two phases (Ni_3P). Hence, the major crystallisation process of the electroless Ni–P deposits involves the crystalline transition of amorphous phase (all deposits), the transformation of metastable to stable phase(s) (the 8–16 wt% P deposits), and the co-development of fcc nickel phase (the 5–12 wt% P deposits). In view of this, the total reaction heat of major crystallisation processes measured at any given time (or temperature) can be regarded as the sum of the reaction heat (heat flow) of individual reactions.

On the other hand, the DSC curves shown in Figs 10.3 and 11.1 might not represent the complete crystallisation process of the deposits, as there could be some amorphous phases remaining even after the major crystallisation process. Besides, any attempt to study individually the crystallisation stages of major crystallisation process of the deposits would face significant difficulties and might lead to incorrect conclusions, because the exact reactions responsible for each of those individual stages might have been severely overlapped.

12.8 Modelling of crystallisation kinetics

In Table 12.3, the n values for the 12–16 wt% P deposits are within a narrow range of 1.35–1.65, while those for the 8 wt% P one are much larger in the range of 2.4–2.8. In contrast, the n values for the 6 wt% P deposit

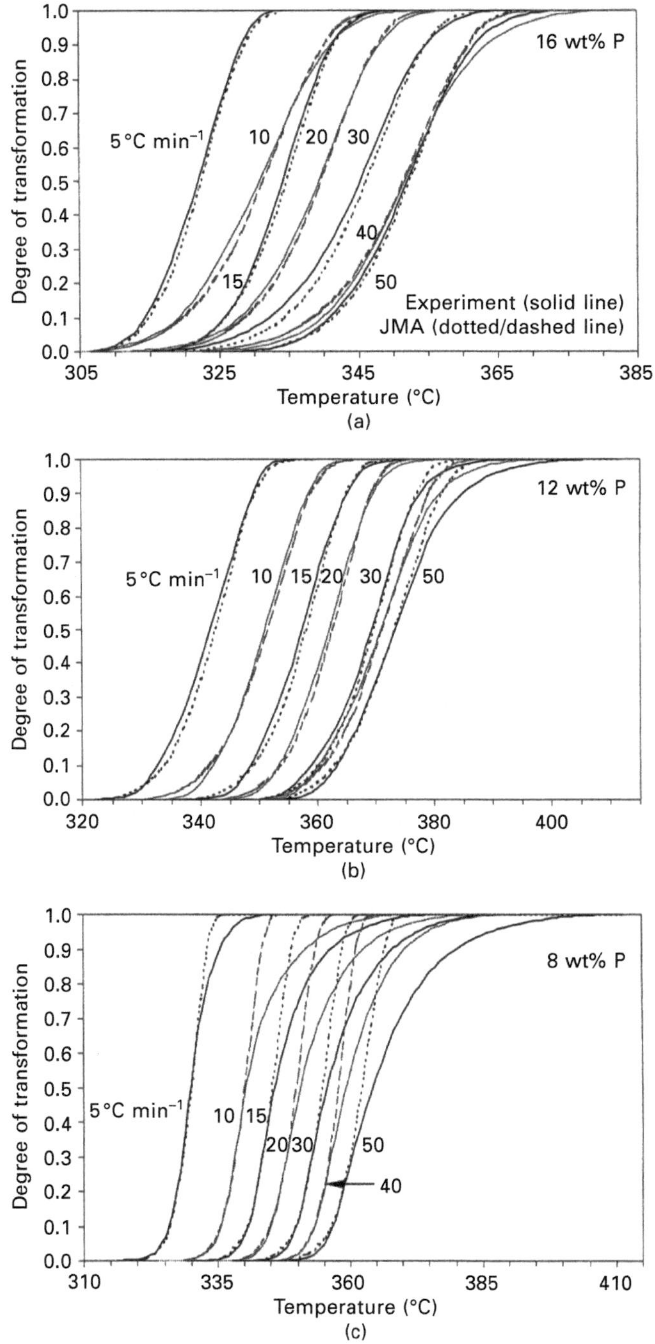

12.14 Degree of transformation, *f*, for the Ni–P deposits (heating at 5–50 °C min^{-1}) calculated from the JMA model and experimental DSC data. (a) 16 wt% P; (b) 12 wt% P; (c) 8 wt% P; (d) 6 wt% P; (e) 5 wt% P; (f) 5 wt% P at 325–400 °C.

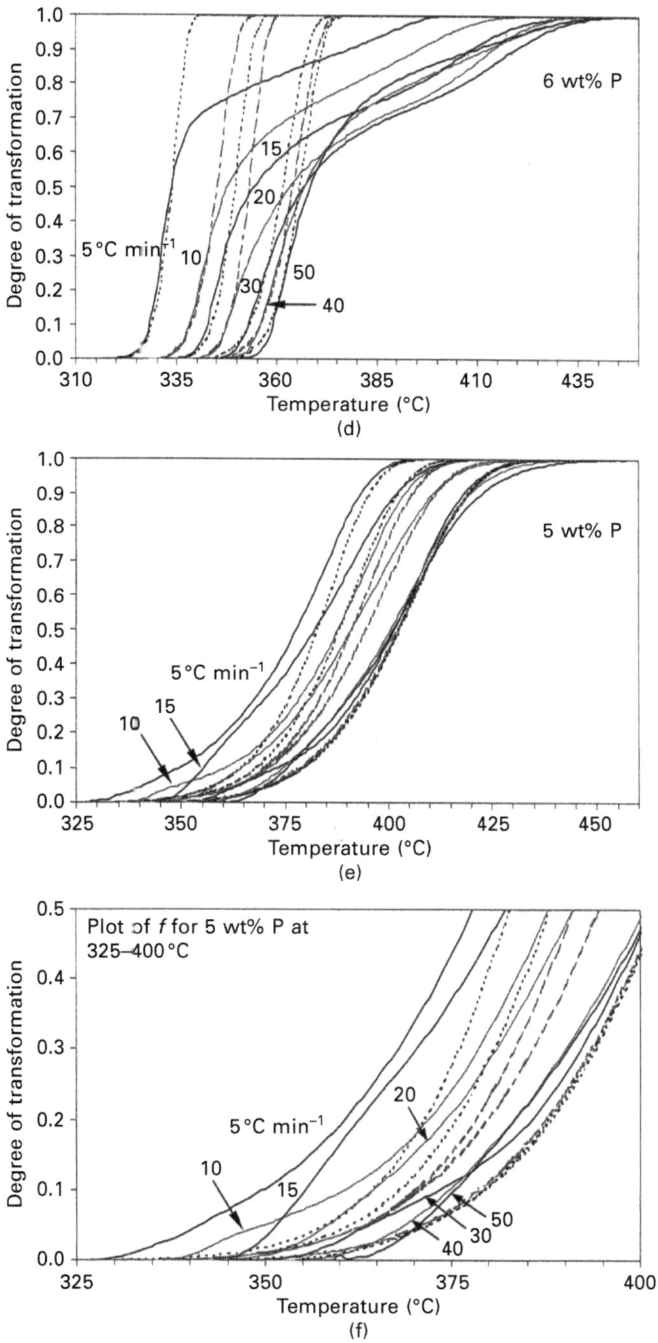

12.14 Continued

Table 12.3 Avrami indexes (n) and pre-exponential factors (k_0) of the deposits corresponding to the JMA simulated degrees of transformation f at different heating rates for the deposits

Deposit	Parameter	5 °C min^{-1}	10 °C min^{-1}	15 °C min^{-1}	20 °C min^{-1}	30 °C min^{-1}	40 °C min^{-1}	50 °C min^{-1}	Average*
16 wt% P	k_0 (× 10^{15})	11.5	6.9	11.5	7.4	6.2	6.6	5.9	6.5±0.7
	n	1.53	1.35	1.55	1.52	1.51	1.50	1.50	1.51±0.01
12 wt% P	k_0 (× 10^{15})	40	37	36.5	38.5	34	41	47	37.8±2.5
	n	1.55	1.50	1.65	1.65	1.60	1.45	1.40	1.57±0.08
8 wt% P	k_0 (× 10^{17})	26.5	21	22	20.4	20	24.5	16	22.4±2.6
	n	2.8	2.7	2.5	2.5	2.5	2.6	2.4	2.6±0.1
6 wt% P	k_0 (× 10^{17})	2.8	2.3	2.6	2.5	1.6	1.9	2.1	–
	n	2.6	2.0	2.2	2.5	1.8	1.8	2.0	–
5 wt% P	k_0 (× 10^{22})	4.5	4.0	9.0	5.7	4.1	6.1	8.2	–
	n	0.90	0.80	0.80	0.70	0.74	0.70	0.70	–

*Average for heating rates 20–50 °C min^{-1} for the 16 wt% P deposit, and for 5–40 °C min^{-1} for the 12 and 8 wt% P deposits.

vary relatively over the heating rates in the range of 1.8–2.6. For the 5 wt% P deposit, the n values do not change significantly over the heating rates. Since the kinetic parameter, n is correlated with the nucleation and growth mechanisms (the geometry of the transformation), the mechanism of transformations for no deposits, except the 6 wt% P deposit, change with the applied heating rates. Hence, the kinetic parameter n for these deposits is independent of time and temperature.

Meanwhile, the variation of pre-exponential factor k_0 values over different heating rates for each of the deposits should also be within a reasonable range, as k_0 is a defined constant in Eq. 12.4(b). The k_0 values for the 16 wt% P deposit, however, are much higher at 5 and 15 °C min^{-1} than at the other heating rates. This could be explained by the fact that this deposit is relatively thin (2.9 μm) in the deposit thickness. Thus, the amount of heat released per unit time during the crystallisation process at lower heating rates could be insufficient for accurate measurement of DSC heat-flow signals. Therefore, the DSC experimental data of the 16 wt% P deposit for the heating rates of 5–15 °C min^{-1} are not reliable, quantitatively for accurate modelling, due to the probable low enthalpy sensitivity problem. In the case of the 12 and 8 wt% P deposits, they do not have the same problem as the 16 wt% P deposit, because the deposits are considerably thicker (51 and 61 μm, respectively) and they could produce sufficient amounts of heat flow for accurate DSC measurements. On the other hand, these deposits could have low temperature sensitivity in their DSC measurements at high heating rate, i.e. 50 °C min^{-1}, due to their larger deposit thicknesses. This phenomenon is evidenced by the corresponding k_0 values that deviate very much from the others at different heating rates (Table 12.3). In the 5–6 wt% P deposits, the k_0 values appear without following a clear tendency in response to the change of heating rates, and thus it is not a constant as defined by Eq. 12.4(b).

Figure 12.14 shows the simulated degree of transformation f (using Eq. 12.4) at different heating rates for the deposits plotted and fitted to those from the experimental data. The overall fittings of f plots for the 16 and 12 wt% P deposits (Figs 12.14(a) and (b)) are considerably satisfactory, though not all of them are completely well fitted. These discrepancies are likely caused by secondary reactions in the major crystallisation process, since the process could involve one or more different reactions as mentioned in Section 12.7. The problem could also be because the experimental DSC curves of the major crystallisation process are not always reproducible at the region between the crystallisation onset (T_S) and end (T_E) temperatures (Fig. 12.15). This is illustrated in the figure by those five DSC curves that have been obtained at a same heating condition, but display apparent peak overlapping at different tendency and temperatures, probably due to the inhomogeneity of deposit compositions and the different thermal contact between specimen and specimen pan. A similar phenomenon is expected

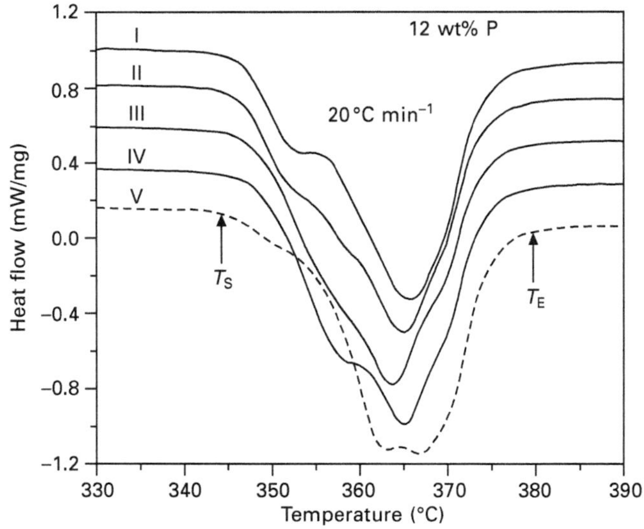

12.15 Experimental DSC curves (I–V) of the 12 wt% P deposit obtained under the same heating condition at 20 °C min^{-1}. For the sake of clarity, the calorimetric curves are shifted arbitrarily on the vertical scale.

in the rest of the DSC results obtained at different heating rates in the both deposits. Nonetheless, the unsatisfactory fits represent only a small portion of the degrees of transformation and they are not considerably significant.

On the other hand, the fittings between the simulated and experimental f plots for the 8 wt% P deposit are not satisfactory as shown in Fig. 12.14(c). Those unsatisfactory results are likely caused by the partially amorphous nature of the deposit at as-deposited condition. From the XRD analysis discussed in Chapter 11 (Fig. 11.3), the 8 wt% P deposit is a mixture of amorphous and nanocrystalline nickel phases at the as-deposited condition. Since different phases could be correlated with different transformation behaviour and mechanism as well as involve different transformation activation energy, the major crystallisation process of the 8 wt% P deposit might not simply correspond to simple crystallisation kinetics as expressed by the JMA model. For a hypoeutectic electroless nickel deposit (with phosphorus level < 11 wt%), the crystallisation process usually begins with the formation of nickel crystallites, followed by the major crystallisation reaction of the remaining amorphous phase into equilibrium phases. Therefore, it is likely that a momentous secondary nucleation process (Málek, 2000) has taken place during the growth of nickel crystal, causing the nucleation and growth processes being overlapped. In the light of this, the nucleation condition as required by the JMA model could have been discounted. At the same time, the crystal-growth induced process could also be responsible for the fast increase of initial crystallisation rate, as indicated by the sharp-start DSC curves of the deposit (Fig. 11.1(a) in Chapter 11). Meanwhile, it is noticeable from Fig. 12.14(c) that the fittings have a tendency of worsening with the increasing heating rates, when the increase of fraction transformed (experimental) over temperatures slows down. The increase of heating rates shortens the time available for crystallisation reactions, and the reactions will end up at much higher temperatures as indicated by the experimental plots. Nevertheless, the reaction(s) that contribute to the overall unsatisfactory fittings have not been exactly known. The result of fittings indicates that the kinetics of the major crystallisation process of the deposit is not fully explicable in the JMA theoretical frame. Regardless of this fact, interpretation of the crystallisation kinetics for the 8 wt% P deposit is further attempted in Section 12.9 using the theory for the purpose of discussion.

Similar to that for the 8 wt% P deposit, the fittings between the simulated and experimental *f* plots for the 6 and 5 wt% P deposits are not satisfactory as indicated by Fig. 12.14(d) and (e), respectively. The sources that have contributed to the unsatisfactory fitting are the same as those being explained for the 8 wt% P deposit. Besides, the presence of significant secondary peaks that overlap with the major exothermic peaks of the 5–6 wt% P deposits (Section 11.1 in Chapter 11) makes it impossible to obtain feasible fittings between the simulated and experimental *f* plots. Accordingly, the interpretation of the crystallisation kinetics for these two deposits using the JMA model is not feasible.

Following the suggestion that the Avrami index n is independent of time and temperature, a second set of *f* values is simulated (using Eq. 12.4) for the 8–16 wt% P deposits using a fixed n value regardless of different heating rates (the last column in Table 12.3). Similarly, a constant pre-exponential factor k_0 value is applied for each of these deposits as conditioned by Eq. 12.4. The discussion on the applied n values in relation to the corresponding nucleation and growth mechanisms is given in the following section. In addition, the calculated DSC curves from the simulated *f* values are plotted for comparison with the experimental data. For the 5–6 wt% P deposits, no such calculations are attempted due to the kinetics of major crystallisation process of the deposits being inexplicable in the JMA theoretical frame (Fig. 12.14).

12.9 Comparison between simulations and experimental data

The fixed values of k_0 and n being used in the JMA modelling for the 8–16 wt% P deposits are averaged from those corresponding to appropriate heating rates (Table 12.3). In the 16 wt% P deposit, the values at 5–15 °C min^{-1} are omitted from the calculation due to the low enthalpy sensitivity of the related DSC measurements. For the 8–12 wt% P deposits, those values corresponding to the high heating rate of 50 °C min^{-1} are omitted due to the inadequate temperature sensitivity in measurements. From such averages, the values of n for the 16, 12 and 8 wt% P deposits are 1.51 ± 0.01, 1.57 ± 0.08 and 2.6 ± 0.1, respectively (the last column in Table 12.3). The value of n does not increase with increasing phosphorus content. In addition, the n value of the 12 wt% P deposit is lower than similar deposits under isothermal conditions, 2.73 for 11.5 wt% P and 2.3 for 11.9 wt% P. Under isothermal conditions, the n value sometimes increases over the crystallisation process. Different heating conditions could produce different crystallisation behaviour that is characterised by the nucleation and growth processes of electroless Ni–P deposits.

According to the classical phase-transformation theory, $n = 1$ indicates one-dimensional growth with grain boundary nucleation after saturation, and $n = 2$ implies two-dimensional growth with grain edge nucleation after saturation. Therefore, the major crystallisation processes of the 12–16 wt% P deposits are one-dimensional growth with decreasing nucleation rate, as implied by their n (≈1.5) values. The process of the 8 wt% P deposit is two-dimensional growth with decreasing nucleation rate. However, the kinetic interpretation for the 8 wt% P deposit by means of the n value is sceptical, due to the poor fittings between the simulated and experimental f plots shown previously in Fig. 12.14(c).

In Figs 12.16–12.18, the degrees of transformation using the averaged k_0 and n values are plotted. The figures also show the calculated (using the simulated f values) and experimental DSC curves derived from those f plots. In the 16 wt% P deposit, considerably good fittings are found between the calculated and experimental DSC curves at different heating rates (Fig. 12.16). The fitting discrepancy near the transformation end in the DSC curves increases with the increase of heating rates. A possible explanation is that the sensitivity of the DSC reduces at higher temperature and higher heating rates. Multiple reactions may also be the cause, for example, the decomposition of intermediate phases (Ni_2P, $Ni_{12}P_5$) into stable phase (Ni_3P). Meanwhile, Fig. 12.17 shows that the fitting for the DSC curves of the 12 wt% P deposit is almost as good as the 16 wt% P deposit, apart from the fact that there are secondary peaks in the experimental curves. The peaks could correspond to the formation of a different stable phase (i.e. Ni) during the major crystallisation process, or are simply caused by the inhomogeneity of deposit compositions and measured heat flow (Fig. 12.15). Although these secondary peaks have not been properly fitted to the JMA simulated DSC curves, they are not considerably substantial due to their small intensity. In Fig. 12.18, the simulated f and DSC curves of the 8 wt% P deposit are presented to show the state of the unsatisfactory fittings. Overall, the simulated DSC curve of the 8 wt% P deposit for each different heating

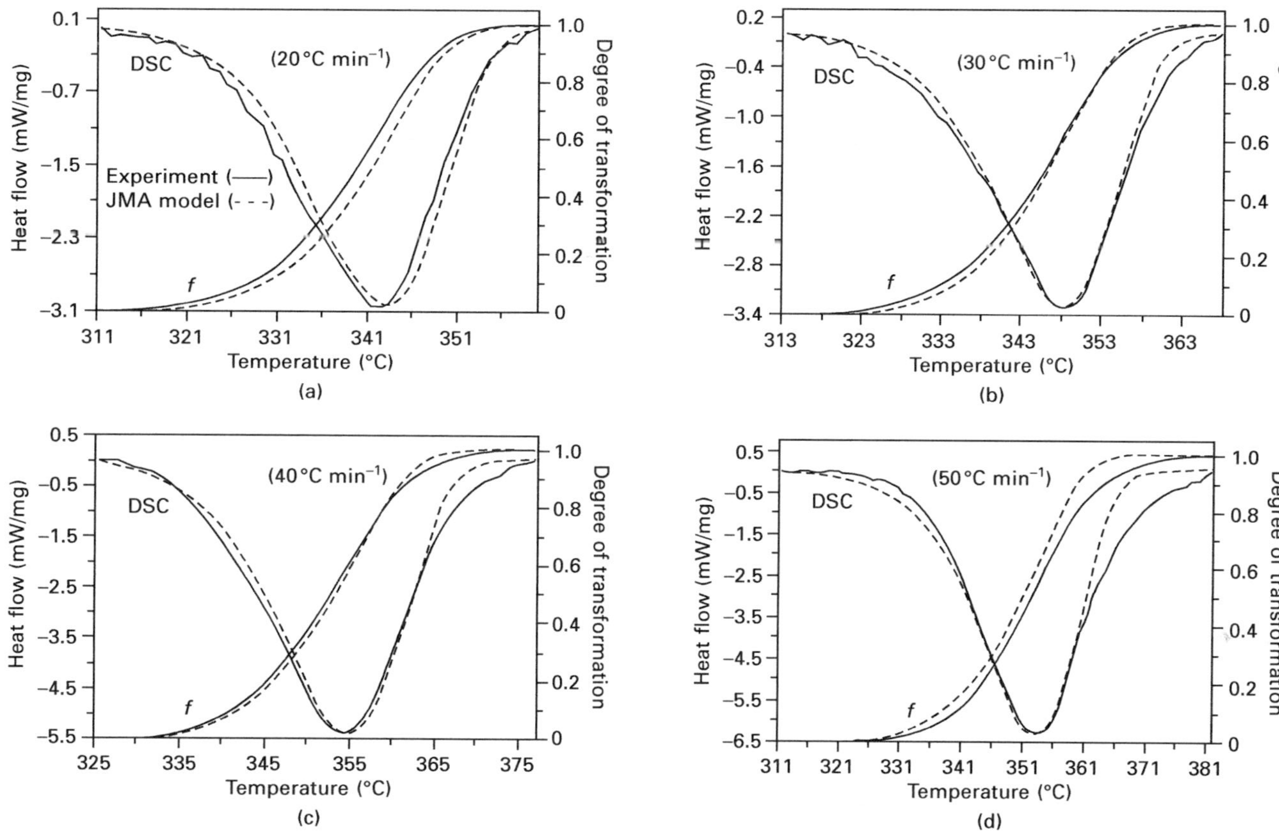

12.16 Comparisons between JMA simulated and experimental results for the degree of transformation *f* and DSC curves of the 16 wt% P deposit, at heating rates of (a) 20 °C, (b) 30 °C, (c) 40 °C and (d) 50 °C min^{-1}.

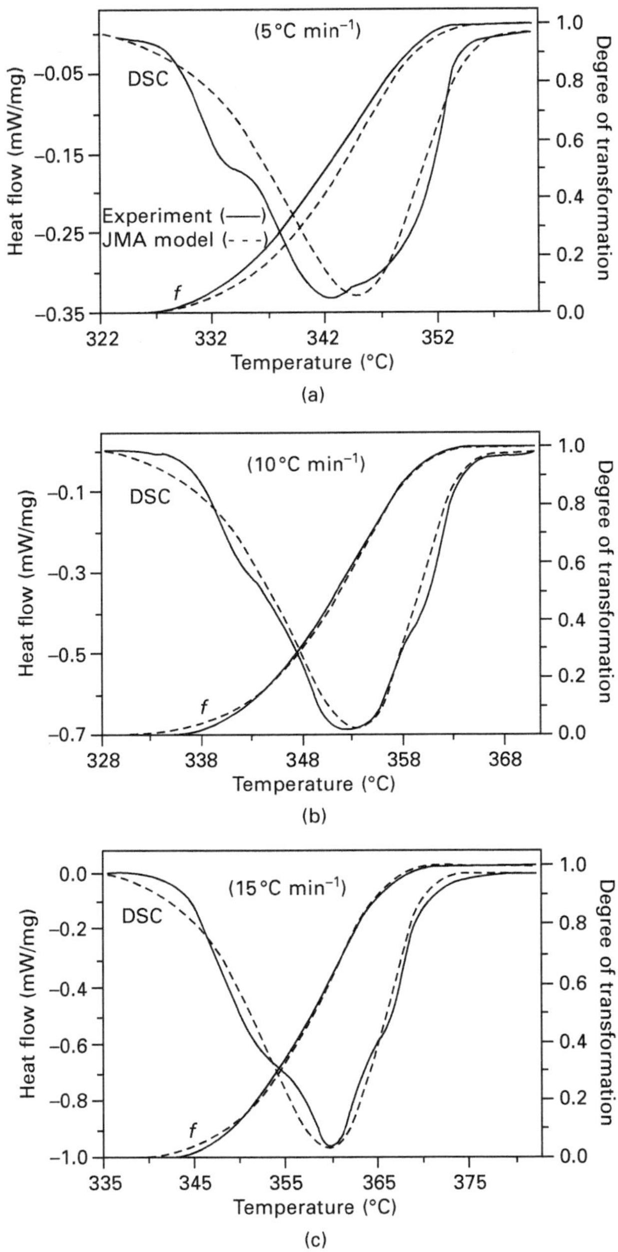

12.17 Comparisons between JMA simulated and experimental results for the degree of transformation *f* and DSC curves of the 12 wt% P deposit, at heating rates of (a) 5 °C, (b) 10 °C, (c) 15 °C, (d) 20 °C, (e) 30 °C and (f) 40 °C min^{-1}.

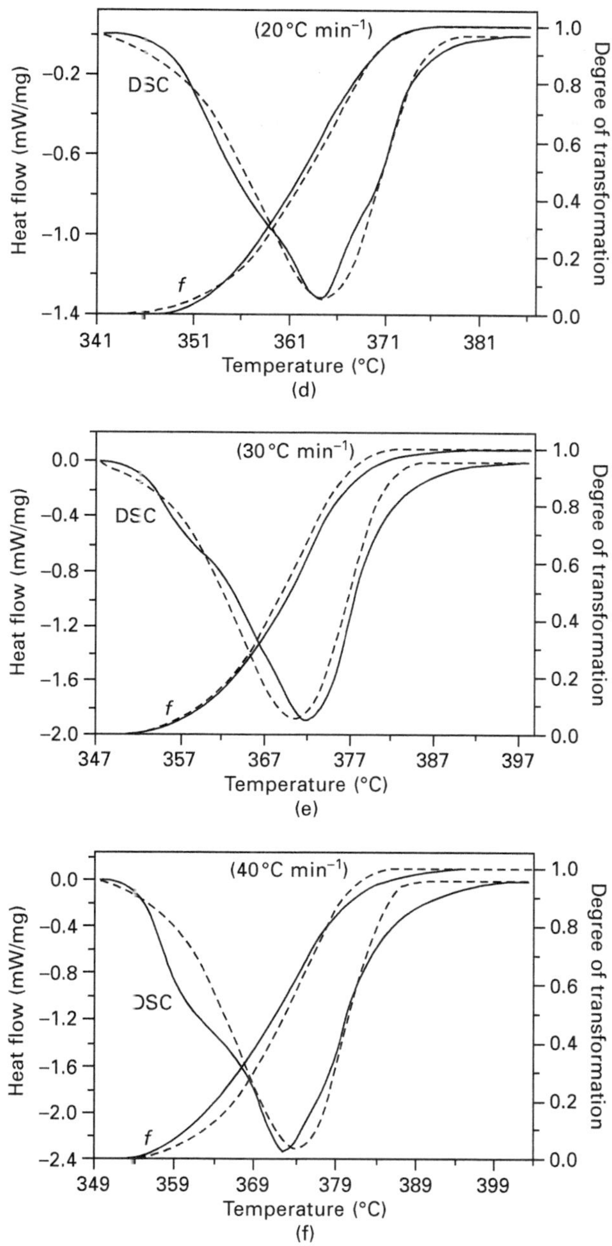

12.17 Continued

rate could not be fitted to those from the experiment, even at the very early stage (transformation start in the DSC curve) of the major crystallisation process. Besides, the fitting discrepancy toward the transformation end in

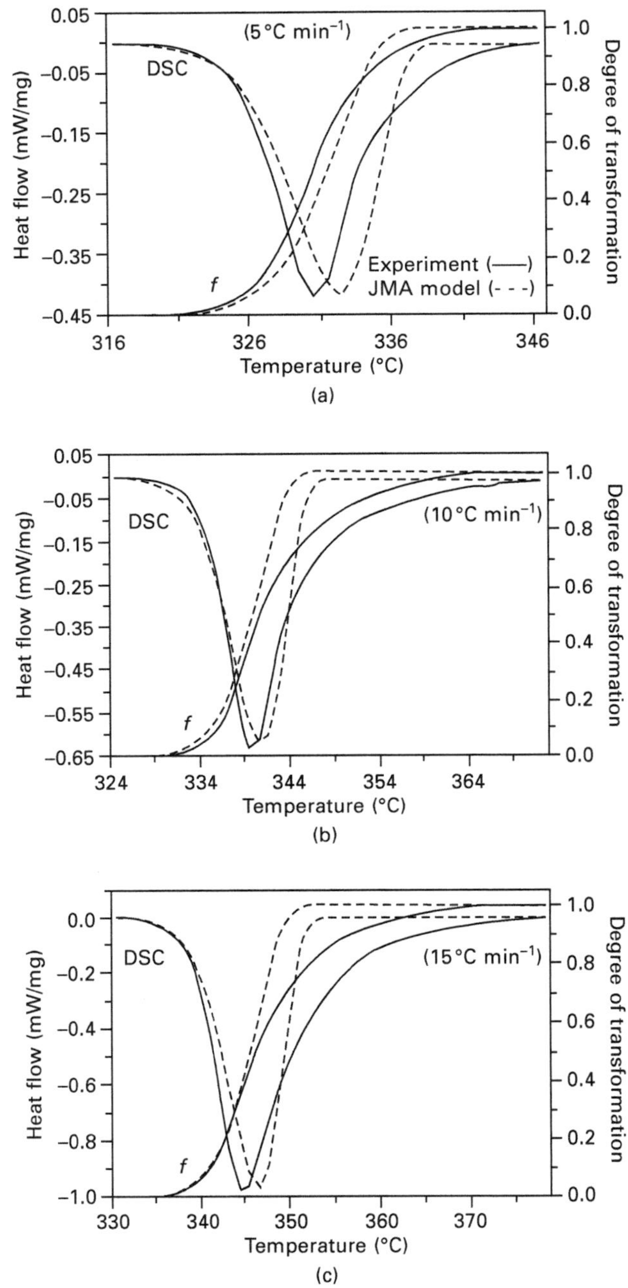

12.18 Comparisons between JMA simulated and experimental results for the degree of transformation *f* and DSC curves of the 8 wt% P deposit, at heating rates of (a) 5 °C, (b) 10 °C, (c) 15 °C, (d) 20 °C, (e) 30 °C and (f) 40 °C min^{-1}.

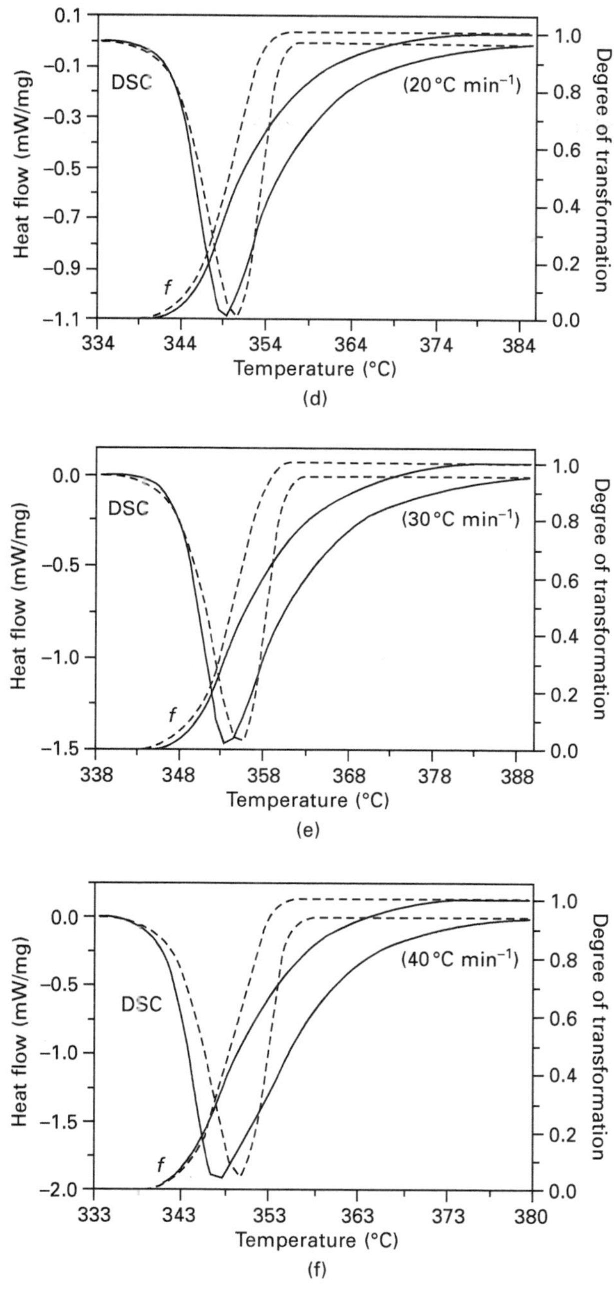

12.18 Continued

the DSC curves is too severe for any feasible good fitting, and probably it is mainly due to the presence of nanocrystalline nickel phase in the deposit at as-deposited condition.

Following the above comparisons between the simulated and experimental DSC curves, JMA modelling could provide reasonable estimations on the crystallisation kinetics of the amorphous electroless Ni–P deposits (12–16 wt% P). Because not all the fittings are good for the entire transformation due to the existence of DSC secondary peak(s) as well as the apparent pushing higher of crystallisation end temperatures (T_E), the JMA estimations can only represent the majority of major crystallisation processes of the deposits. The major crystallisation processes of the electroless Ni–P deposits do involve multiple reactions as mentioned before, so fitting by the JMA models can be either for a single dominant crystallisation reaction, or the sum of multiple reactions of the processes. At present, the kinetic interpretations for the 12–16 wt% P deposits are based on the Avrami index n calculated from the JMA modelling. Therefore, further study to identify any reactions that coexist in the major crystallisation processes of the deposits could be useful in improving the accuracy of the kinetic interpretations. At the same time, the speculation of the involvement of substantial growth-induced and secondary nucleation reaction in the major crystallisation process of the 8 wt% P deposit is perhaps well founded. For this purpose, investigation on the deposits at different stages of crystallisation process is of use, though it could be a difficult task.

12.10 Conclusions

The Johnson–Mehl–Avrami equation can be applied to the crystallisation process in a variety of amorphous solids under isothermal conditions to aid in the understanding of these particular alloy systems. The kinetic parameters derived can be used to interpret the mechanism of the process. The isothermal crystallisation process results in nanometre size grains, therefore largely separating the processes of crystallisation and grain growth, processes that cannot be separated effectively during continuous heating crystallisation. Therefore, the kinetic parameters derived from isothermal experiments are more characteristic of the crystallisation process alone. The main physical insight achieved is in the Avrami parameters derived, which relate to the crystallisation transformation mechanisms.

The JMA modelling on the non-isothermal crystallisation kinetics has shown that the JMA models could make reasonable estimates on the major crystallisation processes of the amorphous electroless Ni–P deposits, namely 12–16 wt% P deposits. However, application of the JMA model on partial crystalline 5–8 wt% P deposits does not give a satisfactory result. The major crystallisation processes of the 12–16 wt% P deposits are one-dimensional growth with decreasing nucleation rate, as implied by their Avrami indexes ($n \approx 1.5$). These n values are constant over the time and temperatures, and thus, independent of the heating rates. Simulations of degree of transformation made by the JMA models can only provide approximations for the kinetic interpretations of the 12–16 wt% P deposits, because not all the fittings with the experimental data are entirely good due to the presence of multiple reactions. Fitting of the DSC curves for the 12–16 wt% P deposits by the JMA models can be regarded as either for a single dominant crystallisation reaction, or the sum of individual reactions of the major crystallisation process.

12.11 References

Málek J (2000), 'Kinetic analysis of crystallization processes in amorphous materials', *Thermochim Acta*, **355**, 239–53.

Mittemeijer E J (1992), 'Analysis of the kinetics of phase transformations', *J Mater Sci*, **27**, 3977–87.

Révész A, Lendvai J, Lóránth J, Pádár J and Bakonyi I (2001a), 'Nanocrystallization studies of an electroless plated Ni–P amorphous alloy', *J Electrochem Soc*, **148**, C715–20.

Révész A, Lendvai J, Cziráki Á, Liebermann H H and Bakonyi I (2001b), 'Formation of nanocrystalline phases during thermal decomposition of amorphous Ni–P alloys by isothermal annealing', *J Nanosci Nanotechnol*, **1**, 191–200.

Révész A, Lendvai J, Varga L K and Bakonyi I (2001c), 'Nanocrystallization studies of a melt-quenched $Fe_{40}Ni_{40}P_{14}B_6$ amorphous alloy', *Mater Sci Forum*, **360–362**, 543–8.

Sha W and Guo Z (2009), *Maraging Steels: Modelling of microstructure, properties and applications*, Woodhead Publishing, Cambridge.

Sha W and Malinov S (2009), *Titanium Alloys: Modelling of microstructure, properties and applications*, Woodhead Publishing, Cambridge.

13
Artificial neural network (ANN) modelling of crystallisation temperatures of nickel–phosphorus deposits

Abstract: A computer model based on a multilayer feedforward artificial neural network is used for the simulation and prediction of temperatures of major crystallisation reaction of the Ni–P based amorphous alloys, as functions of alloy composition, the processing method and heating rate of the heat-treatment process. The model consists of five individual neural networks and is trained using data in Chapters 10 and 11 and other literature.

Key words: neural network, modelling, differential scanning calorimetry curves, crystallisation temperature, nickel–phosphorus.

13.1 Artificial neural network (ANN) modelling

In this chapter, the influence of alloy composition, heating rate and processing parameters on the crystallisation temperatures of the Ni–P based amorphous alloys is studied using a statistical model developed from the artificial neural network (ANN) computing technique. The ANN technique has become one of the very powerful modelling tools. It is based on a non-linear statistical approach, and is particularly suitable for simulation of non-linear and complex correlations that are hard to describe by physical models (Sha and Malinov, 2009). However, a reliable database is needed in developing the model to ensure good performance. In this chapter, the ANN model is developed using the database collected from the published literature as well as data in Chapters 10 and 11.

13.1.1 Basic principles and model description

ANNs were originally developed in the attempt of simulating the biological nervous systems in learning information from the outside environment. They consist of simple synchronous processing elements, which are inspired by biological neural networks. The fundamental unit of ANNs is the neuron (Fig. 13.1). Neurons are arranged in layers and are categorised as the input (N_I), hidden (N_H) and output (N_O) neurons depending on in which layer they are located (Fig. 13.2). Neurons in each layer are linked to each of those in the layers immediately next to it through the connections known as synapses.

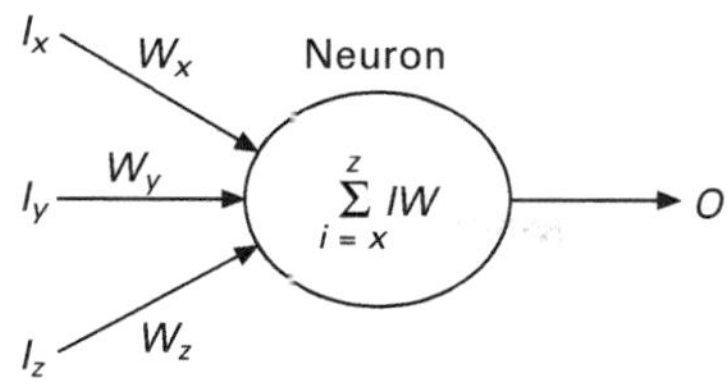

13.1 A single neuron; the basic unit of the ANN.

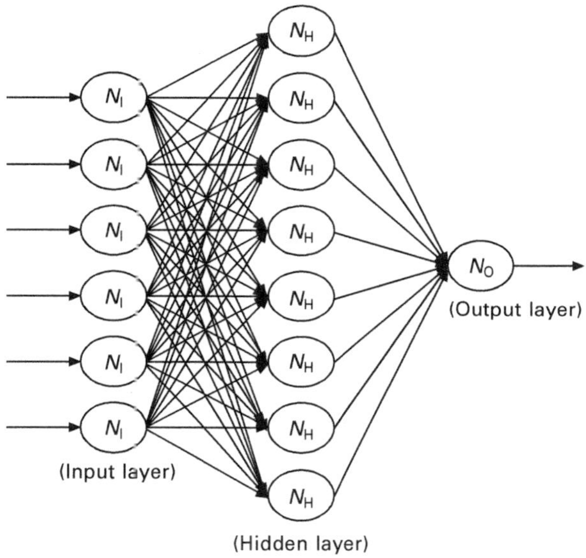

13.2 Schematic structure of the feedforward hierarchical ANN used in the model for the simulation and prediction of crystallisation peak temperature (T_P).

Each synapse is characterised by a weight factor (*W*), which can be adjusted to target the desired output signal (*O*) (Fig. 13.1). The weighted sum of the input signal that a neuron receives is called activation, and the output signal of the neuron is determined by it. As a non-linear statistical networking technique, ANN modelling involves linking up the input and output data through a particular set of non-linear basis functions (Sha and Malinov, 2009). The procedure of ANN modelling is usually the following:

- collecting of data,
- analysis and pre-processing of database,
- training of neural networks,
- testing of the trained neural networks, and
- using the tested neural networks for simulation and prediction (Sha and Malinov, 2009).

These steps are used in developing the model presented in this chapter.

MatLab package (Neural Network Toolbox from The Math Works, Inc.) is widely used to create the model of multilayer feedforward (Section 13.1.2) ANNs in modelling work. The model in this chapter consists of five separate individual neural networks. Each has a three-layer feedforward hierarchical structure, and the flow of information is directed from the input to the output layer. This network shown in Fig. 13.2 consists of six inputs and a single output parameter. The rest of the individual networks in the model are for the simulation and prediction of crystallisation onset (T_{S1}, T_{S2}) and end (T_{E1}, T_{E2}) temperatures (Fig. 13.3). There are eight inputs and a single output parameter in each of these individual neural networks. The reason for using different numbers of input parameters in the neural network for crystallisation peak temperature will be given in Section 13.2. For all neural networks, the numbers of neurons in the input and output layers of the networks are the same as the numbers of the input and output parameters respectively (Fig. 13.2). However, the numbers of neurons in the hidden layer of the neural networks can be different (Table 13.1). These numbers are chosen according to the performance of the neural networks in the training and testing sessions. Discussion on the training and testing of neural networks will be provided later in Section 13.1.5.

Figure 13.4 illustrates the ANN model with the details of its input and output parameters. The reason for having two onset and end crystallisation temperatures (Fig. 13.3) in the output will be discussed in Section 13.1.4.

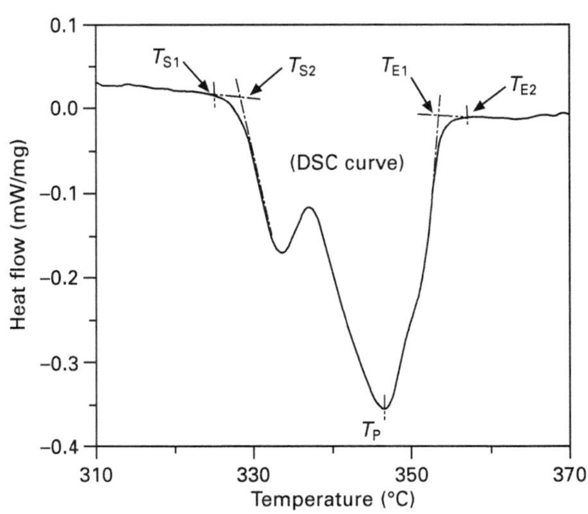

13.3 The definition of crystallisation temperatures (onsets, peak and ends) from the DSC curve.

Table 13.1 Number of neurons in the hidden layer (L_H), and the resulting correlation coefficient (r) values from the training and testing of individual ANNs

Neural network	Neurons in the L_H	$r_{training}$	$r_{testing}$	r_{whole}
T_{S1}	5	0.910	0.914	0.904
T_{S2}	4	0.926	0.906	0.913
T_P*	8	0.958	0.938	0.951
T_{E1}	5	0.966	0.900	0.941
T_{E2}	3	0.959	0.912	0.944

*Only four alloying elements are included for input, P, B, Cu and Fe. For the other four neural networks, two additional elements, Pd and Zr, are included. Ni makes up the balance of the composition.

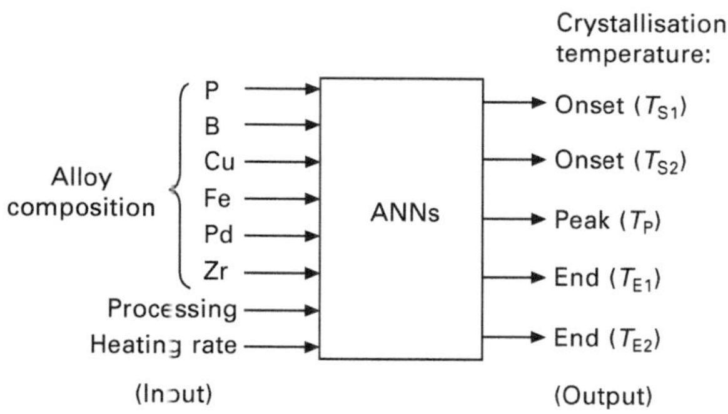

13.4 Schematic diagram of the ANN model for prediction of crystallisation temperatures of Ni–P based amorphous alloys. Only four alloying elements, P, B, Cu and Fe, are included for input for the peak crystallisation–temperature network.

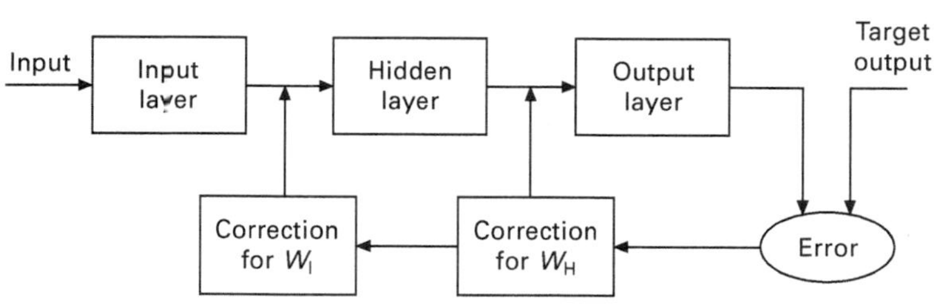

13.5 Multilayer feedforward ANN.

13.1.2 Multilayer feedforward artificial neural networks

A multilayer feedforward ANN is trained until it can approximate a function that can classify the input variables in an appropriate manner as defined by the user (Fig. 13.5).

The computation of a multilayer feedforward network is performed based on a combination of two different modes:

1. mapping mode – information flows forward through the network from input to output layers;
2. learning mode – information flows in turn between forward and backward.

During the computational process, the mapping mode uses the current weight factors (W) to work out the values of the output variables. The result is then compared with the target output variables and the comparison generates the error. Learning mode then is carried out in a manner of backpropagation to adjust the values of W in each different layer. The mapping starts again using the updated values of W and then learning again and so on until an optimum stopping condition (achieved in an automated manner through incorporating Bayesian regularisation (Section 13.1.3)) is reached.

13.1.3 ANN transfer functions, training algorithm and Bayesian regularisation

Consider an elementary neuron with three inputs; with each of them is weighted an appropriate weight factor (W). Thus, the sum of the weighted inputs (ΣIW) forms the input (n) to the transfer function f. Through the applied f, the output signal (O) of the neuron is generated (Fig. 13.6).

The tangent sigmoid (tan-sigmoid or *tansig*) transfer function (Fig. 13.7) is often used in multilayer artificial neural networks, in part, because it is

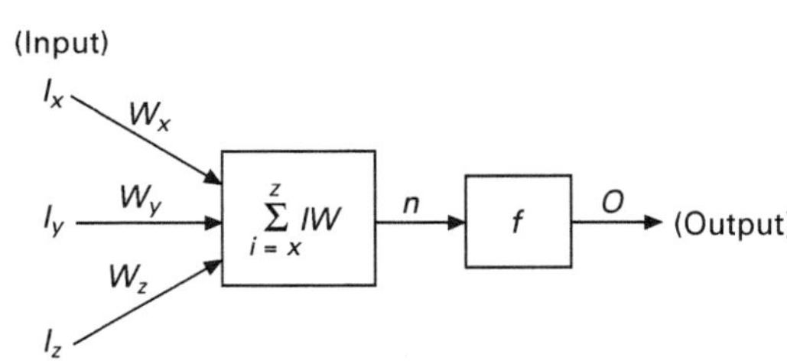

13.6 An elementary neuron.

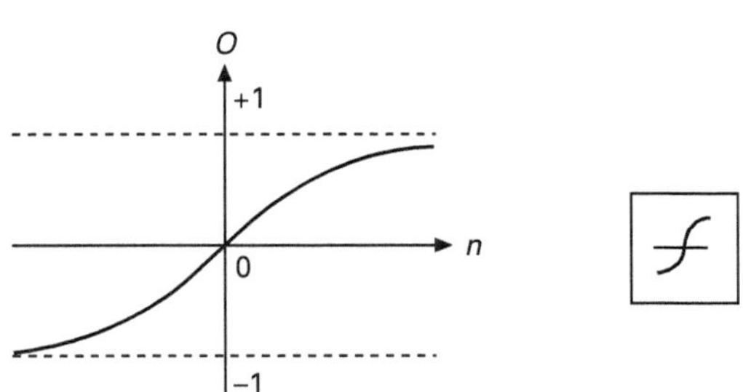

13.7 Tangent sigmoid transfer function.

differentiable. The *tansig* function generates outputs (O) between –1 and 1 as the function's input goes from negative to positive infinity:

$$O = tansig(n) = \frac{2}{1 + e^{-2n}} - 1 \qquad 13.1$$

The linear (*purelin*) transfer function (Fig. 13.8) calculates the neuron output signal O by simply returning the value passed to it. This function is usually used in the output layers:

$$O = purelin(n) = purelin(\Sigma IW) = \Sigma IW \qquad 13.2$$

The Levenberg–Marquardt training algorithm and Bayesian regularisation are two of several different training algorithms. Different algorithms would have different computation and storage requirements, and no single algorithm is best suited to all situations. Thus, the way to find out which training algorithm best suits is through model testing.

By far the Levenberg–Marquardt algorithm is the fastest training algorithm for networks of moderate size, and it has a memory reduction feature for use when the training set is large. However, one problem that can occur when training neural networks is that the network can overfit on the training set and does not generalise well to new data outside the training set. The feature of overfitting is when the error on the training set is driven to very small value, but when new data is presented to the network the error is large. This can be prevented by training with Bayesian regularisation as practised in present neural networks.

Bayesian regularisation is the modified version of Levenberg–Marquardt algorithm. It can produce networks that generalise well. By definition, regularisation is referred to the method for improving generalisation such as that of the Levenberg–Marquardt algorithm. This involves modifying the performance function, which is normally chosen to be the sum of squares of the network errors on the training set. In the framework of Bayesian regularisation, the performance function is improved by adding a term that consists of the mean of the sum of squares of the network weights. This will

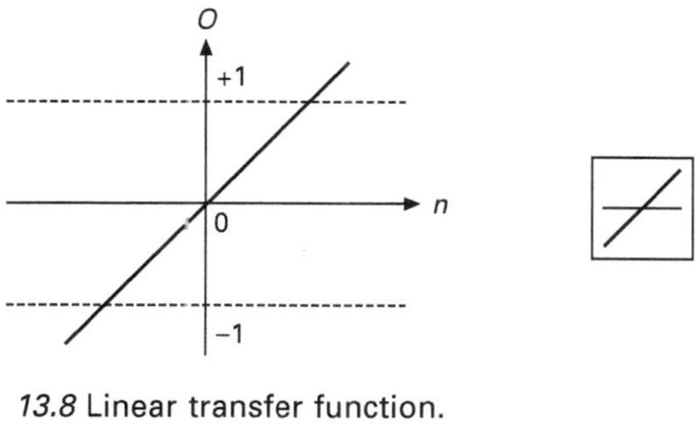

13.8 Linear transfer function.

cause the network to have smaller weights and make it less likely to overfit. Besides, the Bayesian regularisation could determine the optimal regularisation parameters (the number of weights that is effectively in use by the network) in an automated manner by assuming the weights to be random variables with specified distributions. Accordingly, this has the reduced difficulty of determining the optimum network architecture (i.e. numbers of input, hidden and output neurons used).

13.1.4 Database collection and analysis

The datasets used for the construction of ANN model in the modelling work are collected from the available data in published literature (Hentschel *et al.*, 2000; Jiang and Wu, 2001; Martyak and Drake, 2000; Tachev *et al.*, 2000; 2001; Taheri *et al.*, 2001; Ten and Chang, 2001; Wang and Lu, 2000; Zhang and Yi, 2000). Additional data in Chapters 10 and 11 are used, too. Based on the schematic model illustrated in Fig. 13.4, the datasets required are the chemical compositions (P, B, Cu, Fe, Pd and Zr) and processing methods (electroless, electroplated and melt-spun or rapidly quenched) of the Ni–P based amorphous alloys, heating rate (°C min^{-1}) of the continuous heat treatment processes, and the corresponding thermally induced crystallisation temperatures (T). The total numbers of input–output data pairs used in the neural networks are shown in Table 13.2. The majority (about 70%) of the data are for binary Ni–P alloys and the remaining are for ternary and quaternary systems. Many of the datasets for the crystallisation temperatures of alloys are read from the digitised differential scanning calorimetric (DSC) curves. The crystallisation onset and end temperatures can be read either directly from the DSC curve (T_{S1} and T_{E2}) or from the extrapolated lines (T_{S2} and T_{E1}) as shown in Fig. 13.3. The majority of published data regarding the start and end crystallisation temperatures are determined using the latter approach. It

Table 13.2 Alloying elements, and their number and composition range in brackets used in training the ANNs

Neural network	T_{S1}	T_{S2}	T_P	T_{E1}	T_{E2}
P	128 (0–16)	151 (0–16)	129 (0–16)	106 (0–16)	113 (0–16)
B	8 (0–7.1)	18 (0–7.1)	21 (0–7.1)	7 (0–7.1)	7 (0–7.1)
Cu	14 (0–50.3)	25 (0–50.3)	11 (0–50.3)	12 (0–28.5)	12 (0–28.5)
Fe	7 (0–88)	17 (0–88)	24 (0–88)	6 (0–84)	6 (0–84)
Pd	5 (0–61)	13 (0–61)	–	5 (0–61)	5 (0–61)
Zr	5 (0–45.4)	5 (0–45.4)	–	4 (0–23.3)	5 (0–45.4)
Total data pairs	131	164	142	109	116

Balance is Ni. The range of its concentration used in the model is 3.9–98.1 wt%.

should also be noted that the DSC curves (Fig. 13.3) represent only the thermal signal of the transformation of the amorphous to the crystalline phases. Thus, the crystallisation end temperatures (T_{E1} and T_{E2}) might not correspond to the end of the complete phase transformation process of the alloy systems. In addition, the crystallisation peak temperature (T_P) represents only that of the major peak in case of overlapping peaks in DSC curves.

Figure 13.9 and Table 13.2 show the numbers and distributions of the datasets in the training and testing of neural networks. These database analyses highlight the intervals where the neural networks can achieve lower and higher performances. For example, the networks are expected to have the best performance for a binary system, electroless processing method, heating rates in the range of 1–10 °C min^{-1}, and crystallisation temperatures ranged at 301–400 °C.

13.1.5 Training and testing of neural networks

The training of ANNs involves presenting to the networks a sequence of input–output datasets, and targeting the resulting outputs by adjusting the weight factors until they appear as close as possible to those in the training sets. The architecture, transfer function, training algorithms and parameters of the neural networks have to be carefully chosen to ensure the performance of the networks. As shown in Table 13.1, the numbers of neurons in the hidden layer of the neural networks are different. They are determined by conducting several sessions of training and testing of the neural networks with a different number of neurons in the hidden layer at each session. The numbers of neurons that give the best combination of correlation coefficient (r) values in those sessions are chosen. The tangent sigmoid (*tansig*) transfer function is assigned to the hidden layers, whereas a linear (*purelin*) transfer function is used in the output layers. The Levenberg–Marquardt training method with automated regularisation based on Bayesian regularisation is chosen. These are based on trials and findings in the literature on similar types of neural networks (Sha and Malinov, 2009). Descriptions of the ANN transfer functions, Levenberg–Marquardt training method and Bayesian regularisation are given in Section 13.1.3.

In the training and testing sessions, the computer programs of the neural networks are designed to randomly distribute the datasets into two groups: one group with 2/3 of the datasets for training, and the other group with the remaining datasets for testing. In addition, every new session of training and testing starts with new randomly distributed datasets, but the datasets in those groups are still confined to the 2:1 ratio. The performance of neural networks in training and testing is discussed in Section 13.2. The next section, Section 13.1.6, gives an example of the programming used for such training and testing of neural networks (Fig. 13.10).

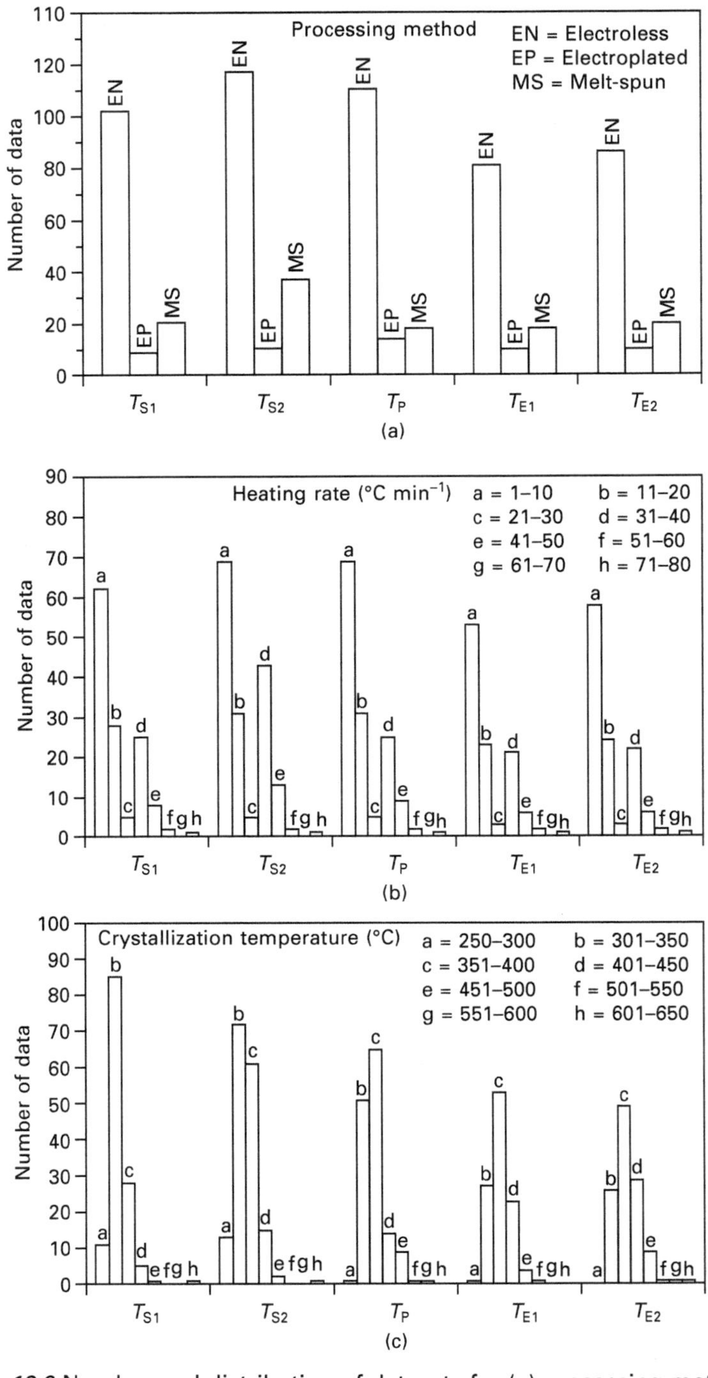

13.9 Number and distribution of datasets for (a) processing method, (b) heating rate and (c) crystallisation temperatures in the training of ANN.

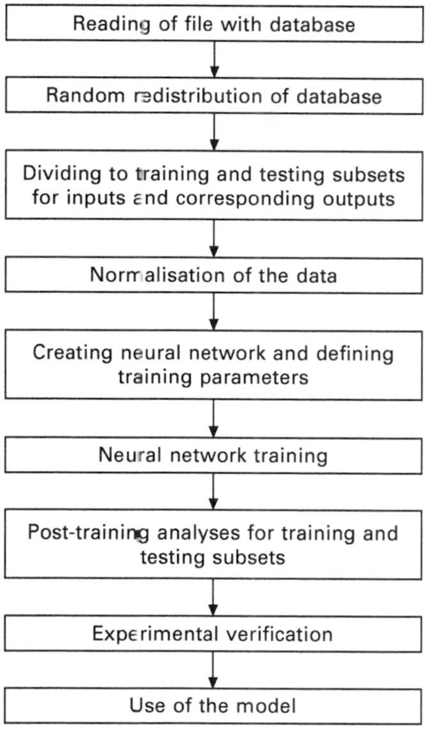

13.10 Flow chart of MatLab programming algorithm for the creation of a neural network model for crystallisation peak temperature (T_P).

13.1.6 MatLab programming of the network for crystallisation peak temperature (T_P)

The program lines are written in regular font style, whereas the description lines are in italic font style all of which start with a '%' symbol.

```
%This program is part from NN for Crystallization Temperature
%Peak Temperature

%The following reads data file and puts it into matrix M
load nn_k_pt.txt;
M = nn_k_pt;
%Matrix dimensions
sz = size(M);
rws = sz(1);
cls = sz(2);
```

```
%Random redistribution of database
rnd1 = rand(rws);
rnd = rnd1(:,1);
M_r = M;
M_r(:,11) = rnd;
M_r = sortrows(M_r,11);
M = M_r(:,1:10);

%The next divide the data into train and test
group
M_all = M(1:rws,:);
M_tr = M(1:101,:);
M_val = M(102:rws,:);
%inputs
PP_all = M_all(:,2:9);
PP_tr = M_tr(:,2:9);
PP_val = M_val(:,2:9);
%outputs
TT_all = M_all(:,10);
TT_tr = M_tr(:,10);
TT_val = M_val(:,10);

%Matrix transpose
p_all = PP_all';
p_tr = PP_tr';
p_val = PP_val';
t_all = TT_all';
t_tr = TT_tr';
t_val = TT_val';

%Normalization
[pn_tr,minp_Rm,maxp_Rm,tn_tr,mint_Rm,maxt_Rm] =
premnmx(p_tr,t_tr);
%min and max of the input
PR = [min(pn_tr'); max(pn_tr')]';

%Neural Network Train
%creating a network
net_Rm=newff(PR,[4,1],{'tansig','purelin'},'train
br');
%train parameters
net_Rm.trainParam.show = 20;
net_Rm.trainParam.lr = 0.01;
```

```
net_Rm.trainParam.epochs = 200;
net_Rm.trainParam.goal = 0.0001;
%train
tic;
net_Rm=train(net_Rm,pn_tr,tn_tr);
toc

%Posttraining analysis
%Train subset
an = sim(net_Rm,pn_tr);
a = postmnmx(an,mint_Rm,maxt_Rm);
figure(1)
[m(1),b(1),r(1)] = postreg(a,t_tr);
title('Train')
ylabel('Experimental Temperature (C)');
xlabel('Neural Network Predictions (C)');

%Test subset
p_val_n = tramnmx(p_val,minp_Rm,maxp_Rm);
a_val_n = sim(net_Rm,p_val_n);
a_val = postmnmx(a_val_n,mint_Rm,maxt_Rm);
figure(2)
[m(2),b(2),r(2)] = postreg(a_val,t_val);
title('Test')
ylabel('Experimental Temperature (C)');
xlabel('Neural Network Predictions (C)');

%The whole subset
p_all_n = tramnmx(p_all,minp_Rm,maxp_Rm);
a_all_n = sim(net_Rm,p_all_n);
a_all = postmnmx(a_all_n,mint_Rm,maxt_Rm);
figure(3)
[m(3),b(3),r(3)] = postreg(a_all,t_all);
title('All')
ylabel('Experimental Temperature (C)');
xlabel('Neural Network Predictions (C)');
```

13.2 Performance of neural networks

In Table 13.1, the individual neural networks with the best performances in training and testing are listed. The performances are judged on the correlation coefficient (r) values of linear regression analyses between ANN predictions and experimental data in the training and testing processes. Among those neural networks, the one used for the simulation and prediction of crystallisation peak temperature (T_P) has the best performance overall. This neural network in the training, testing and the whole datasets shows good fitting by the linear regression analysis between the targeted experimental output given in the datasets and the calculated output (Fig. 13.11). However, this individual network has only four alloying elements for input, as it could not perform well in simulation and prediction if two additional elements Pd and Zr were included. This is believed to be due to the inadequate numbers of datasets containing these two additional elements for the training of the network.

Each of the remaining individual neural networks has six alloying elements for input, and their performances in the training and testing are acceptable although not as good (Table 13.1). The fittings of these neural networks in the training, testing and the whole datasets between the targeted experimental output given in the datasets and the calculated output are shown in Figs 13.12 and 13.13. Those unsatisfactory performances are caused mainly by the errors in some unjustified datasets. Referring to Fig. 13.3, the onset and end temperatures read using either type of definition could have large errors if the DSC curve encountered has a very gradual curvature at the beginning and the end of the peak. Therefore, the trained neural network for the crystallisation peak temperature of Ni–P based amorphous alloys is most reliable and can be used for further simulations and predictions. The remaining neural networks however have to be used with caution. It is also normally understood that the neural networks for all cases could perform well only within the ranges of the given training datasets (Table 13.2 and Fig. 13.9).

13.3 Comparison between calculations and experimental data

The trained ANN model is used to carry out simulations and predictions of the crystallisation peak temperatures (T_P) of Ni–P based amorphous alloys over the influences of alloy composition and heating rate. Comparison is made on some predicted results to the experimental data. Only the crystallisation peak temperatures of the alloys produced from the electroless processing method are shown, because the majority of datasets used in the model are based on the electroless processing method (Fig. 13.9). Predictions on both the onset and the end crystallisation temperatures would not be as reliable.

Figure 13.14(a) shows good agreement with the data from Chapter 10 (the 16 wt% P deposit). Jiang and Wu (2001) reported that the decrease of heating rate in the heat treatment process would result in lower crystallisation temperature of the Ni–P binary alloys. The reason for this is that at a lower heating rate, the thermal ageing effect that could enhance the crystallisation reaction is prolonged. In Section 13.4, the programming used for the

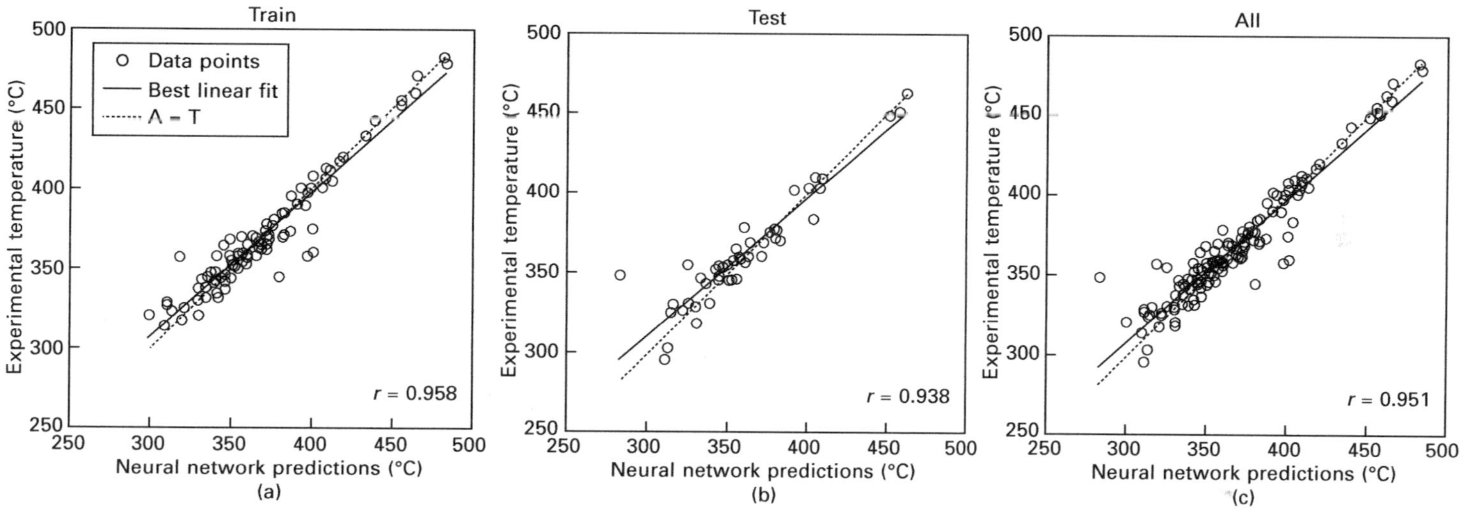

13.11 Performance of the neural network in the training and testing of simulating crystallisation peak temperatures. (a) Training; (b) testing; (c) whole datasets.

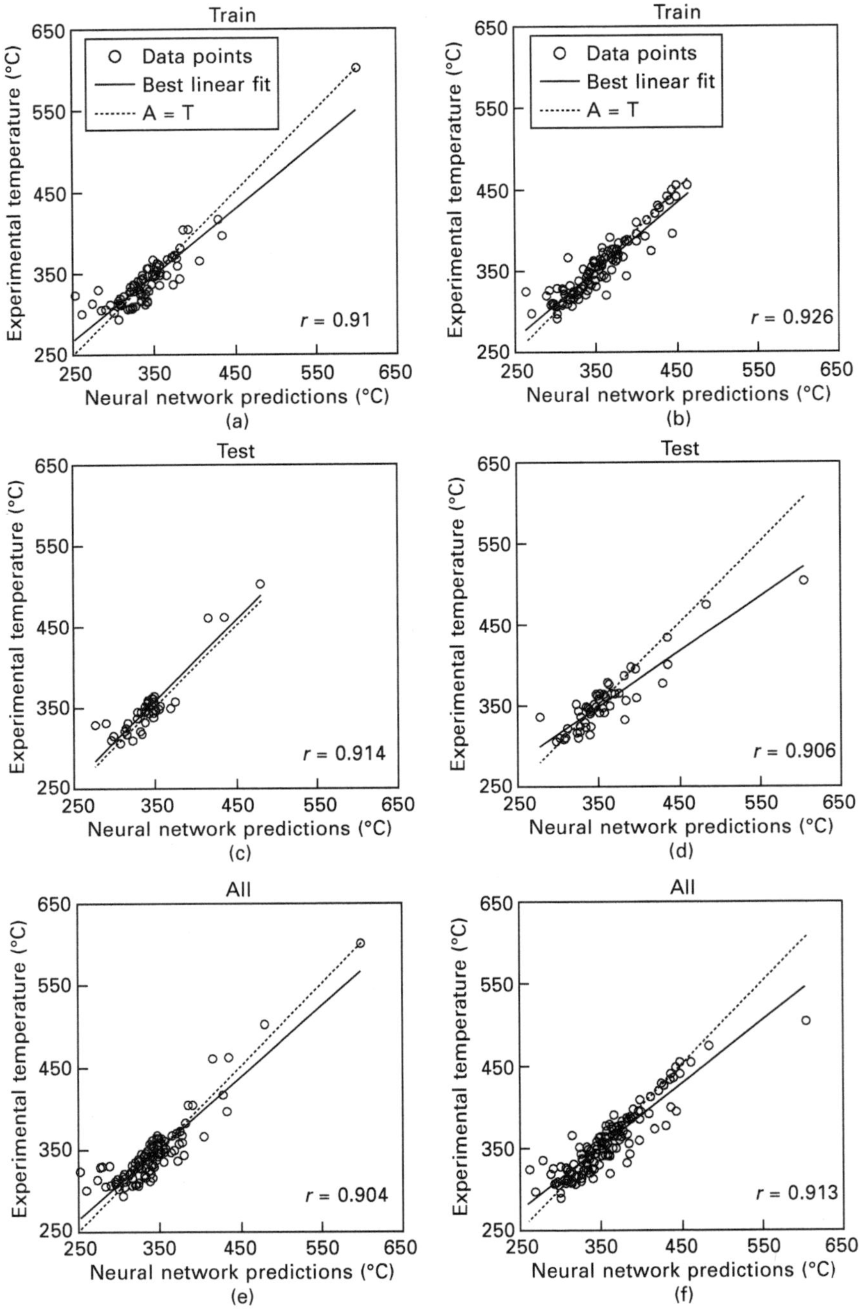

13.12 Performance of the neural network in the training and testing of simulating crystallisation onset temperatures: (a, c, e) T_{S1}; (b, d, f) T_{S2}; (a, b) training; (c, d) testing; (e, f) whole datasets.

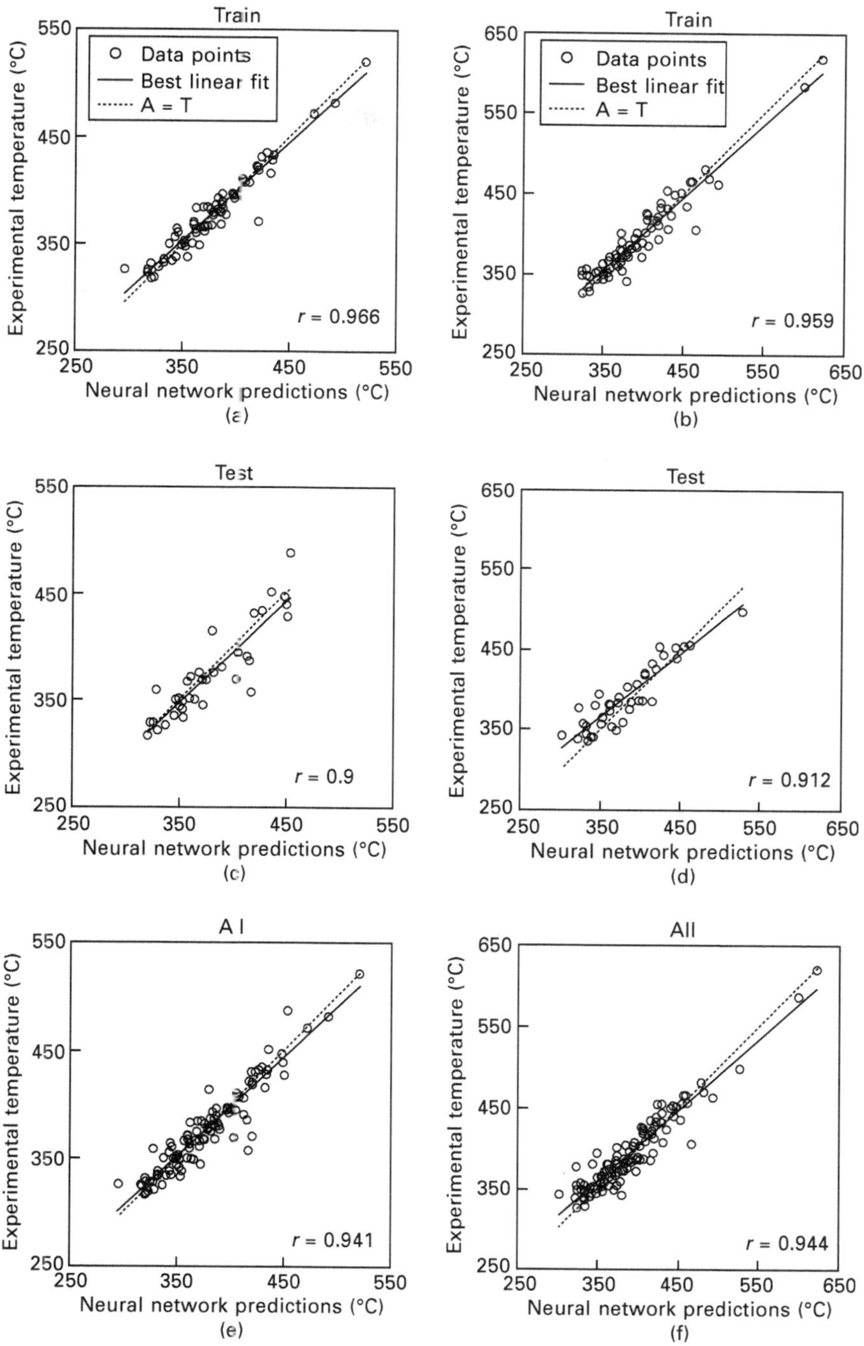

13.13 Performance of the neural network in the training and testing of simulating crystallisation end temperatures: (a, c, e) T_{E1}; (b, d, f) T_{E2}; (a, b) training; (c, d) testing; (e, f) whole datasets.

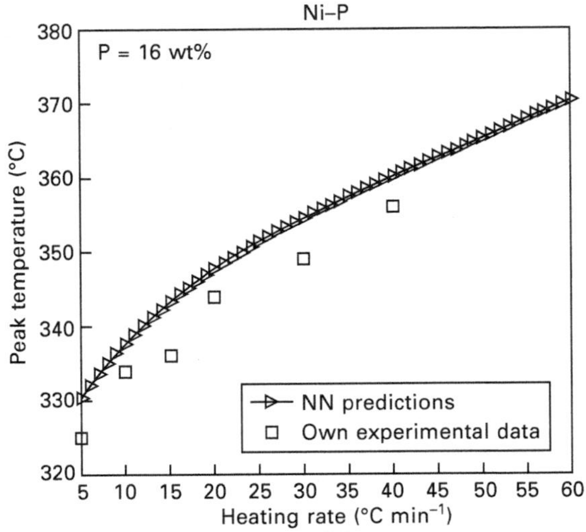

(a)

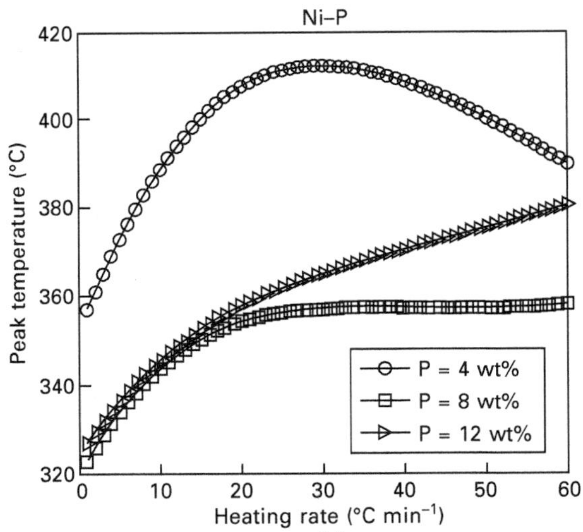

(b)

13.14 Simulation and prediction of the dependency of crystallisation peak temperature of electroless Ni–P binary alloys on the heating rate. Comparison is made in one case with the experimental data. (a) 16 wt% P; (b) 4, 8 and 12 wt% P.

crystallisation in the electroless Ni–P binary alloy (with 16 wt% P) as shown in Fig. 13.14(a) is given.

Similar predictions are made for the electroless Ni–P binary alloys with 4, 8 and 12 wt% phosphorus contents (Fig. 13.14(b)). The low-phosphorus alloy (with 4 wt% P) has the highest crystallisation peak temperature. Xiang *et al.* (2001) have reported an increase of crystallisation temperature with decreasing phosphorus content in electroless Ni–P binary alloys. This tendency is confirmed by the experimental data in Chapters 10 and 11. In general, the occurrence of amorphous structure in Ni–P alloys is related to the phosphorus content. Such structure is thermally unstable but requires activation energy to crystallise. Hence, the decrease of phosphorus level could increase the thermal stability of the alloys and lead to the increase of activation energy for crystallisation, resulting in higher crystallisation temperature. On the other hand, comparison between the medium- and high-phosphorus Ni–P alloys (with 8 and 12 wt% P, respectively) shows contradictory results to these. Nonetheless, the crystallisation peak temperatures of these two alloys have continuously increased with the heating rate with a tendency similar to the alloy with 16 wt% P (Fig. 13.14(a)).

As shown in Fig. 13.14(b), the crystallisation peak temperature of the low-phosphorus alloy increases quite rapidly at the beginning, but starts to decrease after the heating rate reaches 30 °C min^{-1}. The tendency of increase and decrease in the crystallisation peak temperatures has also been found in experimental data (Fig. 13.15) obtained from an alloy with similar phosphorus content (5 wt%). The rise and drop in the crystallisation temperatures could be related with different crystallisation mechanisms. However, the tendency is different with those two alloys with higher phosphorus content, as at different phosphorus contents the effect of high heating rate on the crystallisation temperatures could be different. Overall, the trained ANN model is capable of making acceptable predictions on the crystallisation peak temperatures of electroless Ni–P binary alloys, within the range of the datasets used in the training.

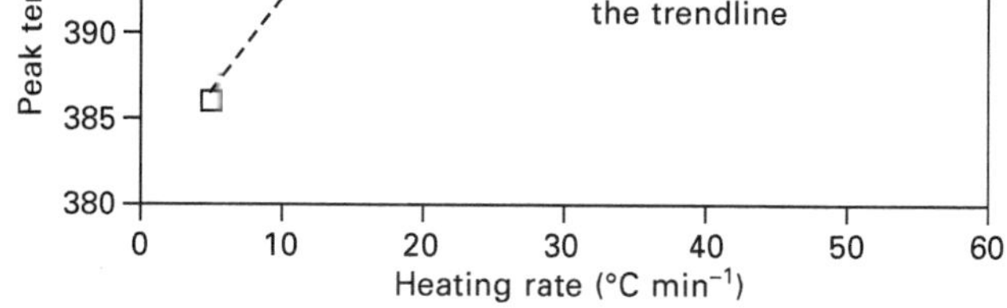

13.15 Crystallisation peak temperatures (T_P) of the 5 wt% P deposit obtained from the experimental DSC curves of the deposit (Fig. 11.1(c)).

In Fig. 13.16(a), there is an acceptable agreement between the model and the experimental data (including the deposits described in Chapters 10 and 11) of the alloys at the heating rate of 20 °C min^{-1}. The prediction agrees with the Xiang *et al.* (2001) in terms of increasing crystallisation temperatures with a decreasing phosphoros level. Besides, there is an obvious

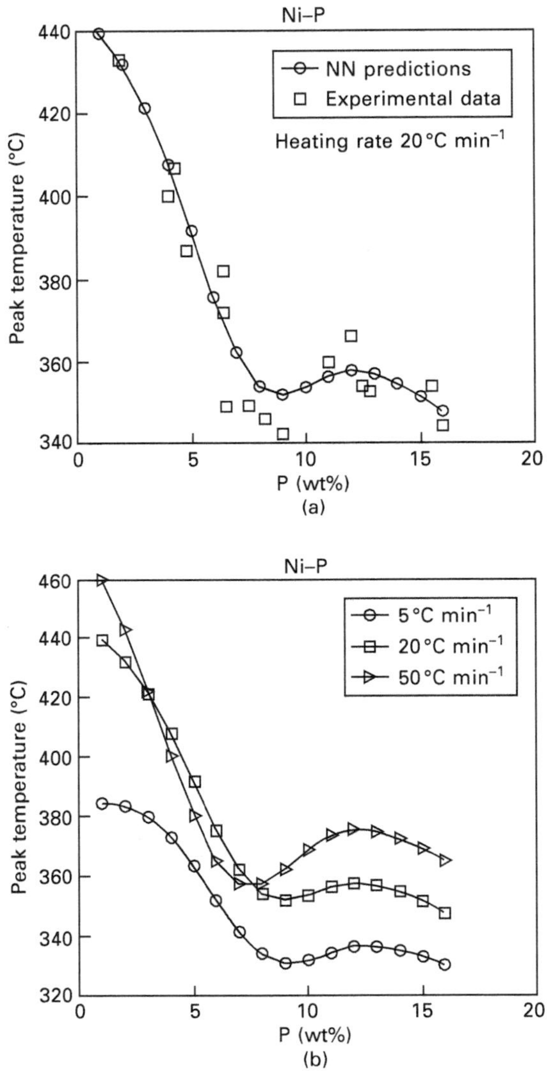

13.16 Simulation and prediction of the effect of phosphorus level on the crystallisation peak temperature of electroless Ni–P binary alloys using the trained ANN model. (a) Heating rate 20 °C min^{-1}; (b) heating rate 5–50 °C min^{-1}. Comparison to the experimental data is made in (a).

temperature peak at the range of phosphorus level between 10 and 15 wt%. The thermal stability of Ni–P amorphous alloys increases as the phosphorus content is close to the eutectic composition (~11 wt% P). Further, two similar predictions are made for different heating rates (5 °C and 50 °C min^{-1}) (Fig. 13.16(b)). Overall, the predictions obtained are very much similar to that at the heating rate of 20 °C min^{-1} regarding the effect of phosphorus levels. In addition, there is the increase of crystallisation peak temperatures with increasing heating rate, which is in agreement with Jiang and Wu (2001). Nevertheless, the effect of heating rate at low phosphorus levels (below 8 wt%) is not monotonous.

13.4 MatLab programming for simulation of peak temperatures versus heating rate

The program lines are written in regular font style, whereas the description lines are in italic font style all of which start with a '%' symbol.

```
%Loading the trained network
load nn_k_pt

%New input
P = 16;
B = 0;
Cu = 0;
Fe = 0;
Proc = 0;

for i = 1:60

HR(i) = i;

pnew = [P B Cu Fe Proc HR(i)]';

%Preprocess of the new input
pnewn_Rm = tramnmx(pnew,minp_Rm,maxp_Rm);

%Simulink
anewn_Rm = sim(net_Rm,pnewn_Rm);

%Postprocessing
anew_Rm = postmnmx(anewn_Rm,mint_Rm,maxt_Rm);

PT(i) = anew_Rm(1);
```

```
end

plot(HR,PT,'k>-')

title('P = 16 wt.%')
xlabel('Heating Rate (C/min)');
ylabel('Peak Temperature (C)');
```

13.5 Prediction using the model

Multicomponent electroless nickel-based alloy systems containing both non-metallic (phosphorus or boron) and metallic (in particular copper, iron and zirconium) elements have been the subject matter of study in search for potential new protective materials. In the following discussion (Figs 13.17 and 13.18), the inclusion of these elements to the systems is at the expense of nickel (wt%), whereas the phosphorus level is kept constant at 8 or 12 wt%. The heating rate of 20 °C min^{-1} is applied to all cases of the predictions using the model.

In Fig. 13.17(a), there is an indication of temperature fluctuation in the prediction before the boron level increases to over 4 wt%. At the boron level of 0–1 wt%, the increase of crystallisation temperature might be related to the change of binary to ternary alloying systems. The crystallisation temperature drops when the boron level further increases up to 4 wt% (approximately the eutectic level for Ni–B binary system). However, the simulation shows the increase of crystallisation peak temperature with increasing boron level when it is over 4 wt%. This is similar to the electroless Ni–P system at the phosphorus level around the eutectic composition (Fig. 13.16).

The dependency of crystallisation peak temperatures of electroless Ni–P–Cu ternary alloys on the copper level is shown in Fig. 13.17(b). The crystallisation peak temperatures increase constantly with the increasing copper level. The same effect of copper level on the crystallisation temperatures was reported by Chang *et al.* (2001) in a study using a similar alloy system. However, in melt-spun Ni–P–Cu alloys, there is a drop in the thermal stability of alloy at the copper level above 5 wt%.

The effect of iron addition on electroless Ni–P alloys will now be examined. The influence on the crystallisation peak temperatures is insignificant (Fig. 13.18). Even the total replacement of nickel with iron (i.e. from Ni–P to Fe–P) modifies the crystallisation peak temperature by 15–20 °C, probably within the experimental error in the measurement. However, a similar alloy system but with the electroplating processing method shows the increase of crystallisation temperatures with increasing iron level in the system.

The heating rate is kept constant at 20 °C min^{-1} in the simulation shown in Fig. 13.19. The temperature differences are high at the low phosphorus

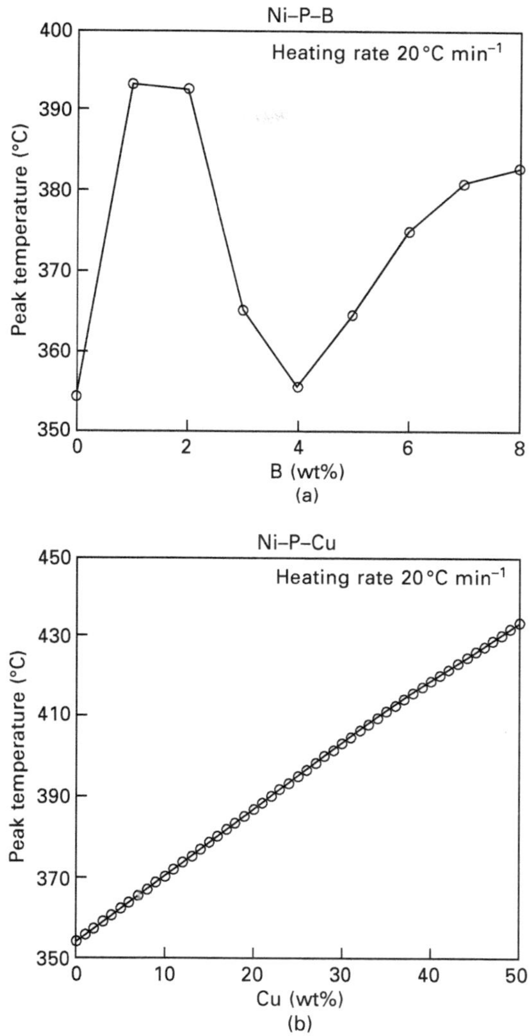

13.17 Influences of the third elements in the 8 wt% P electroless Ni–P ternary alloy systems on the crystallisation peak temperatures simulated using the ANN model. (a) B; (b) Cu.

level. However, the difference decreases with an increasing phosphorus level of up to 12.5 wt%. After that, the temperature difference increases again with an increasing phosphorus level. Accordingly, the prediction from ANN model has shown that the time for crystallisation processes is the shortest at the phosphorus level of 12.5 wt%. The DSC profile of hypoeutectic alloys (<11 wt% P) usually consists of two significant peaks: one related to the formation of nickel crystallites, and the other for major crystallisation reaction. Therefore, the time for complete crystallisation in such alloys is

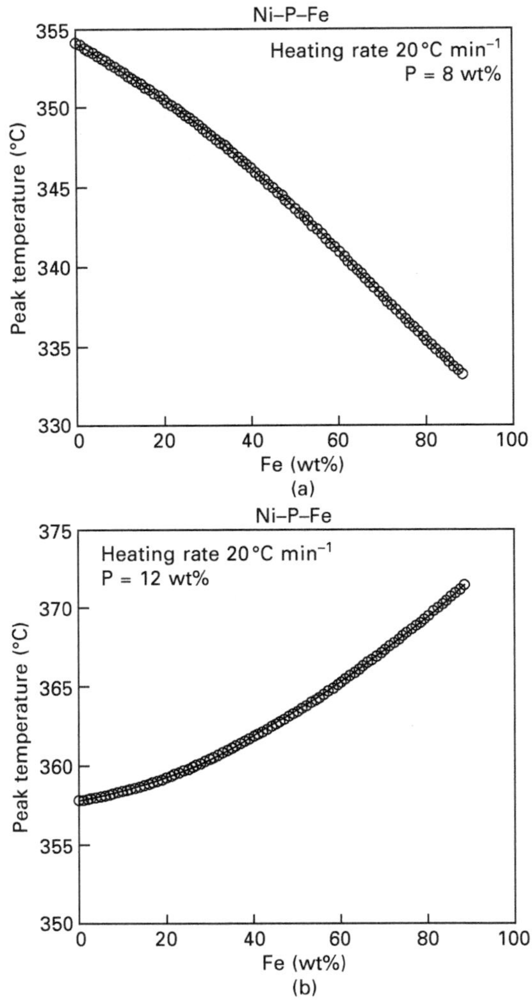

13.18 Influences of the Fe inclusion in the electroless Ni–P ternary alloy systems on the crystallisation peak temperatures simulated using the ANN model. (a) 8 wt% P; (b) 12 wt% P.

long. The DSC profile of the amorphous Ni–P eutectic alloys (~11 wt% P) displayed only one peak, which is for the major crystallisation reaction. The hypereutectic alloys (>11 wt% P), however, have two peaks: one for the major crystallisation reaction and the other related to the decomposition of metastable phases into final equilibrium state. Because of these, the time for complete crystallisation in the eutectic alloy should be the shortest. In the hypereutectic alloys, because the DSC peak of the major crystallisation reaction usually has a gradual ending due to the effect of decomposition

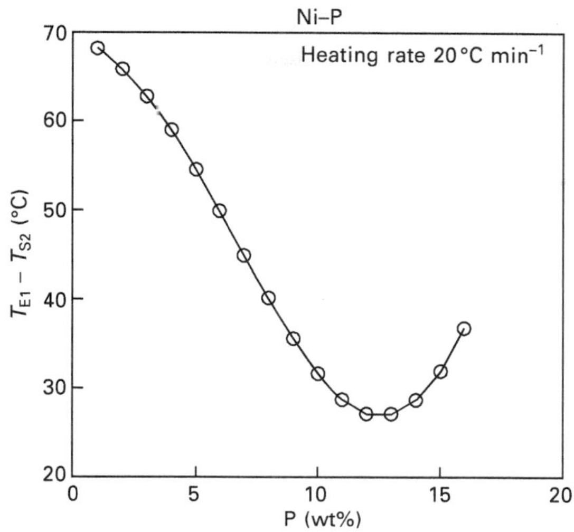

13.19 Model simulation prediction of the difference between the end and onset crystallisation temperatures over the increasing phosphorus level in the electroless Ni–P binary system.

reaction, the difference between crystallisation start and end temperatures of the alloys is larger. Overall, the ANN model can make a reasonable prediction of the difference between the onset and the end temperatures as a function of phosphorus level in the electroless Ni–P binary system.

13.6 Application of the model

Demonstrated by the simulations and predictions made using the ANN model (Figs 13.14 and 13.16–13.19), the model could be a useful tool in assessing the correlations of crystallisation temperatures to the alloy composition, heating rate of heat treatment process and the processing method of Ni–P based alloying systems. Therefore, the model can be used to assess the best combination of parameters in designing new alloys for desired properties.

A graphical user interface (GUI) (Fig. 13.20) makes it easy to carry out simulations and predictions in the further use of the model. The GUI allows users to choose the required input parameters (alloy composition and processing method) and the output parameter (crystallisation onset, peak or end temperature). The resulting simulation shown as a plot of crystallisation temperature versus heating rate will be displayed in the GUI displaying window. In addition, the GUI's 'hold on' function can be used to hold the plots from the previous simulations and display two or more curves in the displaying window.

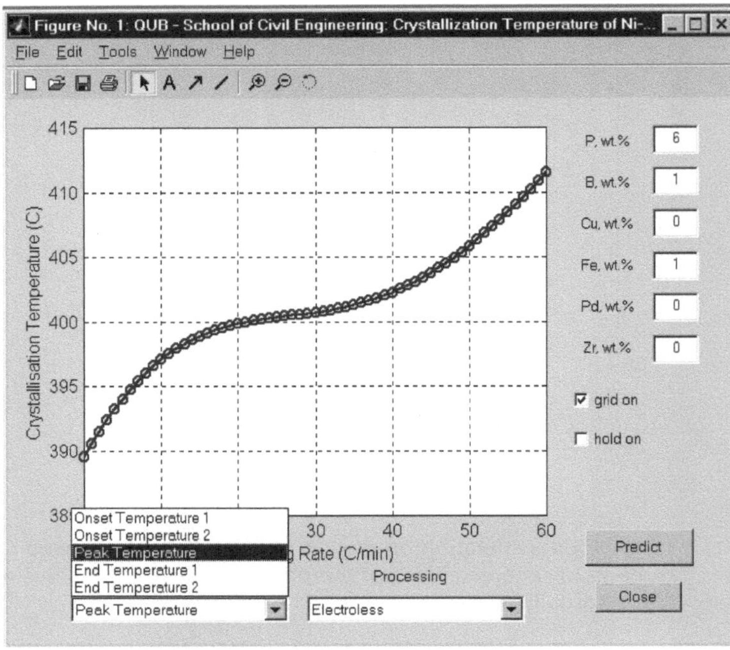

13.20 Use of a GUI in the modelling of crystallisation temperatures as a function of heating rate.

13.7 Conclusions

An artificial neural network model that consists of five individual neural networks can predict the crystallisation temperatures of Ni–P based amorphous alloys under the influences of alloy composition, heating rate of heat treatment process and the processing method. Each of these neural networks can produce one output: crystallisation onset (two definitions), peak, or end (two definitions) temperature. The neural network for predicting crystallisation peak temperatures has the best performance amongst the individual neural networks, indicated by the highest values of correlation coefficient in the linear regression analyses.

The effects of heating rate and phosphorus content on the crystallisation peak temperatures of electroless Ni–P binary alloys predicted using the model are reasonable and satisfactory. Comparison made between the predictions and experimental data gives relatively good agreement in each case. The predictions show that the crystallisation peak temperatures of electroless Ni–P ternary alloy systems are sensitive to the boron and copper level, but insensitive to that of iron.

Users of the model can produce the plots of crystallisation temperature versus heating rate, through the graphical user interface (GUI) for input and output parameters of choice. The model could be a useful tool in the design optimisation of new Ni–P alloys for specific properties as it could weigh up the correlations of crystallisation temperatures with the alloy composition, heating rate of heat treatment processes and the processing method.

13.8 References

Chang Y C, Duh J G and Chen Y I (2001), 'Fabrication and crystallization behaviors of sputtered Ni–Cu–P films on tool steel', *Surf Coat Technol*, **139**, 233–43.

Hentschel T, Isheim D Kirchheim R, Müller F and Kreye H (2000), 'Nanocrystalline Ni–3.6 at.% P and its transformation sequence studied by atom-probe field-ion microscopy', *Acta Mater*, **48**, 933–41.

Jiang T X and Wu H E (2001), 'Studies on microstructure and crystallization behavior of electroless nickel-high phosphorus coatings', *Chem J Chin Univ*, **22**, 976–9.

Martyak N M and Drake K (2000), 'Peak-profile analysis of electroless nickel coatings', *J Alloy Comp*, **312**, 30–40.

Sha W and Malinov S (2009), *Titanium Alloys: Modelling of microstructure, properties and applications*, Woodhead Publishing, Cambridge.

Tachev D, Iorgov D and Armyanov S (2000), 'Magnetothermal investigation of crystallization of Ni–P amorphous alloys', *J Non-Cryst Solids*, **270**, 66–76.

Tachev D, Georgieva J and Armyanov S (2001), 'Magnetothermal study of nanocrystalline particle formation in amorphous electroless Ni–P and Ni–Me–P alloys', *Electrochim Acta*, **47**, 359–69.

Taheri R, Oguocha I N A and Yannacopoulos S (2001), 'Effect of heat treatment on age hardening behaviour of electroless nickel–phosphorus coatings', *Mater Sci Technol*, **17**, 278–84.

Ten S S and Chang F Y (2001), 'Electrical resistivity of electroless nickel coated carbon fibers', *Thin Solid Films*, **388**, 143–9.

Wang Y P and Lu K (2000), 'Crystallization kinetics of amorphous $Ni_{80}P_{20}$ alloy investigated by electrical resistance measurements', *Z Metallkd*, **91**, 285–90.

Xiang Y, Hu W, Liu X and Luo S (2001), 'Evaluation of the phosphorus content of electroless nickel coatings by XRD & DSC', *Plat Surf Finish*, **88**, 96–9.

Zhang B W and Yi G (2000), 'Crystallization kinetics of nano-size amorphous $Fe_{76.9}Ni_{6.2}P_{2.0}B_{14.9}$ powders by differential scanning calorimetry', *Z Metallkd*, **91**, 75–8.

14
Hardness evolution of nickel–phosphorus (Ni–P) deposits with thermal processing

Abstract: Microindentation hardness testing using Vickers or Knoop indenters conducted on electroless nickel–phosphorus (Ni–P) deposits can be used to analyse their hardness evolutions in relation to their deposit phosphorus contents as well as different heating temperatures at isothermal and linear heating conditions. The deposit hardness increases with decreasing phosphorus content at as-deposited condition, and can be enhanced by appropriate heating. The microindentation testing on the cross-sections of the deposits shows variations of hardness across the deposit depth. Scanning electron microscopy (SEM) analysis shows that the lamellar structure in the cross-sections of the as-deposited coatings tends to disappear, and agglomeration occurs when the heat-treating temperature is increased.

Key words: hardness, electroless, nickel–phosphorus, kinetic strength, scanning electron microscopy.

14.1 Concept of kinetic strength

This chapter examines the hardness evolution of electroless nickel–phospohorus (Ni–P) deposits in relation to the deposit phosphorus contents and thermal processing temperatures at isothermal and linear heating conditions. Interpretation of the kinetic energy of increased hardening effects after the isothermal experiments is also attempted. Accordingly, the Vickers indenter is used to obtain the hardness data from the surface as well as cross-sections of the 5–12 wt% P deposits, whereas the Knoop indenter is used to investigate the change of hardness across the depth (distance from the deposit surface towards the deposit/substrate interface) of the deposit cross-sections. Scanning electron microscopy (SEM) and X-ray diffraction (XRD) reveal the microstructures and phase transformations of the deposits at different heating temperatures. Chemical analysis using the electron microprobe technique measures the phosphorus content of the deposits before and after the thermal processing. The empirical kinetic strength concept is adopted to analyse the kinetic energy of increased hardening effects for the deposits at isothermal condition.

In the literature, the hardness data of electroless Ni–P deposits after thermal processing is usually presented as a function of heating temperature (T) or time of heating (t). Therefore, it would be very useful if the combined effects of time and temperature on the hardening of electroless Ni–P deposits could be rationalised into a single and quantifiable parameter. If the calculated parameter, in our case the kinetic energy (Q) of increased hardness, is valid throughout the heating process, it is possible to model the increase of hardness in the electroless Ni–P deposits. For this purpose, the concept of kinetic strength (K_s) that aims to combine the effects of time and temperature for a given heating process could be of use.

Kinetic strength (K_s) is a simple empirical concept that was first proposed from work on time–temperature relations in tempering steel. The concept has been used for investigations of directional recrystallisation of superalloys, hardening of steels, and structure and hardening in welds. According to the concept, the combined effects of time and temperature for a given heating process at isothermal or non-isothermal conditions can be expressed as follows:

$$K_s = t\exp(-Q/RT) \qquad \text{for isothermal} \qquad 14.1$$

and

$$K_s = \int \exp(-Q/RT(t))\,\mathrm{d}t \qquad \text{for non-isothermal} \qquad 14.2$$

where t and T are the time and the temperature of heating respectively, Q is the activation energy, and R is the gas (Boltzmann) constant.

Taking into account the expression in Eq. 14.1, the relationship between the increased hardening effect and the kinetic energy Q of the effect at isothermal treatment can be analysed for electroless nickel–phosphorus deposits. The kinetic energy Q is the slope of the plot of $\ln(\Delta H)$ versus $1000/(RT)$. The denotation for ΔH is the hardness increase, and T is the temperature in unit Kelvin. The calculation of such parameters for the plot of $\ln(\Delta H)$ versus $1000/(RT)$ is illustrated in Table 14.1. The Q value is a constant as it is calculated from linear fitting. Thus, the effectiveness of

Table 14.1 Calculation of kinetic parameters using the Vickers hardness data

Temperature, T (°C)$^+$	Hardness, H ($HV_{0.1}$)*	ΔH*	$1000/(RT)$**	$\ln(\Delta H)$
41	505	0	–	–
97	506	1	0.325	0.10
150	516	11	0.284	2.39
211	534	29	0.249	3.37
266	576	71	0.223	4.27
323	741	236	0.202	5.46
380	978	473	0.184	6.16
404	1032	527	0.178	6.27
440	949	83	0.169	4.42

^+T = isothermal heating temperature (for 1 hour period).
*Decimal places not shown but used when calculating $\ln(\Delta H)$.
**T = temperature in unit Kelvin (K).

adopting the kinetic strength concept, to describe the increased hardening effect in the electroless Ni–P deposits, can be judged from the coefficient of determination (r^2) values for the linear regression analysis of the plots. By knowing that the Q value is a constant, the hardness value at a given heating temperature can be estimated using the relationship in Eq. 14.1 if the Q value is valid for that temperature.

14.2 Thermal processing and phase structure

As described in Chapters 10 and 11, the as-deposited structure of the 12 wt% P deposit is entirely amorphous, whereas those of the 5–8 wt% P deposits are a mixture of amorphous and nanocrystalline nickel phases. When undergoing crystallisation processes during the continuous heating at 20 °C min^{-1}, the phase transformation processes for the 8–12 wt% P deposits follow the following sequence: amorphous (+ nanocrystalline nickel phase) → intermediate phase(s) + fcc nickel + Ni_3P → fcc nickel + Ni_3P stable phases (Chapters 10 and 11). However, the phase transformation processes under the same heating condition for 5–6 wt% P deposits are without the formation of intermediate phases such as Ni_2P and $Ni_{12}P_5$ (Chapter 11).

Comparison of the XRD profiles between Fig. 14.1 and Fig. 11.3 in Chapter 11 indicates that the two different heating conditions produce similar effects on the phase compositions. The sequence of phase transformation and the tendency of increasing intensity for the diffraction peaks in both conditions are comparable. Intermediate phases (symbol '×' in Fig. 14.1) appear at 330–400 °C. On the other hand, an intense diffraction peak (at $2\theta = 47.5°$) belonging to the intermediate phase is absent after heating at 400 °C (Fig. 14.1). Apparently, the intermediate phase decomposes during the isothermal holding for one hour at this temperature. This observation is consistent with what is expected from the theory of phase transformation.

In Fig. 14.2, the XRD profiles showing the texture or composition inhomogeneity are compared with three other XRD profiles from the deposit after heating for the location (2θ) and tendency of their diffractions. The effect of texture inhomogeneity in the as-deposited state should not affect significantly the overall phase transformation process of the deposit. The average of those two as-deposited profiles is used for the 8 wt% P deposit in Fig. 14.1. As previously discussed in Chapter 11 (Section 11.2), the inhomogeneity in the 8 wt% P deposit discussed based upon Fig. 14.2 would be expected in other deposits.

14.3 Vickers surface hardness

Figure 14.3 shows the dependence as a function of temperature after the continuous heating process. Hardness of these deposits depends on the

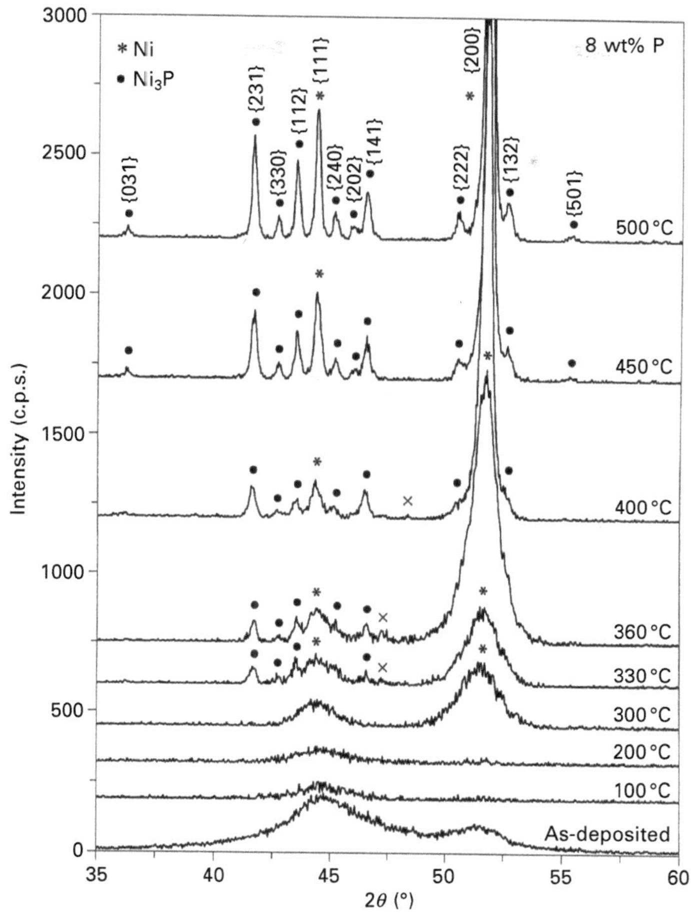

14.1 XRD profiles of the 8 wt% P deposit at as-deposited condition and after heating at 100–500 °C for one hour at isothermal conditions. For the sake of clarity, the profiles are shifted arbitrarily on the vertical scale.

phosphorus content and heating temperature with several tendencies:

1. The as-deposited hardness shows a trend of increase with decreasing phosphorus content.
2. The hardness increases with heating temperature and reaches a maximum at 400–450 °C. Further heating to higher temperatures will decrease the hardness values.
3. The degree of hardness loss (after achieving the maximum hardness) increases as the phosphorus content decreases.

Taheri *et al.* (2001) also reported the increase of maximum hardness values with decreasing phosphorus content in annealed electroless Ni–P deposits.

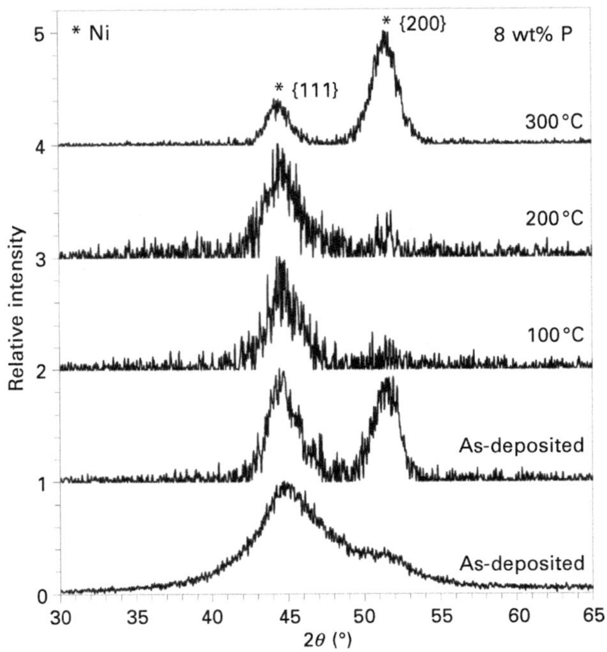

14.2 XRD profiles (with normalised intensity) of the 8 wt% P deposit at as-deposited condition and after isothermal heating for one hour. Two profiles from different specimens at the as-deposited condition are presented to show the specimen-to-specimen variation in diffraction intensity distribution at that state.

In the case of non-isothermal annealing, the high annealing rate normally can reduce the time needed to reach the maximum hardness value. In the case of isothermal annealing, the time taken to achieve maximum hardness becomes shorter when the phosphorus content is reduced and the annealing temperature is below or around that for the maximum hardness. The hardness normally increases with annealing time to the maximum, and then decreases with the further increase of annealing time if the annealing temperature is not much higher than that for achieving maximum hardness.

Different phosphorus content of the electroless Ni–P deposits causes different as-deposited microstructure that responds to heating temperature resulting in the hardness evolutions of the deposits. At as-deposited state, the reason for the lower phosphorus deposits to have higher hardness could be the high internal stress. During thermal processing, the increase of hardness is related to the amount of Ni_3P stable phase formed in the deposits. The massive formation of Ni_3P phase happens at the heating end temperature of 400 °C. Besides, the Ni_3P phase present at this temperature could be accompanied by high internal stresses revealed by its short and wide XRD

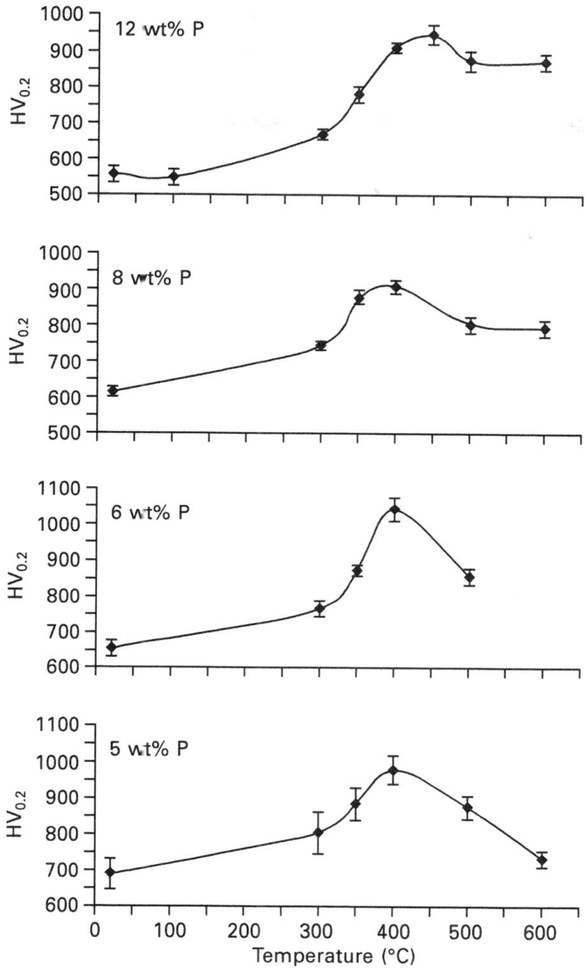

14.3 Vickers surface hardness of Ni–P deposits after linear heating to different temperatures at 20 °C min^{-1}.

peaks. These stresses are caused by the lattice mismatch between Ni_3P and nickel matrix, because in the early stages of transformation the interfaces are likely to be coherent. Additionally, there could be distortion caused by the non-equilibrium locations of the nickel and phosphorus atoms at this stage. Accordingly, the hardness of the deposits increases with increasing heating temperatures and reaches the maximum values at around 400 °C. On the other hand, the hardness drops begin when thermal processing continues to higher heating end temperatures when grain and particle coarsening is significant. Meanwhile, the atoms of nickel and phosphorus have perhaps moved to more stable locations in their lattice arrangements. The grain growth and coarsening of Ni_3P precipitate impose not as much effect for high-phosphorus alloys at heating to above the end temperatures of maximum hardness. The degree of hardness loss for the deposit with high-phosphorus content is therefore somewhat gentle because of Ni_3P dominance. Since the lower phosphorus deposits are dominated by nickel precipitate, their coarsening process produces a larger size but smaller number of grains mainly of the nickel phase. The subsequent rate of hardness loss is higher.

Also shown in Fig. 14.3, the deviations (± limit) of hardness value for some data points corresponding to different heating end temperatures are somewhat large. Vander Voort (2000) has discussed several possible reasons for such deviation, including the experimental errors related to hardness measurement and instrumental limitation, and the errors caused by material factors, such as uneven surface condition and inhomogeneity of the tested specimen. The electroless Ni–P deposits have the uneven surface condition as shown in Fig. 14.4. The effect of deposit inhomogeneity due to the presence of more than one phase during thermal processing is not clear though, as considerable hardness deviations have already been found in the as-deposited state. Any probable variation of hardness among the deposits caused by the indentation size effect (Factor and Roman, 2000) can be eliminated if the surface hardness measurements for all deposits are made at a same load.

14.4 Vickers hardness of the cross-sections

Comparison of Fig. 14.5 (the data also given in Table 14.2, with error values included) with Fig. 14.3 suggests that the behaviours of these deposits, in terms of the hardness change relating to phosphorus content and heating temperature, are largely consistent with those of the deposits heated at 20 °C min^{-1}. The result of comparison is analogous to that between the XRD profiles for the two conditions where similar phase compositions and transformation behaviours are evident (Figs 14.1 and 11.3). Nevertheless, the comparison between Figs 14.3 and 14.5 does indicate some differences, such as that in the tendencies (rates) of increasing and decreasing hardness. These differences are however rather insignificant considering that they are mostly within the variations (± limits) of hardness data as indicated in Fig. 14.3 and Table 14.2. The hardness data presented in the two figures are measured with different indentation loads from different deposit specimens. Hence, any comparison made between the hardness data of these two figures could be at risk of being interfered by the indentation-size effect factor.

The hardness of the deposit cross-sections increases with increasing heating temperature and reaches peak values at 400–450 °C. The 12 wt% P deposit attains its maximum hardness at a temperature of 450 °C which is higher than for the other deposits. This may indicate that the temperature for attaining maximum hardness increases with increasing phosphorus content. The

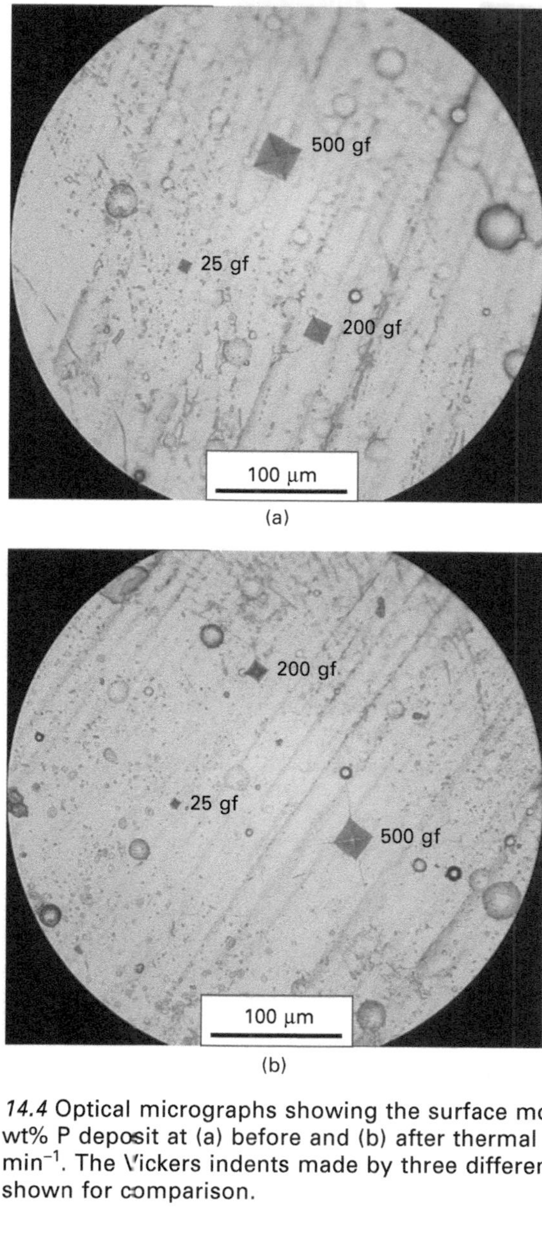

14.4 Optical micrographs showing the surface morphology of the 12 wt% P deposit at (a) before and (b) after thermal processing at 20 °C min^{-1}. The Vickers indents made by three different applied loads are shown for comparison.

reason for the high-phosphorus electroless Ni–P deposit attaining maximum hardness at a higher temperature could be the as-deposited structure of the deposit. During the crystallisation process, the hardening effects in the fully amorphous deposit are controlled by the transformation of amorphous phase to the Ni_3P stable phase, whereas those in the partially amorphous deposits

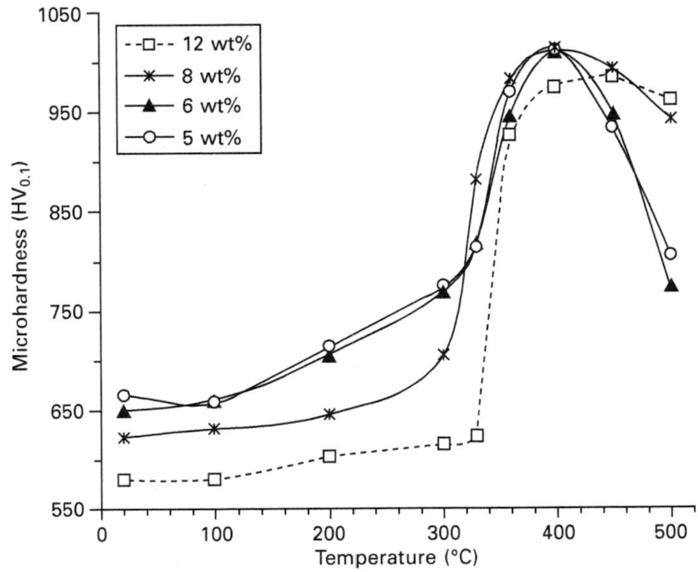

14.5 Vickers hardness data measured along the centre line of the cross-sections after heating at different isothermal temperatures for one hour.

Table 14.2 Vickers hardness ($HV_{0.1}$) measured along the centre line of deposit cross-sections

Isothermal heating temperature (°C)	12 wt% P	8 wt% P	6 wt% P	5 wt% P
No heating	579±11	623±4	650±8	664±6
100	581±3	631±5	660±4	658±13
200	603±7	645±4	704±7	712±6
300	614±4	702±8	768±19	774±11
330	623±5	881±9	817±19	814±9
360	926±5	982±8	945±10	970±9
400	973±4	1010±4	1009±10	1011±9
450	983±6	992±6	946±10	932±9
500	961±5	942±9	773±36	803±5

would have to depend on the interaction with the existing nanocrystalline nickel phase. Because the grain growth process of these nickel precipitates could have started at lower heating temperature, so does the subsequent attenuation process of hardening effect.

The errors for the hardness data measured from the deposit cross-sections are relatively small when compared with those from the deposit surface (Table 14.2 and Fig. 14.3). The most reasonable explanation is that the deposit cross-sections have been polished before the measurements, while no surface preparation is performed on the uneven deposit surface (Fig. 14.4).

As a result, the deposit cross-sections would have a better surface/indenter contact condition than the deposit surface for making indentation testing.

14.5 Knoop hardness over the depth of cross-sections

Figure 14.6 shows the hardness values as a function of the deposit depth (distance from the deposit surface towards the deposit/substrate interface). As indicated by the horizontal axis of this figure, the depth-distances at which the indentations are made are not drawn to scale. Each of the depth-distances is allocated an equal length separating it from the adjacent distances by vertical lines. There are nine numbers of Knoop hardness data plotted for each depth-distance representing the hardness values of the deposit cross-sections heated at different isothermal temperatures (100–500 °C).

The narrow Knoop indenter makes it feasible for testing specimens with potential hardness gradient such as electroless Ni–P deposit cross-sections (Vander Voort, 2000). The Knoop indenter is used with the long diagonal parallel to the substrate in order to detect possible hardness change across the depth (Fig. 14.7). This alignment also accommodates the limited thickness of the deposit cross-sections. In Fig. 14.6(a), the hardness increase with the depth in the 12 wt% P deposit is obvious and outside the deviations of the data points. The tendency for the increase is similar for all specimens corresponding to different heating temperatures. However, the change of hardness across the depth for the other deposits (Fig. 14.6(b)–14.6(d)) is not monotonous and varies with heating temperature. Added to that, the deviations in many of the hardness data are substantial, and therefore it is difficult to clearly define the tendency of increasing or decreasing hardness with depth. By considering the overall tendencies of hardness change for the three deposits, however, it seems that the hardness is somewhat lower at depths closer to the surface of the deposit cross-sections.

Several possible sources could contribute to the relatively large standard deviations of the hardness data. First, the phase compositions and microstructural properties of the electroless Ni–P deposits might be different across the depth. Metallographic examination of the deposit cross-sections by SEM reveals a laminated structure for the Ni–P deposits at as-deposited condition. The appearance of this structure is illustrated in Fig. 14.8(a) and (c) by the SEM micrographs. The formation of such a structure is associated with a periodic change of deposition conditions leading to slight periodic variation of microstructure and phosphorus content (Peeters *et al.*, 2001). The laminations can be subsequently eliminated through thermal processing at relatively high temperatures due to the agglomeration of constituent phases. The deposits, after being heated at 450 °C for one hour, show sign of less obvious lamination with increasing agglomeration (Fig. 14.8(b) and (d)).

Nevertheless, the gradual disappearance of the laminated structure has not resulted in the decrease of hardness variations, perhaps because the effect on the hardness variations of the structure is minor. On the other hand, it is

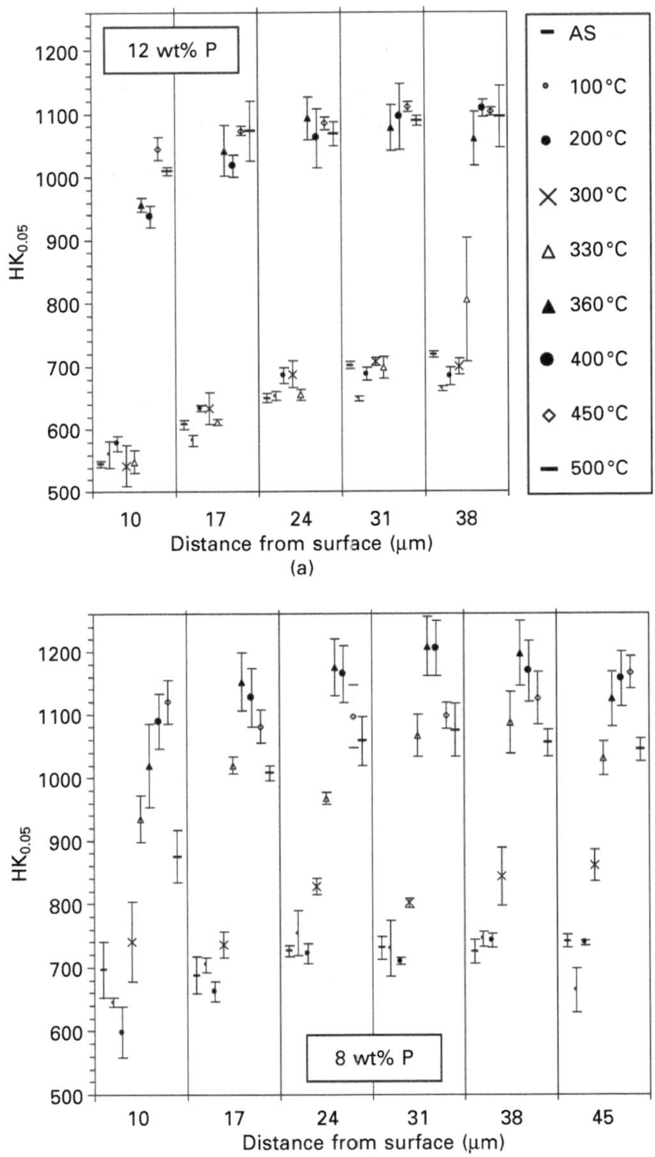

14.6 Variations of Knoop hardness across the depth of deposit cross-sections. There are one as-deposited (AS) and eight heat-treated (100–500 °C for one hour) specimens measured for each deposit. (a) 12 wt% P; (b) 8 wt% P; (c) 6 wt% P; (d) 5 wt% P.

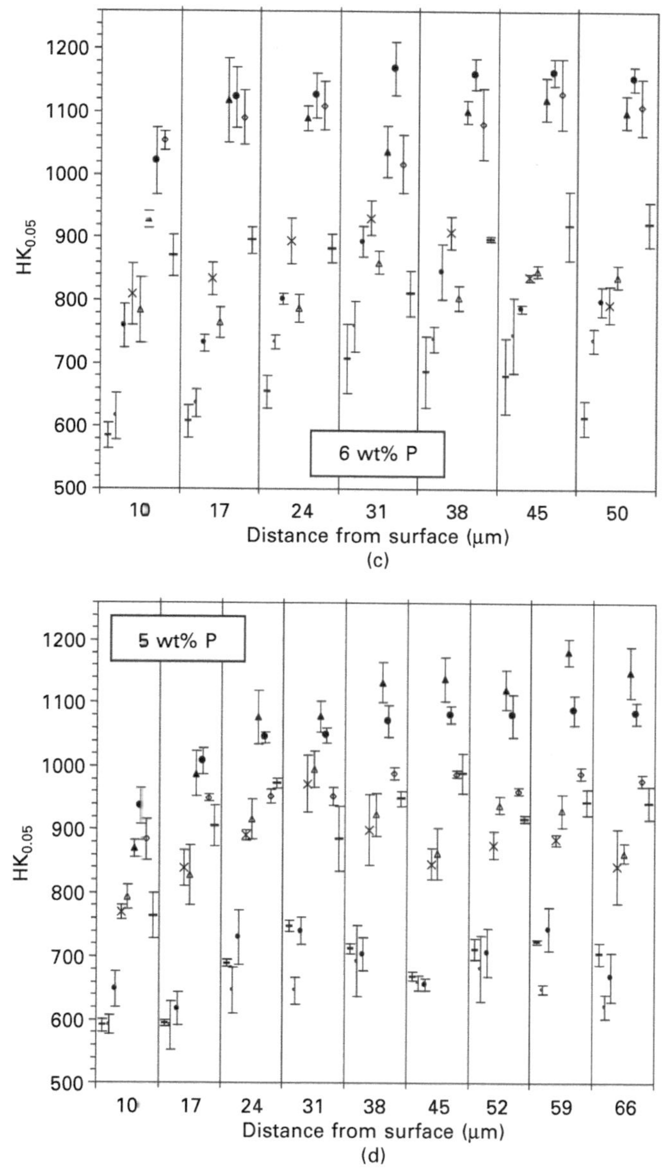

14.6 Continued

also possible that the decrease of such variations is offset by the effect of crystallisation processes.

Meanwhile, the formation of different phases might not be distributed evenly across the entire plating during the crystalline transition processes. Therefore, the hardness values obtained from different points at a given

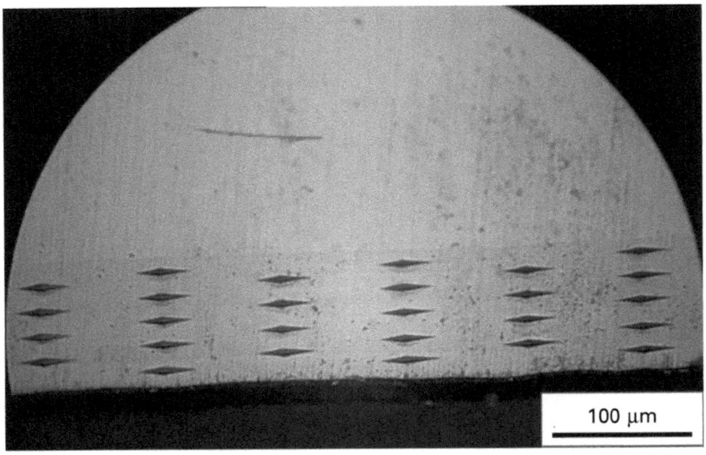

14.7 Indents made by the Knoop indenter across the depth (varying distance from the deposit surface towards the deposit/substrate interface) of the coated deposit (only one side is shown) on the deposit cross-section.

deposit depth might be different. In Fig. 14.6, the hardness variations of the 12 wt% P deposit are less significant than in the other deposits. The reason for this could be that the 12 wt% P deposit consists of mainly the Ni_3P phase, whereas the other deposits have significant amount of nickel phase in addition to the Ni_3P phase (Fig. 14.1). Thus, phase distribution is not as critical in the 12 wt% P deposit as that in the other deposits. Nonetheless, chemical analysis at different points across the depths of the deposits after being heated at 330–500 °C does not show large fluctuation in phosphorus content.

Another source for the presence of relatively large standard deviations in the electroless Ni–P deposits could be that the Knoop microindentation is more sensitive to the surface conditions, such as defects and anisotropies of material, because the depth of indentation is shallower (roughly half) than that of the Vickers measurement. The defects might include the curved surface and tilting of sample cross-sections caused by the difficulty in maintaining specimen flatness during polishing. Such difficulty prevails especially when polishing hard deposits coated on a softer substrate and mounted within a softer mounting media (Factor and Roman, 2000). Besides, the anisotropies of deposit microstructures are obvious as indicated by the presence of laminated structure and different phases.

14.6 Kinetics of increased hardening effects

Using the isothermal hardness data in Table 14.1, the plot of Fig. 14.9 has been made using the data points corresponding to the increasing hardening effects. The slope of the plot therefore represents the activation energy of increased hardening, that is Q = 40.9 kJ mol^{-1}. Based on the value of coefficient of determination (r^2), the result of kinetic modelling has been satisfactory.

Figure 14.10 uses the isothermal hardness data measured from the deposit cross-sections. The plot for the 6 wt% P deposit is the most satisfactory. A sudden increase of relative hardness (ΔH) value between the two heating

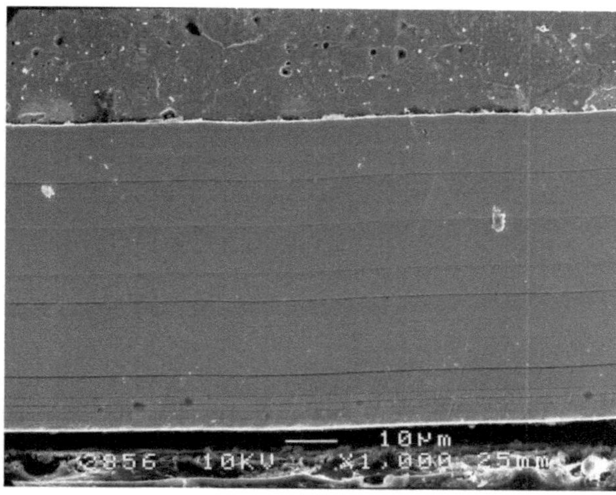

(a)

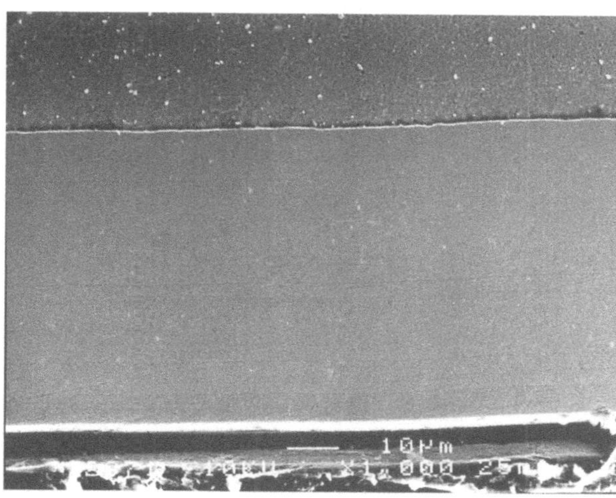

(b)

14.8 SEM micrographs showing the effect of agglomeration by thermal processing (450 °C for one hour) of the laminated structure of cross-sections. (a) 8 wt% P, as-deposited, (b) 8 wt% P, heat treated; (c) 5 wt% P, as-deposited, (d) 5 wt% P, heat treated.

(c)

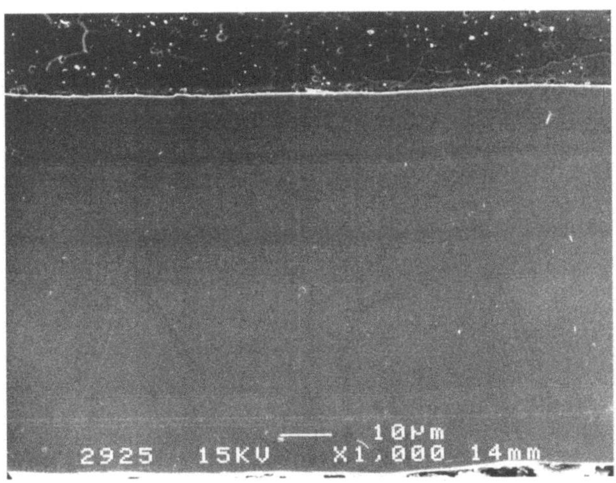

(d)

14.8 Continued

temperatures (300 °C and 330 °C for the 8 wt% P deposit, and 330 °C and 360 °C for the 12 and 5 wt% P deposits) is noticeable. Because of this, the values of the coefficient of determination for these plots are lower. This signifies that the kinetic strength theory may not be applicable to the increased hardening. The increased hardening effects for the electroless Ni–P deposits might have involved more than one stage with different kinetic energies. More satisfactory r^2 values are probable because the kinetic energies involved at different stages are somewhat similar. As a whole, the concept of kinetic strength could be used to provide approximations for the kinetic interpretation of increased hardening effects for the electroless Ni–P deposits, if such effects do not involve stages with very different kinetic energies.

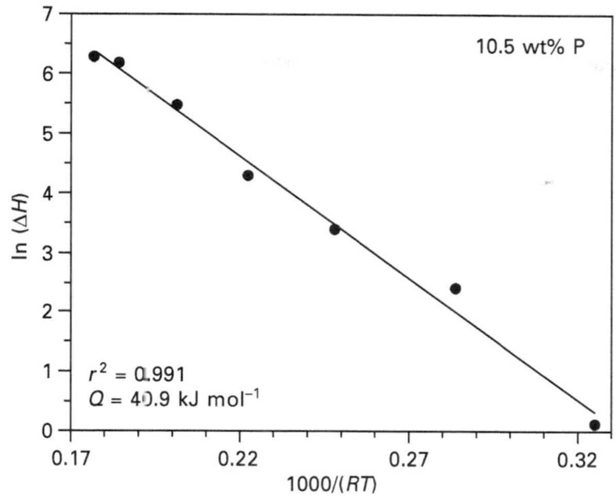

14.9 Plot of $\ln(\Delta H)$ versus $1000/(RT)$ using the Vickers hardness ($HV_{0.1}$) data in Table 14.1.

14.7 Microstructure and hardness of alloy coatings containing tin and tungsten

Electroless nickel alloys are widely used in many industries such as the chemical, petroleum, printing, textiles, automotive, electronics, as well as the aviation and aerospace sector due to their desirable properties. Among these properties are high hardness, high wear resistance, high corrosion resistance, high melting point and thermal resistance. There has been an interest within the scientific community to further improve these properties by co-depositing one or more metallic or non-metallic element into the alloy matrix.

One direction is aimed at furthering the research and development of nickel–phosphorus alloy coatings with an additional element to make a ternary deposit with tungsten or tin, as well as the quaternary deposit with the addition of both the elements. Tungsten has high hardness, a high melting point and high tensile strength, which makes it a favourable additional element. Tin on the other hand is non-toxic and has good corrosion resistance and solderability.

Electroless plating of binary Ni–P, ternary Ni–Sn–P and Ni–W–P as well as quaternary Ni–W–Sn–P alloy coatings can be made in alkali-citrate baths. After plating, several kinds of test can determine improvements of the characteristics and properties due to the additional elements as well as the change in behaviour when heat treatment is applied to these coatings. These include X-ray diffraction analysis to identify whether the coatings are amorphous. Interesting surface morphology features can be revealed using scanning electron microscopy. For the as-coated deposits, addition of a third element improves the hardness. Heat-treating to 200 °C leads to increases in hardness for all the coatings, and heat-treating to 400 °C will improve the hardness further for the ternary alloys. Heat-treating at 600 °C reduces hardness.

14.8 Conclusions

The Ni–P deposits have similar microstructural and phase transformation characteristics regardless of difference in the heating conditions (isothermal and linear heating). The evolution of Vickers hardness in response to phosphorus

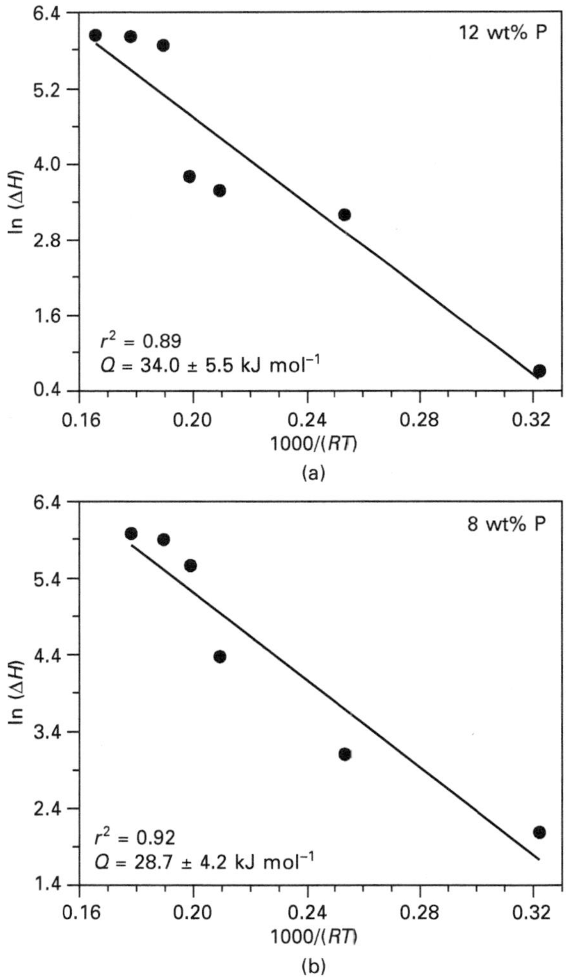

14.10 Plots of $\ln(\Delta H)$ versus $1000/(RT)$ using the Vickers hardness data of the Ni–P deposits (Fig. 14.5).

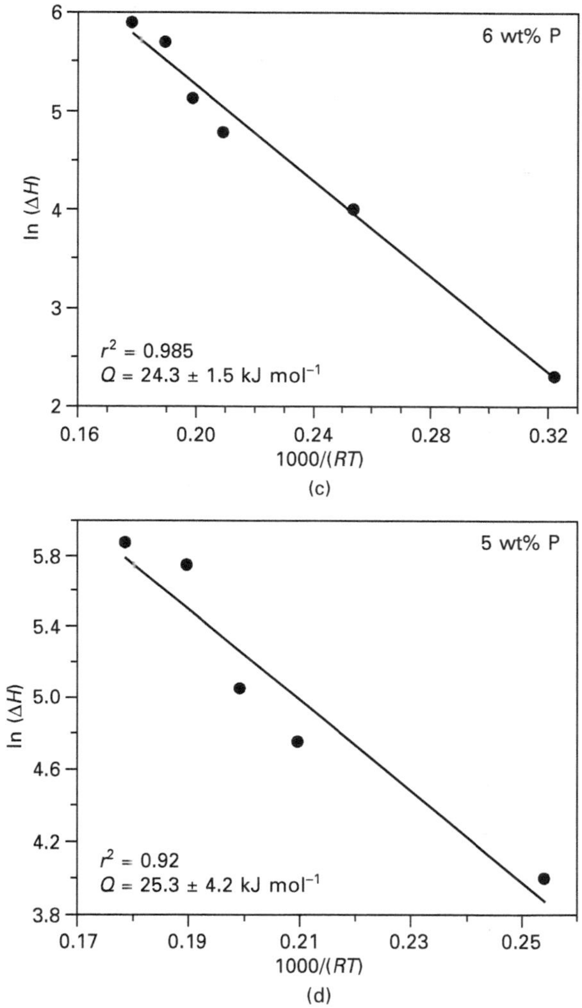

14.10 Continued

content and heating temperature, between those measurements taken from the deposit surface and cross-sections, are similar.

The deposits have an increase of Vickers as-deposited hardness with decreasing phosphorus content. The behaviour of hardness evolutions during thermal processing is also affected by the phosphorus content, in that the temperature for attaining maximum hardness increases with increasing phosphorus content. Besides, the degree of hardness loss (after achieving the maximum hardness) increases as the phosphorus content decreases.

Vickers hardness of the deposits increases with heating temperature and reaches a maximum at 400–450 °C. Further heating at the higher temperatures would decrease the hardness values.

Ni_3P stable phase formed during the thermal processing acts as a function of precipitation hardening, whereas the grain growth and coarsening of nickel and Ni_3P precipitates are responsible for the subsequent decreasing hardness at further heating.

Hardness varies across the depth on the cross-sections, but the tendencies of change are usually not clear. The concept of kinetic strength could not make accurate quantification of the kinetics of increased hardening effects for the deposits, probably due to the presence of different stages in such effects, which are related to different kinetic energies.

14.9 References

Factor M and Roman I (2000), 'Vickers microindentation of WC–12%Co thermal spray coating: Part 1: statistical analysis of microhardness data', *Surf Coat Technol*, **132**, 181–93.

Peeters P, Hoorn G V D, Daenen T, Kurowski A and Staikov G (2001), 'Properties of electroless and electroplated Ni–P and its application in microgalvanics', *Electrochim Acta*, **47**, 161–9.

Taheri R, Oguocha I N A and Yannacopoulos S (2001), 'The tribological characteristics of electroless Ni–P coatings', *Wear*, **249**, 389–96.

Vander Voort G F (2000), 'Microindentation hardness testing', in *ASM Handbook*, Volume 8, *Mechanical Testing and Evaluation*, ASM International, Materials Park, OH, 221–31.

15
Applications of electroless nickel–phosphorus (Ni–P) plating

Abstract: Data for electroless plating will be useful for design of processing parameters and optimising control parameters which will advance the development of surface engineering technology and contribute to industry. A comparison is made with other commonly used engineering deposits such as electroplated nickel and hard chromium. The advantages, limitations, and development of electroless nickel–phosphorus (Ni–P) are outlined and the use of heat treatment to tailor properties for specific industrial applications is described. Finally, the challenges this technology faces in the future are considered. Research is proposed on friction and wear of an electroless nickel–phosphorus–silicon carbide (Ni–P–SiC) coated aluminium cylinder liner in a case study.

Key words: electroless plating, nickel-phosphorus, applications, electroplated nickel, hard chromium.

15.1 Comparisons with other common engineering deposits

Major advance in electroless nickel–phosphorus (Ni–P) technology was achieved when high-phosphorus nickel coatings were developed. It is known that high-phosphorus electroless nickel coatings are good in wear and corrosion resistances due to their high hardness and amorphous structure, respectively. Considering the significance of wear and corrosion problems that the various industries have encountered, such coatings are finding increasing demand and application.

Electroplated nickel deposits for engineering use usually contain about 99 wt% nickel and are sulphur-free. Most deposits are plated from Watts and nickel sulfamate solutions. Other plating solutions are used also, but they tend to be reserved for plating deposits having specific engineering requirements. The deposits are used in various engineering applications to enhance surface properties, such as hardness, corrosion and wear resistance, and magnetic properties. For example, turbine blade manufacturing in the aerospace industry uses electroplated nickel to prevent surface oxidation during high temperature forging. The deposits are also used as a corrosion undercoat to hard chromium plated on hydraulic components operating in extreme conditions. Apart from those, electroplated nickel may be specified to salvage or build up worn or undersized components. Some mechanical properties of electroplated nickel deposits are listed in Table 15.1, and can be compared to those for electroless Ni–P deposits. On the other hand, electroplated nickel deposits could have certain disadvantages, such as uneven deposit thickness due to the effect of irregular plating current density distributions and substrate configurations, and the deposit's crystalline structure means that there is limited scope to modify the material properties through heat treatments.

Hard chromium is also known as industrial, functional or engineering chromium and it differs from decorative chromium in many technical aspects. Introduced to the industries in the 1920s, it can be electroplated from the plating solution containing both chromic acid (CrO_3) and catalytic anions. Because of their extremely high hardness (800–1000 $HV_{0.1}$) and corrosion resistance, hard chromium deposits can be used for applications in areas where excellent wear and corrosion resistance is required. However, the deposits are very brittle and usually consist of relatively high internal tensile stresses usually greater than 200 MPa. This could cause cracks in the deposits and result in fissures that propagate into the substrate. Accordingly, the corrosion resistance is severely reduced as cracks offer pathways for a corrosion environment to reach the substrate. Also because of high brittleness, hard chromium deposits are incapable of withstanding deformation and are preferable for use only on hard substrates. During the electroplating processes, the massive amount of hydrogen produced causes the deposit and substrate to suffer from hydrogen embitterment. This problem, however, can be solved by proper heat-treatment at 100–200 °C after the deposition processes, but this might have adverse effects on the hard chromium's properties and so to overall performance. For example, heat treatment at low temperatures increases the tensile stress in deposits and results in decreasing fatigue strength of substrates. The effects of heat treatment on the hardness of hard chromium deposits vary with temperatures. Heating at 400 °C or lower has little effect on the hardness, but heating at higher temperatures leads to significant reduction. The thickness of hard chromium deposits varies significantly depending upon the substrate's configuration and distance to the anodes. Compared to electroless Ni–P deposits, hard chromium has poor thickness uniformity. Table 15.2 lists and compares the properties of hard

Table 15.1 Mechanical properties of electroplated nickel from Watts and nickel sulfamate solutions

Mechanical properties	Watt nickel	Nickel sulfamate
Tensile strength (MPa)	345–485	415–610
Elongation (%)	10–30	5–30
Hardness ($HV_{0.1}$)	130–200	170–230
Internal tensile stress (MPa)	125–210	0–55

Table 15.2 Properties of electroless Ni–P deposit and hard chromium

Properties	Electroless Ni–P	Hard chromium
Melting point (°C)	890	1610
Internal stress on steel (MPa)	<7	200–300
Electrical resistivity (μΩ·m)	0.90	0.14–0.66
Thermal conductivity (W/m·K)	8	67
Modulus of elasticity (GPa)	200	100–200
Elongation (%)	1–1.5	<<0.1
Coefficient of thermal expansion (10^{-6}/°C)	12	6
Adhesion strength (MPa)	300–400	Good
Tensile strength (MPa)	>700	<200
Hardness ($HV_{0.1}$)	480–500*	800–1000
Coefficient of friction versus steel (lubricated)	0.13	0.16
Taber wear resistance (mg/1000 cycles)	15–20^{+}	2–3

*1100 $HV_{0.1}$ after heat treatment.
$^{+}$2–9 mg/1000 cycles after heat treatment.

chromium and electroless Ni–P (10–11 wt% P) deposits. In general, hard chromium deposits have a low coefficient of friction, good anti-sticking property and high adhesion strength to substrates, but they are not as good as electroless Ni–P deposits. In comparison, hard chromium deposits also have much lower ductility than electroless Ni–P deposits. However, their high thermal conductivity enables the rapid dissipation of high frictional heat caused by localised frictional effects.

Overall, the principal industrial uses of hard chromium deposits are for wear-resistance applications, improvement of equipment's life and performance, and part salvage. Because of the increasing environmental concerns in recent years, the use of hard chromium deposits has faced mounting challenges from many potential substitutions. However, none of these exhibits all of the desirable qualities of hard chromium deposits.

15.2 Advantages of electroless nickel–phosphorus (Ni–P) deposits

Compared to some other engineering deposits like electroplated nickel and hard chromium, electroless Ni–P deposits exhibit numerous advantages owing to their unique material properties. Notable among these are:

- Flexibility in tailoring material properties – most of the material properties being adjustable; in particular through plating and heat-treatment processes. This gives the deposits flexibility in meeting specific performance and requirements.
- High hardness – as-deposited hardness between 500 and 700 $HV_{0.1}$ depending on phosphorus content and up to 1100 $HV_{0.1}$ (equivalent to hard

chromium) after being heat-treated at 400 °C for one hour (Fig. 15.1).

- Excellent corrosion resistance – by forming a barrier between the substrate and the environment. Amorphous high-phosphorus deposits perform better in neutral and acidic environments, whereas low-phosphorus deposits are better in alkaline environment. In general, because of their amorphous nature and passivity, the corrosion resistance is superior to that of electroplated nickel and chromium alloys in many environments.
- Excellent wear resistance – because of their high hardness. The deposits have remarkable resistance to wear and abrasion. The Taber wear indexes (TWI) for electroplated nickel, electroless 9 wt% P as-deposited and heat-treated Ni–P, and hard chromium deposits are 25, 17, 4 and 2 mg/1000 cycles, respectively. However, the inclusion in Ni–P of hard particles like silicon carbide or diamond increases the wear resistances to the equivalent range of hard chromium.
- Inherent lubricity – better than that of hard chromium, stainless steel, steel and aluminium. The deposits have another advantage over hard chromium because of the environment concerns. In addition, excellent lubricity can be obtained by the co-deposition of solid particles within deposits.
- High solderability – the deposits can be easily soldered on light metal such as aluminium. Because of this, they are used extensively in electronics applications.
- Adjustable magnetic and electrical properties – these unique properties in response to the annealing effect have made the deposits a vital material in making items such as the underlayer in thin film magnetic disks, and thin film resistors.
- Excellent plating thickness uniformity even in recessed areas – this

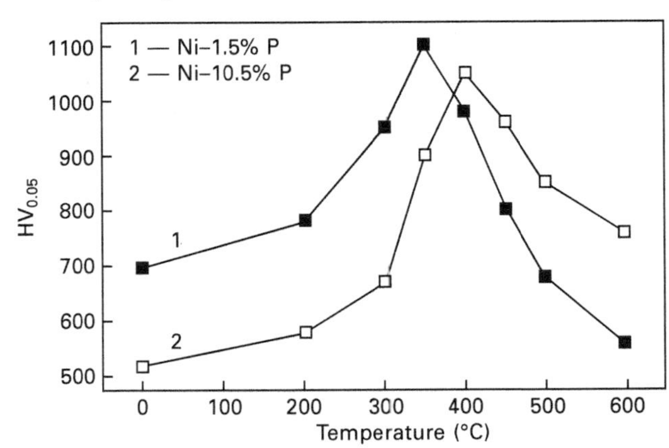

15.1 Effect of phosphorus content and heat-treatment temperature on the hardness of electroless Ni–P deposits (Zhang *et al.*, 1998). (Reproduced by permission of the Deputy Chief–editor of *Transactions of Nonferrous Society of China*.)

property is well known especially when compared with electroplated nickel deposits (Fig. 15.2). In addition, the plating property is free from the effect of the substrate's configuration.

- Ability to be plated on a conductor and non-conductor – as the plating process does not use electrical current, plating can be done on any material if it is catalytic to the chemical reduction of nickel.

15.3 Enhancement through heat-treatment processes

The industrial sectors that have widely used electroless Ni–P deposits include those listed in Table 15.3.

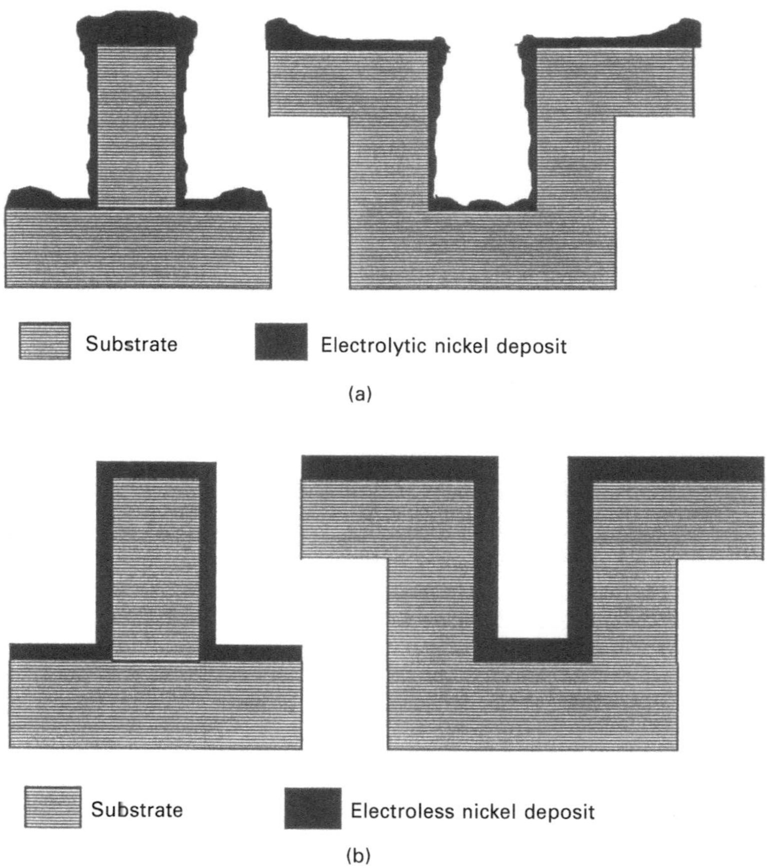

15.2 Coverage and thickness uniformity of (a) electroplate nickel and (b) electroless Ni–P deposits on the substrates (With kind permission from Lea Manufacturing Company: Electroless Nickel Technology Training Course Notes, Chapter 8: Properties and Applications, 1999, Figs 5 and 6).

Table 15.3 Industrial sectors in which electroless Ni–P deposits have found broad applications

Industrial sectors		
Aerospace	Material handling	Railroad
Automotive	Marine	Printing
Chemical and oil	Medical and pharmaceutical	Textiles
Electronics	Military	Wood and paper
Firearms	Mining	
Food	Moulds and dies	

Table 15.4 Recommended heat treatments for achieving maximum hardness in electroless Ni–P deposits (ASTM, 2000)

Deposit type (wt% P)	Temperature (°C)	Time (hour)
1–3	350–380	1
2–4	360–390	1
5–9	365–400	1
10 and above	375–400	1

If it is necessary, for example, to increase the hardness of deposits for a specific industrial application, appropriate heat treatment will be chosen and applied to the deposits in order to fulfil the requirement. At the heating temperatures lower than crystallisation temperature, structure changes can be insignificant. On the other hand, several structure changes due to crystalline transition are expected in the deposits heat-treated at higher temperatures. In addition, the process of heat treatment and the onset of crystallisation reaction are time-dependent. Furthermore, factors such as deposit phosphorus content, thickness of deposit, and type and nature of substrate can undermine the outcomes of heat-treatment process. Due to these reasons, the selection of heat-treatment processes has to depend on the deposit conditions and requirement of applications.

Several examples of industrial application that involve electroless Ni–P deposits and heat treatment processes are given and briefly described.

1. In automotive component manufacturing, brake pistons plated with 13 μm thick medium-phosphorus deposit are heat-treated at 400 °C for one hour to improve the deposit hardness and the wear resistance. The improved hardness extends the durability of the disk brake when compared to electroplated nickel and hard chromium finishes. The recommendations for achieving maximum hardness in the deposits at different ranges of phosphorus content are listed in Table 15.4 (ASTM, 2000).
2. In the fabrication of computer memory disks, high-phosphorus deposits have the important function of protecting the aluminium substrate and providing a non-magnetic polishable base for the thin film magnetic

medium. The plated non-magnetic high-phosphorus deposits are heat-treated to increase hardness for giving better protection to the substrate. In addition, the deposits create a barrier between the substrate and the sputtered magnetic surface, thus preventing electronic crossover from one side of the disk to the other. Because of this, appropriate heat-treatment usually at low temperatures is necessary to prevent the deposits from developing ferromagnetic properties, which may affect the overall magnetic performance of the memory disks.

3. In general, the metallic components plated with electroless Ni–P deposits can be heat-treated to improve the adhesion of deposits on the components. Heat-treatment processes conducted at relatively low temperatures (<200 °C) normally could prevent the hardening process caused by rapid precipitation of nickel phosphide. Table 15.5 shows the heat-treatment conditions recommended by the ASTM B733 Standard for improving adhesion bond with various substrates (ASTM, 2000). The very small extent of inter-diffusion at the deposit-substrate interface helps to improve the bond strength between deposit and component. With proper pre-treatment on the components, heat-treatment processes may have only minimum effect on the adhesion strength.
4. Heat-treatment is also used to reduce hydrogen embrittlement in metallic components. For example, steel components can be heat-treated at about 200 °C to aid the absorbed hydrogen to diffuse out of the steel. Table 15.6 gives the ASTM B733 Standard recommended heat-treatment conditions for embrittlement relief of steels on different strength levels (ASTM, 2000). Hydrogen can be introduced into components such as through electro cleaning and electroless deposition processes. The absorbed hydrogen can cause cracking in the plated components, especially in those made of high-strength steels.
5. Heat-treatment processes can be used for providing coloured finishing of electroless Ni–P deposits. Coloured finishes can aid component identification and they can be used for decorative purposes. In addition, some components such as baking pans, and optical and solar devices are

Table 15.5 Recommended heat treatments for improving deposit's bond strength with different substrates (ASTM, 2000)

Substrate	Temperature (°C)	Time (hours)
Precipitation hardened Al alloys	120–130	1–6
Carburised steel	120–130	1–6
Aluminium and beryllium alloys	140–150	1–2
Copper and copper alloys	140–150	1–2
Carbon and alloy steel	180–200	2–4
Nickel and nickel alloy	200–300	1–4
Titanium alloys	300–320	1–4

Table 15.6 Recommended heat treatments for hydrogen embrittlement relief of steels on different strength levels (ASTM, 2000)

Maximum specified tensile strength (MPa)	Time (hour) at 190–210 °C
1050 or lower	Not required
1051–1450	2
1451–1800	18
above 1800	24

coloured black in order to be light and heat absorbent. Different coloured finishes of yellow to black can be produced in an air or an oxidising atmosphere at the temperatures between 350 °C and 950 °C. The oxide layer formed on the surface gives colour to the deposit.

15.4 Electroless nickel–phosphorus–silicon carbide (Ni–P–SiC) coated aluminium cylinder liner: a research case study

A research goal is to develop surfaces to withstand wear, corrosion, fatigue and oxidation at high temperatures by the available surface modification techniques. To achieve this goal, we will need the following three research steps. First, we need to identify suitable surface modification techniques from the available processes for such service conditions and conduct studies. Fundamental understanding of the processes is very important to successful implementation. Second, we need to extend our specialisation in the techniques to developing new coating systems. Finally, we need to study the mechanical properties of the coatings, characterisation and implementation.

Current trends of coating techniques involve composite coatings, such as multilayer or multiphase coatings, which are expected to have tailor-made properties for specific applications. Electroless coatings are used for the following reasons:

- uniform coating thickness
- unique structural features and morphology
- excellent chemical, mechanical and magnetic properties
- possibility for improvement of properties for specific applications
- absence of power supply and electrical contacts during plating
- ability to deposit even on non-conducting surfaces.

The mechanical and tribological properties of these coatings can be further improved by the incorporation of hard particles (SiC, B_4C, Al_2O_3 and diamond) and dry lubricants (PTFE, MoS_2 and graphite), resulting in a film with self-lubricating and excellent anti-sticking characteristics. Among the particles, SiC is most widely used.

Cylinder components contribute to around 30% of total friction in an engine. The most important requirements for satisfactory service life of the cylinder component in an engine are long life, high temperature resistance and good mechanical properties. Most of the parts inside the cylinder should undergo surface treatment. The choice of the process depends on the functional efficiency and cost. Generally, cast iron is used for cylinder components, which exhibits outstanding wear characteristics, and high strength properties even at high temperatures, but the disadvantages are high weight and excess fuel consumption. Aluminium engine blocks have the advantage of being lightweight but have the disadvantage of having poor wear and scuff resistance between the piston and rings and the mating cylinder wall. Hence, there is a need to investigate the techniques to improve the performance of aluminium alloys to enable practical applications.

To improve wear and scuff resistance, several techniques have been used in aluminium engine blocks. The installation of cast iron cylinder liners is one technique; however, extensive machining is required of both the engine block and cylinder liner so that they fit together properly. It is also possible to cast the aluminium block around a cast iron liner but this adds complexity to the casting process. Additionally, cast iron liners have the disadvantage of adding weight to the aluminium engine block. Another technique is to cast the entire aluminium block out of a high-silicon aluminium alloy. This material has excellent wear resistance but is difficult to machine and difficult to cast.

Another technique is to cast the aluminium block out of a lower-silicon content aluminium alloy and apply plating to the bore of the block or aluminium alloy liner to improve wear resistance. The plating is typically a nickel alloy with a controlled fine dispersion of silicon carbide or boron nitride particles distributed uniformly in the metal matrix. Electroless Ni–P–SiC coatings have a high potential to be used for cylinder liners with high wear resistance and hardness.

The properties and microstructure of electroless nickel coatings depend on the post-deposition heat treatment, which is frequently used to improve adhesion or to modify properties in order to satisfy the needs of a particular application. Electroless nickel coatings can be a good alternative to the chromium coatings without the negative environmental impact.

Research is needed to study:

- the effect of process parameters on the properties of the coatings
- the effect of size and fraction of SiC particles incorporated on hardness, friction coefficient and wear resistance
- both mechanical (hardness) and tribological (friction resistance and wear) properties of the coatings
- the effect of heat treatment on the microstructure of the coatings

- the microstructural changes and the phases formed in the coatings responsible for the hardness by X-ray diffraction (XRD) analysis
- wear of coated cylinder liners with different surface roughness and the difference of the wear performance between aluminium and cast iron bores
- wear surfaces and debris by using scanning electron microscopy (SEM).

A summary of methodology is as follows:

- electroless Ni–P–SiC coating on aluminium: experimental setup
- annealing/heat treatment: furnace
- microhardness measurements: microhardness tester
- microstructural studies: image analyser
- identification of phases formed: XRD
- wear behaviour: pin–on–disc machine
- characterisation of the coated layer formed: SEM.

This research is a continuation of the wear studies of Ni–P coating on cylinder liners against aluminium and cast iron cylinder bores. Among the anticipated outcomes, maximum hardness can be achieved after heat treatment depending on the phosphorus content. This is attributed to fine Ni crystallites and hard intermetallic Ni_3P particles precipitated during the crystallisation of the amorphous phase. The research will lead to the development of cylinder liner coating with high hardness and wear resistance to replace the cast iron cylinder liner in automotive engine blocks. Electroless Ni–P–SiC composite plating on aluminium should produce high wear resistance and sound corrosion resistant coatings. The hardness of electroless Ni–P–SiC composite coatings can be increased to HV800, which has a high potential to be used as cylinder liners in the automotive industry.

15.5 Development and future trends

The development of electroless Ni–P deposits over the past 60 years has been remarkable (Longfield *et al.*, 2001). In the 1950s and 1960s, the deposits were mainly used for railroad tank cars. The growing demand came only in the late 1960s and 1970s when metal finishing companies started introducing their own labelled plating solutions, as well as providing technical support to the plating operators, and therefore making them much easier to use. In the gas and oil industry where the deposits were used on valves and other components for wear and corrosion protection, the demand was particularly high. The growth however was halted by the crash of the gas and oil market in the mid 1980s. Because of this adverse event, the deposits have, in fact, been diverted and introduced into various industries for different applications. The deposits are replacing coatings that are more traditional in certain critical and difficult applications owing to their unique combination of material properties. In addition, with the progression of research on material properties and areas for potential application, the deposits have found steady growth in the market since recovering from the slump in the mid 1980s. However, it is very difficult to estimate precise growth statistics of the markets as the majority of published information is concerned with technical, rather than commercial, aspects. In the United Kingdom, the consumption of sodium hypophosphite is around 400 metric tonnes per annum, and this probably could give us some insight into the electroless Ni–P deposit market in this country.

The unique material properties of electroless Ni–P deposits have warranted their selection in many applications. Under appropriate plating conditions, and using proper post-plating options such as heat-treatment processes in particular, many of the material properties can be further enhanced giving the deposits flexibility in meeting specific performance requirements. Accordingly, maintaining continuous research in the material aspects and the correlations between properties and the alloy compositions, plating and thermal processing conditions, and crystallisation and phase transformation behaviour of deposits is important. Such correlations have strong implications in both processing and service of electroless Ni–P deposits.

It is undoubted that the success of electroless Ni–P deposits also relies on the cost and environmental considerations. By considering practical, cost and environmental factors, the electroless Ni–P industry is expected to face several challenges at present or in the immediate future:

- First, it is known that the largest single use of electroless Ni–P deposits is in the manufacturing of aluminium memory disks (Longfield *et al.*, 2001). Hence, the discovering of any novel technology eliminating electroless Ni–P deposits from this application might bring serious effect to the deposit's present and future markets. For this reason, continuous exploration for diverse new applications is essential.
- Secondly, it is very important to ensure that the existing related technologies are continuously improved and are capable of meeting future high performance need and new requirements. For example, the improvements in recording technology allowing for increases in recording densities have resulted in the need for continual quality improvements in both the electroless Ni–P plating and the pre-treatment processes. The new technologies using alloys, composites and special 'hard' electroless Ni–P deposits are being used for high performance requirements.
- Thirdly, because of environmental concerns, it is highly demanded that the plating industry be able to effectively reduce the amount of waste produced and to ensure that the disposed waste is dealt with safely and legally. In the United Kingdom, for example, a spent plating solution containing 10 g/l or more nickel metal is defined as special waste that has to be dealt with under specific environmental direction. For this

reason, technical knowledge for proper control of plating parameters and bath maintenance for electroless Ni–P deposits is primarily important in preventing unnecessary waste of chemical use in the plating processes.

- Finally, it is vital for the industry to be able to meet the deposit's quality and performance requirements at competitive costs. The cost factor can be important when competing with alternative options. As a result, continuous advancement in the technologies such as in the plating operation, and pre- and post-treatments processes is necessary. With more advanced technologies, the industry may be able to more effectively optimise both performance and costs, while minimising adverse impact on the environment (Longfield *et al.*, 2001).

15.6 Conclusions

Owing to their unique combination of material properties, electroless Ni–P deposits are used in various industrial sectors in particular as functional and protective coatings. It has been shown that thermal processing can be very useful in modifying and improving many of the material properties, like hardness, wear and corrosion resistances, and magnetic and electrical behaviours. This gives the deposits flexibility that many other engineering deposits may not have in meeting specific performance requirements in various applications. Overall, the unique characteristics of electroless Ni–P deposits can be summarised as below:

- amorphous or crystalline at as-deposited state with the presence of weak laminations
- microstructural changes in deposit taking place at elevated temperature due to crystallisation of amorphous structure making the deposits heat-treatable for further enhancement of material properties
- high hardness but limited ductility
- excellent corrosion and wear resistance
- inherent lubricity and high solderability
- adjustable magnetic and electrical properties
- excellent plating thickness uniformity on either conductor or non-conductor.

15.7 References

ASTM (2000), *Annual Book of ASTM Standards*, 02.05 (Standard B733), Philadelphia, PA, 456–66.

Longfield P D, Rauppius R and Richtering W (2001), 'Novel long-life electroless nickel with continuous regeneration by electrodialysis', *Met Finish* **99**(5), 36, 38–43.

Zhang Y Z, Zhang K, Fan J Z, Zhang S M and Yao M (1998), 'Hardness and wear resistance of electroless nickel deposits', *Trans Nonferrous Met Soc China* **8**, 642–7.

Index

Kolibri 360 GmbH
Ferdinand-Jühlke-Straße 7
99095 Erfurt, Deutschland
produktsicherheit@kolibri360.de